9th Curtin University of Technology Science and Engineering International Conference 2014

(CUTSE2014)

IOP Conference Series: Materials Science and Engineering Volume 78

Sarawak, Malaysia
3 – 4 December 2014

ISBN: 978-1-5108-0149-3
ISSN: 1757-8981

Printed from e-media with permission by:

Curran Associates, Inc.
57 Morehouse Lane
Red Hook, NY 12571

Some format issues inherent in the e-media version may also appear in this print version.

Printed by Curran Associates, Inc. (2015)

Institute of Physics
Dirac House, Temple Back
Bristol BS1 6BE UK

Phone: 44 1 17 929 7481
Fax: 44 1 17 920 0979

techtracking@iop.org

Additional copies of this publication are available from:

Curran Associates, Inc.
57 Morehouse Lane
Red Hook, NY 12571 USA
Phone: 845-758-0400
Fax: 845-758-2634
Email: curran@proceedings.com
Web: www.proceedings.com

TABLE OF CONTENTS

WIRE ROPE ISOLATORS FOR VIBRATION ISOLATION OF EQUIPMENT AND STRUCTURES – A REVIEW 1
P S Balaji, M E Rahman, Leblouba Moussa, H H Lau

STUDY ON FLEXIBLE PAVEMENT FAILURES IN SOFT SOIL TROPICAL REGIONS 11
M Jayakumar, Lee Chee Soon

REVIEW ON METALLIC AND PLASTIC FLEXIBLE DYE SENSITIZED SOLAR CELL 23
A R Yugis, R F Mansa, C S Sipaut

EFFECT OF VARYING ACID HYDROLYSIS CONDITION IN GRACILARIA SP. FERMENTATION USING SASAD 30
H Mansuit, M D C Samsuri, C S Sipaut, C F Yee, S M Yasir, R Mansa

SOL – GEL SYNTHESIS AND CHARACTERIZATION OF MAGNESIUM PEROXIDE NANOPARTICLES 38
J Jaison, C Ashok Raja, S Balakumar, Y S Chan

OPTIMIZATION AND PERFORMANCE EVALUATION FOR NUTRIENT REMOVAL FROM PALM OIL MILL EFFLUENT WASTEWATER USING MICROALGAE 46
Raheek I Ibrahim, Z H Wong, A W Mohammad

EXPERIMENTAL INVESTIGATION ON SANDSTONE ROCK PERMEABILITY OF PAKISTAN GAS FIELDS 53
Arshad Raza, Chua Han Bing, Ramasamy Nagarajan, Mohamed Ali Hamid

COMPARISON OF BUFFER EFFECT OF DIFFERENT ACIDS DURING SANDSTONE ACIDIZING 62
Mian Umer Shafiq, Hisham Khaled Ben Mahmud, Mohamed Ali Hamid

CUTTINGS-LIQUID FRICTIONAL PRESSURE LOSS MODEL FOR HORIZONTAL NARROW ANNULAR FLOW WITH ROTATING DRILLPIPE 70
T N Ofei, S Irawan, W Pao

VOLTAGE HARMONIC ELIMINATION WITH RLC BASED INTERFACE SMOOTHING FILTER 80
K Chandrasekaran, V K Ramachandaramurthy

CMOS SPDT SWITCH FOR WLAN APPLICATIONS 88
M A S Bhuiyan, M B I Reaz, L F Rahman, K N Minhad

AN ANALYTICAL STUDY OF VARIOUS TELECOMMINICATION NETWORKS USING MARKOV MODELS 94
M Ramakrishnan, E Jayamani, P Ezhumalai

TRANSMISSION LINE SERIES COMPENSATION FOR WIND ENERGY TRANSMISSION 105
C Palanichamy, Y C Wong

ENHANCED ELECTRIC POWER TRANSMISSION BY HYBRID COMPENSATION TECHNIQUE 115
C Palanichamy, G Q Kiu

ADAPTIVE PRECODED MIMO FOR LTE WIRELESS COMMUNICATION 128
A F Nabilla, T C Tiong

ADAPTIVE MODULATION AND CODING FOR LTE WIRELESS COMMUNICATION 134
S S Hadi, T C Tiong

PERFORMANCE ANALYSIS FOR LTE WIRELESS COMMUNICATION 140
S Tholhath, T C Tiong

OPTIMIZING ELECTRO-THERMO HELDS FOR SOOT OXIDATION USING MICROWAVE HEATING AND METAL 148
Haitham B Al-Wakeel, Z A Abdul Karim, Hussain H Al-Kayiem

COMPARATIVE PERFORMANCE ANALYSIS OF A DUAL-SOLENOID MECHANICAL OSCILLATOR 161
V C C Lee, H V Lee, H G Harno, K C Woo

IMPROVING OF THE PHOTOVOLTAIC / THERMAL SYSTEM PERFORMANCE USING WATER COOLING TECHNIQUE 167
Hashim A Hussien, Ali H Numan, Abdulmunem R Abdulmunem

ON RADIAL FLOW BETWEEN PARALLEL DISKS 176
A Y L Wee, A Gorin

TEMPERATURE AND VOLTAGE RESPONSES OF A MOLTEN CARBONATE FUEL CELL IN THE PRESENCE OF A HYDROGEN FUEL LEAKAGE 182
M C Law, G V Y Liang, V C C Lee, S K Wee

INFLUENCE OF CARBON NANO TUBES ON THE THERMO-MECHANICAL PROPERTIES OF UNSATURATED POLYESTER NANOCOMPOSITE 191
A K M Moshiul Alam, M D H Beg, Rosli Mohd Yunus

THIN-WALLED COMPOSITE TUBES USING FILLERS SUBJECTED TO QUASISTATIC AXIAL COMPRESSION 197
Haidar F Al-Qrimli, Fadhil A Mahdi, Firas B Ismail, Ibrahim S Alzorqi

EXPERIMENTAL INVESTIGATION AND 3D FINITE ELEMENT PREDICTION OF TEMPERATURE DISTRIBUTION DURING TRAVELLING HEAT SOURCED FROM OXYACETYLENE FLAME 209
Adam Umar Alkali, Turnad Lenggo Ginta, Ahmad Majdi Abdul-Rani

BAD ACTORS CRITICALITY ASSESSMENT FOR PIPELINE SYSTEM 216
Meseret Nasir, Kit Wee Chong, Sabtuni Osman, Wee Siaw Khur

PROBABILISTIC LIFE CYCLE COST MODEL FOR REPAIRABLE SYSTEM 225
Meseret Nasir, H Y Chong, Sabtuni Osman

WIND CANNOT BE DIRECTED BUT SAILS CAN BE ADJUSTED FOR MALAYSIAN RENEWABLE ENERGY PROGRESS 233
C Palanichamy, Meseret Nasir, S Veeramani

ROBUST H_∞ STABILIZATION OF A HARD DISK DRIVE SYSTEM WITH A SINGLE-STAGE ACTUATOR 241
Hendra G Harno, Raymond Song Kiin Woon

EFFECT OF MACHINING VELOCITY IN NANOSCALE MACHINING OPERATIONS 250
Sumaiya Islam, Raafat Ibrahim, Noman Khondoker

SOME ADVANTAGES OF IMPLEMENTING AN ADAPTIVE MOVING MESH FOR THE SOLUTION TO THE BURGERS EQUATION 257
Sudi Mungkasi

PHENOL BIODEGRADATION BY FREE AND IMMOBILIZED CANDIDA TROPICALIS RETL-CRL ON COCONUT HUSK AND LOOFAH PACKED IN BIOFILTER COLUMN 263
D Shazryenna, R Ruzanna, M S Jessica, M T Piakong

ISOLATION AND CHARACTERIZATION OF SURFACE AND SUBSURFACE BACTERIA IN SEAWATER OF MANTANANI ISLAND, KOTA BELUD, SABAH BY DIRECT AND ENRICHMENT TECHNIQUES 270
L D Benard, P M Tuah, E G Suadin, N Jamian

REVIEW PAPER ON CELL MEMBRANE ELECTROPORATION OF MICROALGAE USING ELECTRIC FIELD TREATMENT METHOD FOR MICROALGAE LIPID EXTRACTION 279
C Joannes, C S Sipaut, J Dayou, S M Yasir, R F Mansa

A FIRST SURVEY ON THE ABUNDANCE OF PLASTICS FRAGMENTS AND PARTICLES ON TWO SANDY BEACHES IN KUCHING, SARAWAK, MALAYSIA 287
V James Noik, P Mohd Tuah

ISOLATION, CHARACTERIZATION AND SCREENING OF RHIZOSPHERIC BACTERIA OF PITTOSFERUM RESINIFERUM HEMSL. 299
R Hanirah, M T Piakong, L Syaufi

EXPERIMENTAL INVESTIGATION OF WOOD FIBRE CEMENT COMPOSITE WALL PANEL UNDER AXIAL LOADING 305
Mst Sadia Mahzabin, Roszilah Hamid

THE EFFECTS OF CONCENTRATION AND SALINITY ON POLYMER ADSORPTION ISOTHERM AT SANDSTONE ROCK SURFACE 311
M Ali, H Ben Mahmud

DESIGN AND DEVELOPMENT OF LOW-COST WIND TUNNEL FOR EDUCATIONAL PURPOSE 318
T H Yong, S S Dol

DESIGN AND DEVELOPMENT OF LOW-COST WATER TUNNEL FOR EDUCATIONAL PURPOSE 329
M Zahari, S S Dol

THE EFFECTS OF SPRING STIFFNESS ON VORTEX-INDUCED VIBRATION FOR ENERGY GENERATION 337
M Zahari, H B Chan, T H Yong, S S Dol

THE FEASIBILITY OF WIND AND SOLAR ENERGY APPLICATION FOR OIL AND GAS OFFSHORE PLATFORM 347
Y K Tiong, M A Zahari, S F Wong, S S Dol

Author Index

IOP Conf. Series: Materials Science and Engineering **78** (2015) 011001 doi:10.1088/1757-899X/78/1/011001

Foreword

A very warm welcome to all participants of the 9th Curtin University Technology, Science and Engineering (CUTSE) Conference 2014.

This annual conference dates back to 2006 when the first Curtin University of Technology Science and Engineering (CUTSE) Conference was held in Curtin University, Miri Sarawak. CUTSE Conference was initially intended for Curtin's undergraduates such that they are able to experience the presentation of their work in a conference environment. As time passes and following the urge of knowledge dissemination, CUTSE Conference is hence open to public. This year the Department of Mechanical Engineering has been given the honour to organize the 9th CUTSE Conference. It has been a pleasure to watch CUTSE grow from strength to strength over the years. This year, our theme is "Discovering, Innovating and Engineering". We hope that it is in this spirit that CUTSE participants may align their respective work, such that we all aim for a greater and better implementation of "Discovering, Innovating and Engineering".

The 9th CUTSE Conference 2014 is an excellent avenue for researchers, engineers, scientists, academicians, professionals from industry and students to share their research findings and initiate further collaborations in their respective fields. Parallel sessions in Mechanical, Electrical, Computer, Civil and Chemical engineering as well as the sciences will be hosted over a period of two days. Each year, the conference attracts participation from a number of countries in addition to Malaysia and Australia. In addition, student participants will get the opportunity to present their research projects and gain valuable feedback from industry professionals. This year the Conference will be organised by the Department of Mechanical Engineering of Curtin Sarawak's School of Engineering and Science in collaboration with The Institute of Engineers Malaysia, Miri Branch.

On behalf of the organizing committee, I would like to thank this year's sponsors and supporters. We appreciate your support for CUTSE 2014 and in research and development, and your foresight in nurturing cutting edge research into industrial applications. CUTSE 2014 would not be possible without the dedicated work and efforts of the organizing committee, who worked tirelessly in all aspects of the conference organization. I thank you for your hard work and commitment towards making CUTSE 2014 a success.

Selamat Datai (Welcome) and enjoy the conference.

Dr Vincent Lee Chieng Chen
Organizing Chairperson,
9th CUTSE Conference 2014

IOP Conf. Series: Materials Science and Engineering **78** (2015) 011002 doi:10.1088/1757-899X/78/1/011002

CUTSE2014 Organizing Committee

Chairperson
Dr Vincent Lee Chieng Chen

Deputy Chairperson
Dr Wee Siaw Khur

Secretariat
Dr Meseret Nasir Reshid
Ms Vivian Dayong
Ms Alicia Lyndie Rogers
Ms Dora Cecelia Indin
Ms Jacqueline Martin

Treasury
Dr Wee Siaw Khur
Dr Stephanie Chan Yen San

Programme and Logistics
Dr Hendra Gunawan Harno (Committee Chair)

Dr Mohan Reddy Moola (Programme Coordinator)
Dr Rajamohan Ganesan
Dr Mesfin Gizaw Zewge
Ms Chloe Wong Siew Fan
Mr Chan Hiang Bin
Mr Yong Tshun Howe

Mr Pang Po Ken (Logistics Coordinator)
Mr Donovan Mujah Bernard Lium
Mr Jason Chia Keh Leong
Mr Kamaroizan Mohamad

Proceedings and Publication
Associate Professor Dr Sujan Debnath (Committee Chair)

Dr M. E. Rahman (Proceedings Coordinator)
Dr Stephanie Chan Yen San
Dr Law Ming Chiat
Dr Charles Loo Chin Moy
Eur Ing Tiong Teck Chai
Mr OmidNabinejad

Dr Zinedine Khatir (Publication Coordinator)
Dr B. M. Kumar

Sponsorship and Publicity
Associate Professor Dr Sharul Sham bin Dol (Committee Chair)
Dr Khor Ee Huey
Ir Lim Chye Ing
Ms Cynthia Paul Sebli
Ms Wee Yee Boon
Ms Nor Hasimah Mohamad Zain
Dr Edwin Jong Nyon Tchan

IOP Conf. Series: Materials Science and Engineering **78** (2015) 011003 doi:10.1088/1757-899X/78/1/011003

Scientific Committee

Peer review statement

All papers published in this volume of *IOP Conference Series: Materials Science and Engineering* have been peer reviewed through processes administered by the proceedings Editors. Reviews were conducted by expert referees to the professional and scientific standards expected of a proceedings journal published by IOP Publishing.

Wire rope isolators for vibration isolation of equipment and structures – A review

P.S.Balaji[1*], M.E.Rahman[1], Leblouba Moussa[2], H. H. Lau[1]

[1]Department of Civil and Construction Engineering, Curtin University Sarawak, Malaysia
[2]Department of Civil & Environmental Engineering, University of Sharjah, UAE

*Email: balaji.palani@postgrad.curtin.edu.my

Abstract. Vibrations and shocks are studied using various techniques and analyzed to predict their detrimental effect on the equipment and structures. In cases, where the effects of vibration become unacceptable, it may cause structural damage and affect the operation of the equipment. Hence, adding a discrete system to isolate the vibration from source becomes necessary. The Wire Rope Isolator (WRI) can be used to effectively isolate the system from disturbing vibrations. The WRI is a type of passive isolator that exhibits nonlinear behavior. It consists of stranded wire rope held between two metal retainer bars and the metal wire rope is made up of individual wire strands that are in frictional contact with each other, hence, it is a kind of friction–type isolator. This paper compiles the research work on wire rope isolators. This paper presents the research work under two categories, namely monotonic and cyclic loading behaviors of WRI. The review also discusses the different terminologies associated with vibration isolation system and highlights the comparison between various isolation systems.

1. Introduction

Vibrations and shocks are studied using various techniques and analyzed to predict their detrimental effect on the equipment and structures. In many cases, the vibrations are unavoidable, but it will be within tolerable limits. In other cases where the vibration becomes intolerable, it is required to analyse the system for the effects of vibrations and improve the mechanical properties or in a place where are system design restricted in improving mechanical properties, it is then required to add an isolation system to counter the vibrations [1]. The application of vibration isolation system requires an understanding of the vibration control components, namely, source, path and receiver of the vibration [2].

The source of the vibration can be either natural or man-made. The natural sources include earthquake, wind, ocean waves, etc. and man-made sources include vibrations due to the operation of heavy machineries, construction works, roads and railways, etc. The path is the medium through which the vibration is transmitted, such as building components, pipe ducts, etc. The receiver refers to the building or equipment which receives the vibration from the source. The vibration control is employed based on the level of vibrations that a receiver can withstand without undergoing structural damage or affecting the functionality. The cases in which the level of vibrations is unacceptable, the isolation system is generally applied to cut off the path of the vibration to enhance the safety of the receiver.

In order to understand the isolation system, first it is required to understand the terms, namely, transmissibility and natural frequency. Transmissibility is a ratio of the vibrational force being measured in a system to the vibrational force entering a system (Figure 1). The transmissibility is a direct measure of the amount of vibrations transmitted by the isolation system, for example, if an isolation system has the transmissibility of 0.8 then it indicates that 80% of the vibration forces have been transmitted through the isolation system. The isolation efficiency provides the direct measurement of efficiency of isolation system and it is given by,

$$\text{Isolation efficiency} = 1 - \text{Transmissibility} \quad (1)$$

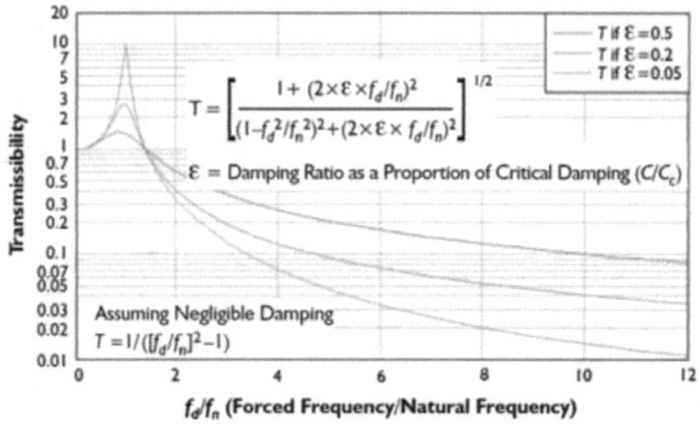

Figure 1. Transmissibility curve with respect to frequency ratio [2]

Natural frequency is the frequency at which the system will vibrate in the absence of any driving force or damping. Theoretically, when there is no external force is applied against the system vibration, then the system vibrates forever as per Newton's first law. However, in the real world, there exist the force in different forms which act against the vibration and bring the system to rest and such phenomenon is called as damping. It is absolutely required to prevent the system vibrate at its natural frequency, as this results in the larger magnitude of displacement and further to failure of components or structures in a phenomenon called resonance.

The concept of isolation and damping are often misunderstood, however, both are different from each other. It is required to understand the exact difference between them in order to proceed further. Damping refers to the absorption of vibrational energy which has entered the system and dissipating it by converting the kinetic energy of vibration into a different form of energy. On the other hand, isolation refers to the prevention of vibrations from entering a system. The transmissibility curve shown in figure 2 depicts the region of isolation and damping. Damping is primarily used to prevent the system from vibrating in a larger amplitude at its natural frequency, however; the isolation refers to the region, which are having lower values of transmissibility [1-3]. However, both isolation and damping are used in conjunction to achieve the desirable performance [2].

A general isolation system has two main components, namely, stiffness and damping. Firstly, the stiffness of the isolation system controls the static deflection required for different levels of isolation. The lower value of transmissibility can be achieved by increasing the static deflection. The amount of transmissibility required for an application differs in each case and depends mainly on the criticalness of the application. In general, the transmissibility of about 3% for critical applications and 5 % for sensitive applications and 10% for non-sensitive are applied [2]. The values indicated are only for reference and the exact amount of transmissibility required has to be calculated for the required applications.

Secondly, the damping component of isolator enables the energy dissipation of the external excitation to suppress it. There are three types of damping: viscous, coulomb, and structural/solid damping. Viscous damping is dependent on the velocity of the movement. Coulomb damping is dependent on surface friction and the pressure between surfaces. Structural damping, also known as hysteresis, is internal friction within the material; all materials exhibit some amount of hysteresis. The amount of damping required is subjective and varies with respect to applications. Most of the structures have less than 5% inherent structural damping. According to the previous study [3], the general guidelines for amount of damping can be taken as 20-25 % of damping for standard automobile suspension, 30% damping is typical for a heavy duty suspension and 40% damping is typically used for high performance suspense and damping levels above 40% proves very uncomfortable to human and tends to loose structural joints.

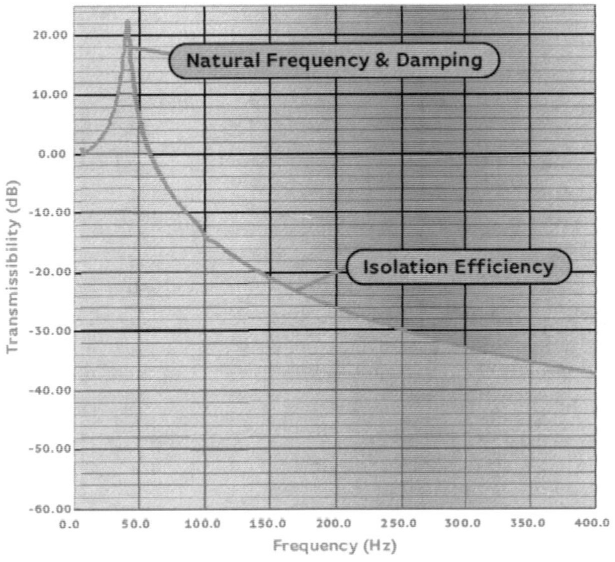

Figure 2. Transmissibility curve with region of damping and isolation [1]

2. Types of isolation system

The isolation system is categorized into three types, namely, Passive system, Active system and Semi-Active system. The active vibration isolation system generally can achieve high performance in suppressing the structural vibrations [4-6]. Semi-active vibrations isolations, also called as adaptive–passive system, contains along with the spring and the dampers, a feedback circuit which consists of a sensor, controller, and an actuator (such as Magneto-rheological (MR) or shape memory alloys). Feedback circuit monitors real time vibration excitation using sensors and controls the actuators to dissipate the energy and achieves relatively better isolation than passive isolation methods [5, 7]. However, a fully active isolation system (figure 3) isolates the vibration directly using actuator systems without having additional passive dampers [8-10].

Many researchers have studied the performance of various control methods for damping using MR fluids [11-19] and stated that the damping property of the MR fluid actuators can be controlled by varying the viscosity of the MR fluid using a magnetic field. However, active systems are very design intensive and require sensors and processors to provide real time data to the isolator. In addition, it requires a large amount of power to operate. These necessary features of active isolation systems make it the most expensive isolation design. As a consequence of the expense and the large power requirement, active isolation systems are very uncommon in major industries, except in critical applications that require high level isolation such as sensitive testing equipment [20].

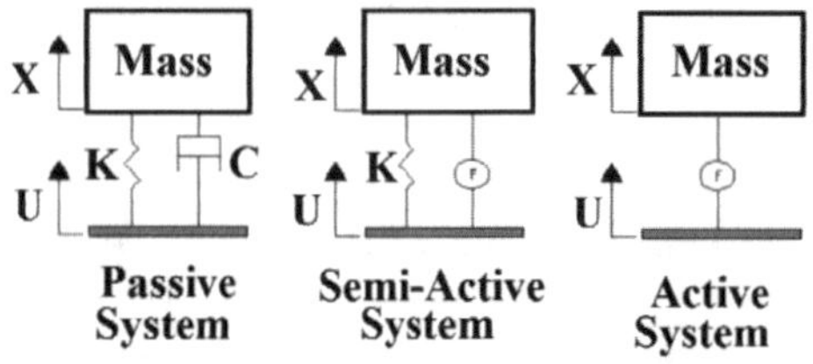

Figure 3. Types of isolation system [9]

Passive vibration isolation, on the other hand, refers to vibration isolation or mitigation of vibrations using passive techniques such as rubber pads or mechanical springs. In its simplest form, it is represented as a combination of a spring having stiffness (K) and a damper having damping coefficient (C), as shown in the figure 3. The spring provides the elasticity to the structure and energy dissipation is provided by the damper. Presently, passive energy dissipation devices are the most widely used vibration protection systems in industries and buildings because of its simple design and low cost [20]. These systems encompass a range of materials and devices for enhancing the damping, stiffness and strength, and can be used both for natural hazard mitigation and for rehabilitation of aging or deficient structures. Contrary to semi-active systems, passive energy dissipation systems do not need an external supply of power in order to operate. Moreover, since energy dissipation devices are not an integral part of the supporting structure, they can easily be replaced for maintenance without compromising the structural integrity [21].

The conventionally designed passive vibration Isolators exhibit linear behavior in both elastic stiffness and damping, which restricts the energy dissipation capability [22]. The design of conventional passive isolators needs a trade-off with stiffness and damping. The low transmissibility over a wide range of frequency can be obtained by reducing the elastic stiffness of the isolator as small as possible, however, such case will lead to large static and quasi-static displacements which are likely to be detrimental to the equipment supported [2].

On the other hand, damping of the isolator needs to be increased to reduce the transmissibility at the resonance which may cause deterioration to the transmissibility over the high frequency range. Moreover, the linear passive isolators are useful only if the excitation frequencies are well above the natural frequency of isolators and thus they are limited to cases which has moderate environmental disturbance. However, under cases such as shock, random ground motions or impact loads the spectrum contains low frequencies which are dangerous to the structures or equipment. Hence the limitations of linear can be overcome by developing a passive vibration isolator which exhibits non-linear behavior [22].

Several authors have developed different types of nonlinear vibration isolators and have investigated the unique dynamic behaviors [23-25]. A comprehensive survey of recent developments of nonlinear vibration isolators has been performed by Ibrahim [22], in which many cited studies [26-31] reveal that the introduction of nonlinear damping and stiffness are of great benefit in vibration isolation. Recently Wire Rope Isolators (WRI), a type of passive isolator which exhibits non-linear behavior in both elastic stiffness and damping, has become the subject of intensive studies [32, 33].

3. Wire rope isolators

Wire rope isolators consist of wire rope strands held between two metal retainer bars in the form of helix shape (Figure 4 (a)) or polycal shape (Figure 4 (b)). As the name implies, wire rope isolators use metal wire rope made up of individual wire strands that are in frictional contact with each other. WRI can provide flexibility in all directions and possess inherent damping characteristic derived from rubbing and sliding friction between the intertwined cables. It is therefore a kind of friction-type damper which adopts stranded wire rope as the elastic component and utilities friction damping (Coulomb damping) between the individual wire strands. This friction dominant property causes the viscous damping in WRI to be relatively insignificant [34, 35].

The advantages of WRI include wide temperature range operations between -100°C to +250°C [36] and less susceptible to the detrimental effects of environmental conditions like extreme temperature, salt, fog, grease, radiation, dust and low manufacturing cost [37]. The polycal type of WRI is primarily used for micro mechanical and electronic applications [37]. For heavy machinery applications the helical WRI is used [38].

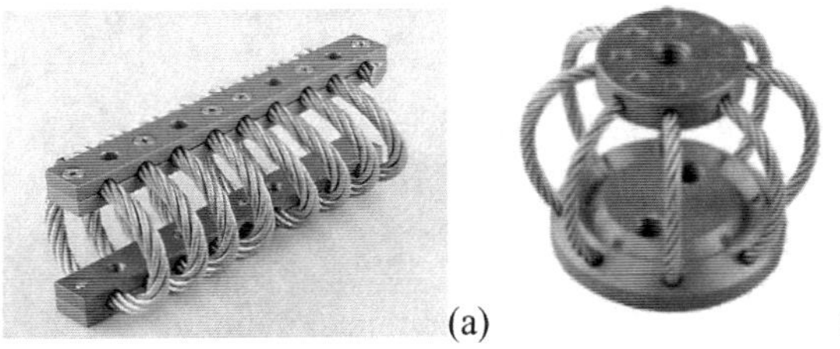

(a) (b)

Figure 4. (a) Helical Isolator [36] (b) Polycal Isolator [36]

4. Characteristics of WRI

The geometric characteristic of the WRI is shown in figure 5. As WRI is a type of passive isolator, it can also be represented in a simplified form as a spring-damper element with static elastic stiffness (K) and damping coefficient (C) (Figure 6(a)) [39]. The static elastic stiffness can be determined from the load-displacement curve under monotonic loading and the damping coefficient can be determined using the hysteresis curve obtained from cyclic loading. Monotonic loading refers to loading in one direction and cyclic loading refers to a set of loading-unloading sequence. There are three loading modes (Figure 6) in which WRI can be used- tension/compression load (along Z), shear load (along X) and roll load (along Y) [32, 39].

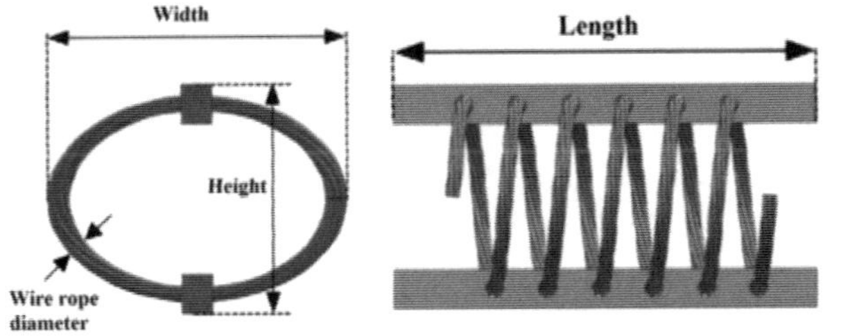

Figure 5. Geometric characteristics of the WRI

5. Selection of WRI

ITT Inc [38] develops a series of vibration and shock insulators of all types and sizes, including WRI and has provided the possible orientations of WRI that can be used in practical applications as shown in figure 7. The orientation of the WRI depends mainly on the supporting structure to fix the WRI. In the majority of the cases, the tension/compression loading mode is preferred. However based on the availability of the supporting structures, the WRI can also be used in the shear and the roll mode. ITT Inc [38] provides the procedure for selection of WRI required for applications. The procedure mainly relies on the calculation of static elastic stiffness (K) of the WRI. Upon estimating the K, based on the weight of the equipment to be supported and disturbing frequency, it is then required to identify a suitable WRI model from the catalogue.

The catalogue contains the values of the static elastic stiffness values of all WRI provided by ITT Inc., which have been experimentally determined under monotonic loading. The static elastic stiffness in all different loading modes are available in the catalogue. The WRI, which possess static elastic stiffness in the loading mode as required for the application will be selected. The selection procedure of WRI is fully based on the monotonic loading response and it should be noted that the behavior under cyclic loading is not considered during the selection process due to a lack of information for this loading type. The research work on the behavior of WRI under monotonic loading and cyclic loading is presented in the next section.

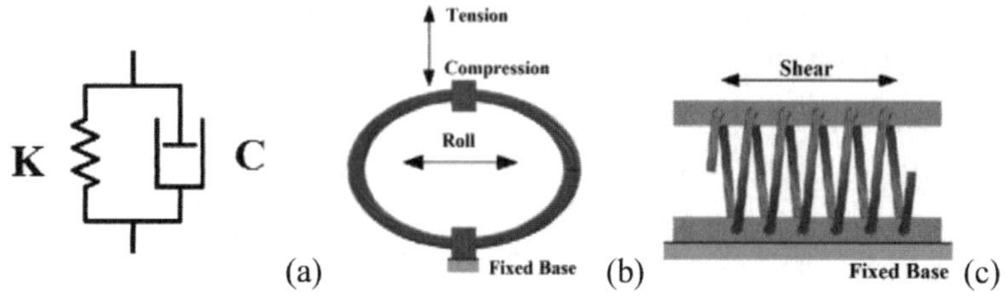

Figure 6. (a) Spring-Damper element (b) Tension/compression and roll, (c) Shear

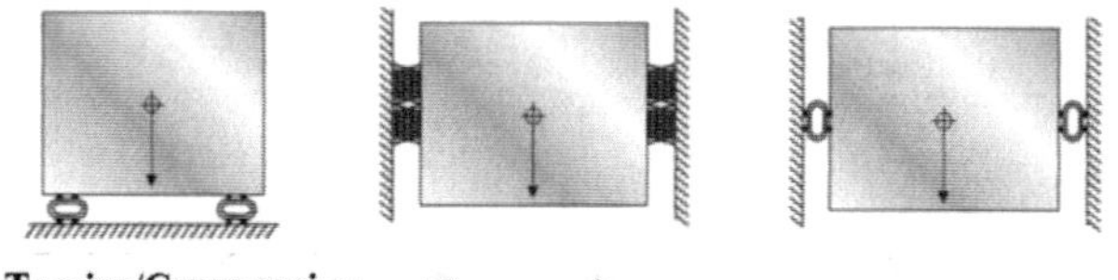

Figure 7. Possible orientations of WRI as suggested by ITT Inc [38]

6. Behavior under monotonic loading

The study of WRI behavior under monotonic loading provides the knowledge on the static elastic stiffness. ITT Inc [38] and Devflex Co [36] develops a variety of WRI in different sizes and stiffness values which facilitates the easy selection of WRI for a particular application. The selection procedure relies mainly on the static elastic stiffness of the WRI and presently the experimental methods are used for its determination. Tinker et al [34, 35] preformed a monotonic shear loading on WRI for the static stiffness (Figure 8). However, a comprehensive study on the static elastic stiffness of WRI is not available in the literatures. All available literatures on WRI were on the cyclic loading behavior. Though the selection procedure relies mainly on static stiffness, presently experimental methods are used to estimate the stiffness. However, an analytical model for the static elastic stiffness will avoid the time consuming experimental work.

Knowledge of the behavior of WRI's post elastic limits enables the prediction of the WRI's capability to sustain further loadings. However, no analytical research on the static elastic stiffness and the post elastic behavior of WRI is available. An accurate analytical model, including all the influencing parameters will greatly assist the appropriate selection and the design of WRI.

7. Behavior under cyclic loading

The study of WRI behavior under cyclic loading can provide the knowledge on isolation capabilities. WRI exhibits hysteresis behavior under cyclic loading due to energy dissipation through coulomb damping. The literature shows only limited research on the hysteresis behavior of WRI using experimental methods [32-35, 39]. A study has been conducted by Demetriades et al. [39] on the hysteresis behavior of the WRI using a cyclic loading test. They show that WRI exhibits symmetrical hysteresis curve for roll and shear loadings as shown in figure 9(a) and asymmetrical hysteresis curve (Figure 9(b)) for tension/compression loading. They also reported that WRI provides a 10% damping for large deformation and about 20% to 30% for small deformation. Tinker and Cutchin [35] studied the damping phenomenon, the stiffness and the hysteresis behavior and reported that the damping in WRI arises from the friction between the individual wire strands.

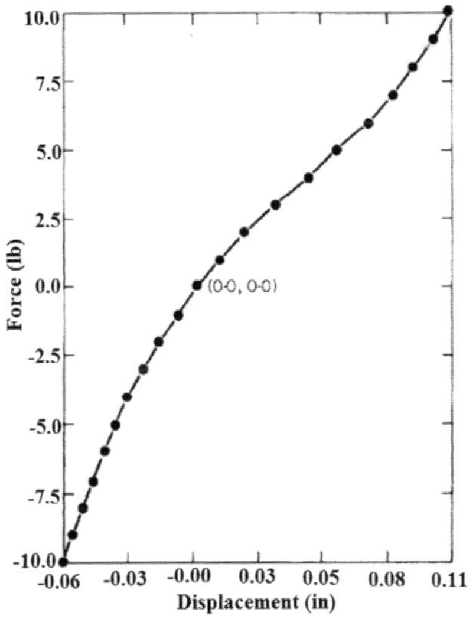

Figure 8. Experimental stiffness curve of WRI under shear load [35]

Schwanen [40] reported that the asymmetric hysteresis curve during tension/compression is due to hardening and softening of the wire rope spring and found that the wire rope spring undergoes softening in compression and hardening in tension. The softening is due to the decrease in the contact points between the wire strands under compression load. The hardening in the tension is due to the increase in the contact points under tension load, which results in increased friction between the wire strands. Massa et al. [33] introduced a ball bearing in the polycal WRI to increase its vertical stiffness. Ball bearing provides additional stiffness in the vertical direction to support the normal load of the equipment and hence increases the load carrying capacity. Paolacci and Giannini [32] conducted a study on the effectiveness of steel cable dampers for the seismic protection of electrical equipment. They developed a numerical model for electrical equipment supported by WRI and subjected it to the seismic load of the 1980 Irpinia earthquake (Italy). The study shows the effectiveness and potential of WRI as a base isolation system

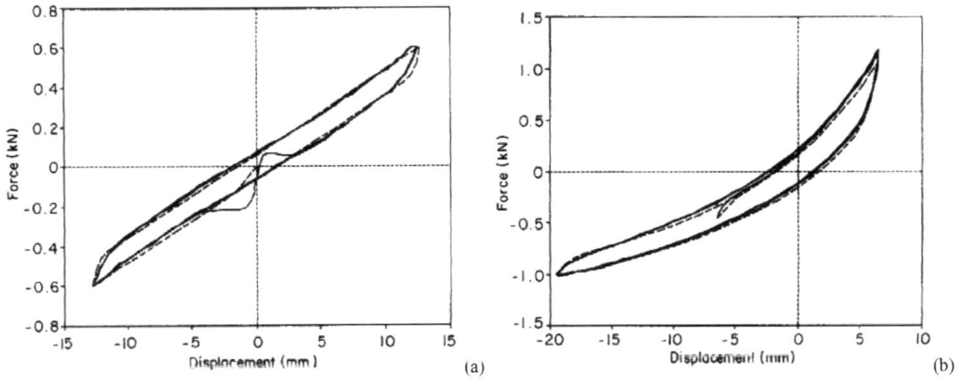

Figure 9. Hysteresis behaviour (a) under shear loading (b) under tension/compression loading [39]

7.1. Bouc-Wen Model

The hysteresis behaviour of the WRI depends on their geometrical properties including wire rope diameter, length and width and the number of turns and loading directions [39]. The area under the hysteresis curves is the energy dissipated by the WRI during the cyclic loading test [34, 35]. A well-developed mathematical model of hysteresis behavior would facilitate the prediction of energy dissipation capabilities and avoids the time-consuming experimental work. The generalized Bouc-Wen (BW) model is commonly used to model the symmetric hysteretic behavior. Many researchers have used the BW model to perform mathematical modelling for the hysteresis system in their area of research work such as magetorhelogical dampers [13, 42, 43] and piezo electric actuators [41]. The generalized Bouc-Wen model of the hysteresis curve is given by

$$Q(x,\dot{x},t) = g(x,\dot{x}) + h(x) \tag{2}$$

where Q is the restoring force , g the non-hysteretic component (Equation 3), an algebraic function of the instantaneous displacement x and the velocity $x^{\cdot}$ and h the hysteretic component (Equation 4), a function of the time history of x [42].

$$g = c\dot{x} + \alpha kx \tag{3}$$

$$h = (1-\alpha)kz \tag{4}$$

where k is the pre-yielding stiffness and α the ratio of post-yielding to pre-yielding stiffness. z is described by the non-linear differential equations of Equation.5,

$$\dot{z} = \left(\frac{1}{\eta}\right)\left[A\dot{x} - \upsilon\left(\beta|\dot{x}||z|^{n-1}z - \vartheta|\dot{x}||z|^{n}\right)\right] \tag{5}$$

In Eq.5 the model parameters A, υ, β, ϑ, η and n govern the amplitude, the shape of the hysteresis loop, and the smoothness of transition from the elastic to the inelastic region. Different choices of model parameters control the shape of the hysteresis curve. When using the Bouc-Wen model for a practical application, it is necessary to perform the identification of the model parameters using system identification techniques. Many researchers have used techniques like the least square method [43], Kalman filter method [44], genetic algorithm based identification [45] to identify the system parameters, so that the output of the model matches as accurate as possible with the experimental data. The symmetrical hysteresis behavior of WRI under shear and roll loading can be modelled using the BW model. However, under tension/compression loading it exhibits asymmetrical behavior and the BW model has to be modified suitably for accurate prediction of this asymmetric behavior.

8. Conclusion

The conventional passive isolators exhibits linear behavior and it needs an improvement to enhance the safety of the isolated system. WRI is a non-linear passive isolator and can provide a better isolation than the conventional passive isolators. The major advantage of WRI is that, it can provide isolation in all three planes and in all orientations. The behavior of WRI can be studied under monotonic and cyclic loading to understand its stiffness and damping behavior. WRI is a newly developed isolation system relatively, when compared with the linear vibration devices hence literature has only few research work. This survey presented the major research work carried on WRI. This paper also highlighted the area for the future scope of work in the field of wire rope Isolators.

Acknowledgment

This paper is based upon the work supported by the Ministry of Higher Education (MOHE), Malaysia under the ERGS grant scheme.

References

[1]. A. R. Klembczyk. 25/04/2014). Introduction to Shock and Vibration Isolation and Damping Systems. website: http://taylordevices.com/Tech-Paper-archives/literature-pdf/74-Introduction_ShockandVibration.pdf

[2]. R. Simmons 2007 Vibration isolation *ASHRAE* **49** 30-40.

[3]. D. P. Taylor 2014 The Application Of Energy Dissipating Damping Devices To An Engineered Structure Or Mechanism. Website: http://taylordevices.com/Tech-Paper-archives/literature-pdf/74-Introduction_ShockandVibration.pdf Accessed on 20/11/2014.

[4]. M. E. Hoque, T. Mizuno, D. Kishita, M. Takasaki, and Y. Ishino, 2010 Development of an Active Vibration Isolation System Using Linearized Zero-Power Control With Weight Support Springs *Journal of Vibration and Acoustics* **132** 041006 01-06 2010.

[5]. A. Preumont, Vibration Control of Active Structures 2002: An Introduction: Springer London, Limited.

[6]. N. Tanaka and Y. Kikushima 1989 On the Hybrid Vibration Isolation Method *Journal of Vibration and Acoustics* **111** 61-70.

[7]. S. Dong, K.-Q. Lu, J. Q. Sun, and K. Rudolph 2005 Rehabilitation device with variable resistance and intelligent control *Medical engineering & physics* 27 249-255.

[8]. K. V.G and T. Ya. V 2011 Active Damping of the resonant vibrations of a flexible cylindrical panel with sensors and actuators *International Applied Mechanics* **47** 720-726.

[9]. D. Karnopp 1995 Active and Semi-Active Vibration Isolation *Journal of Mechanical Design* **117** 177-185.

[10]. J. Alanoly and S. Sankar 1987 A New Concept in Semi-Active Vibration Isolation *Journal of Mechanical Design* 109 242-247.

[11]. W Zhu and X Rui 2014 Semiactive Vibration Control Using a Magnetorheological Damper and a Magnetorheological Elastomer Based on the Bouc-Wen Model *Shock and Vibration* **2014** 1-10.

[12]. Y. Li and J.Li 2014 Dynamic characteristics of a magnetorheological pin joint for civil structures *Frontiers of Mechanical Engineering* **9** 15-33.

[13]. Bhowmik and Subrata 2013 Experimental calibration of forward and inverse neural networks for rotary type magnetorheological damper *Structural Engineering and Mechanics* **46** 673-693.

[14]. A. Spaggiari, E. Dragoni 2012 Efficient dynamic modelling and characterization of a magnetorheological damper *Meccanica* **47** 2041-2054.

[15]. A.M.Tusset and J.M.Balthazar 2012 On the Chaotic Suppression of Both Ideal and Non-ideal Duffing Based Vibrating Systems Using a Magnetorheological Damper *Differential Equations and Dynamical Systems,* **21** 105-121.

[16]. H. Metered, P. Bonello and S.O. Oyadiji 2010 The experimental identification of magnetorheological dampers and evaluation of their controllers *Mechanical Systems and Signal Processing* **24** 976-994.

[17]. H.Xiang, Q.Fang, Z.Gong and Hao Wu 2008 Experimental investigation into magnetorheological damper subjected to impact loads *Transactions of Tianjin University* **14** 540-544.

[18]. A. Dominguez, R. Sedaghati, and I. Stiharu 2008 Modeling and application of MR dampers in semi-adaptive structures *Computers & Structures* 86 407-415.

[19]. S.Zhao,H.deng, J.Zhang, L.Yu and S.Sun 2013 A Novel MR Device with Variable Stiffness and Damping Capability *International Journal of Aerospace and Lightweight Structures* 3 325-334.

[20]. E. I. Rivin, 2003 Passive Vibration Isolation: Professional Engineering Publishing.

[21]. G. Li and H.-n. Li 2009 Pushover analysis method for asymmetric structure with passive energy dissipation devices," **728824** 728824-8.

[22]. R. A. Ibrahim 2008 Recent advances in nonlinear passive vibration isolators *Journal of Sound and Vibration* 314 371-452.
[23]. G. Popov and S. Sankar 1995 Modelling and analysis of non-linear orifice type damping in vibration isolators Journal of Sound and Vibration 183 751-764.
[24]. A.K. Mallik, V. Kher, M. Puri, and H. Hatwal 1999 on the modelling of non-linear elastomeric vibration isolators *Journal of Sound and Vibration* **219** 239-253.
[25]. C.Alhan and H. Gavin 2004 A parametric study of linear and non-linear passively damped seismic isolation systems for buildings Engineering Structures **26** 485-497.
[26]. D.Xu, Q. Yu, J. Zhou, and S. R. Bishop 2013 Theoretical and experimental analyses of a nonlinear magnetic vibration isolator with quasi-zero-stiffness characteristic *Journal of Sound and Vibration* 332 3377-3389.
[27]. L. Jiang, D. Stredulinsky, J. Szabo, and M. W. Chernuka 2002 Numerical characterisation of nonlinear stiffness properties of pre-stressed vibration isolation mounts **30** 70-71.
[28]. M. S. Hundal 1979 Response of a base excited system with Coulomb and viscous friction," *Journal of Sound and Vibration* **64** 371-378 1979.
[29]. S. M. Metwalli 1986 Optimum Nonlinear Suspension Systems *Journal of Mechanical Design 108* 197-202.
[30]. T. A. Nayfeh, E. Emaci, and A. F. Vakakis 1997 Application of Nonlinear Localization to the Optimization of a Vibration Isolation System *AIAA Journal* **35** 1378-1386.
[31]. A.Ravindra and A. K. Mallik 1994 Performance of Non-linear Vibration Isolators Under Harmonic Excitation *Journal of Sound and Vibration* **170** 325-337.
[32]. P. Paolacci and R. Giannini Study of The Effectiveness of Steel Cable Dampers For The Seismic Protection of Electrical Equipment *The 14th World Conference on Earthquake Engineering*, Beijing, China, 2008.
[33]. G. Massa, S. Pagano, E. Rocca, and S. Strano 2013 Sensitive equipments on WRS-BTU isolators Meccanica **48** 1777-1790 2013.
[34]. T. Michael Loyd 1989 Damping Phenomena in a Wire Rope Vibration Isolation System Doctor of Philosophy Aerospace Engineering *Auburn University* Auburn.
[35]. M. L. Tinker and M. A. Cutchins 1992 Damping phenomena in a wire rope vibration isolation system *Journal of Sound and Vibration* **157** 7-18.
[36]. "GGT Series Wire Rope Isolator, WUXI HONGYUAN DEVFLEX CO.LTD, China," www.dpflex.com, Ed., ed, 2014.
[37]. Shock & Vibration Control Wire rope Isolators L. WUXI HONGYUAN DEVFLEX Co., Ed., ed. China, 2014.
[38]. Wire rope Isolators, ITT Enidine Inc, New York, www.enidine.com, 2014.
[39]. G. F. Demetriades, M. C. Constantinou, and A. M. Reinhorn, 1993 Study of wire rope systems for seismic protection of equipment in buildings *Engineering Structures* **15** 321-334.
[40]. W. Schwanen 2004 Modelling and identification of the dynamic behavior of a wire rope spring Master Thesis, Department of Mechanical Engineering, *Technische Universiteit Eindhoven*, The Netherlands, Eindhoven.
[41]. C.-J. Lin and S.-R. Yang 2006 Precise positioning of piezo-actuated stages using hysteresis-observer based control *Mechatronics* 16 417-426.
[42]. Y. K. Wen 1989 Methods of Random Vibration for Inelastic Structures *Applied Mechanics Reviews* **42** 39-52.
[43]. C.-H. Loh and S.-T. Chung 1993A three-stage identification approach for hysteretic systems Earthquake Engineering & Structural Dynamics **22** 129-150.
[44]. L. Jeen-Shang and Z. Yigong 1994 Nonlinear structural identification using extended kalman filter *Computers & Structures* **52** 757-764 1994.
[45]. G. A. Ortiz, D. A. Alvarez, and D. Bedoya-Ruíz Identification of Bouc–Wen type models using multi-objective optimization algorithms Computers & Structures **114-115** 121-132.

Study on Flexible Pavement Failures in Soft Soil Tropical Regions

M. Jayakumar [1]
Department of Civil and Construction Engineering
Curtin University Sarawak Miri, Malaysia

E-mail: m.jayakumar@curtin.edu.my

Lee Chee Soon [2]
Inovatif Jurutera Perunding Sdn Bhd
Kota Kinabalu, Sabah, Malaysia

Email: cheesoon05@gmail.com

Abstract. Road network system experienced rapid upgrowth since ages ago and it started developing in Malaysia during the colonization of British due to its significant impacts in transportation field. Flexible pavement, the major road network in Malaysia, has been deteriorating by various types of distresses which cause descending serviceability of the pavement structure. This paper discusses the pavement condition assessment carried out in Sarawak and Sabah, Malaysia to have design solutions for flexible pavement failures. Field tests were conducted to examine the subgrade strength of existing roads in Sarawak at various failure locations, to assess the impact of subgrade strength on pavement failures. Research outcomes from field condition assessment and subgrade testing showed that the critical causes of pavement failures are inadequate design and maintenance of drainage system and shoulder cross fall, along with inadequate pavement thickness provided by may be assuming the conservative value of soil strength at optimum moisture content, whereas the exiting and expected subgrade strengths at equilibrium moisture content are far below. Our further research shows that stabilized existing recycled asphalt and base materials to use as a sub-base along with bitumen stabilized open graded base in the pavement composition may be a viable solution for pavement failures.

1. Introduction

Surface transport system plays an important role on rapid development in agriculture and industrialization of any countries. A high quality of road network system maximises its benefits to the society and facilitates to achieve the economic growth of the country. Flexible pavement, an asphaltic-pavement that flexes or deflects under vehicular loading, is the major road network in Malaysia. The bituminous road network, which has an approximate total length of 14,757 Km, is excellently bringing conveniences for road users to travel from places to places [1].
In general, the service life period of flexible pavement in bearing traffic loads without suffering any failures is designed as ten to fifteen years [2]. However, incremental of vehicles from time to time has

[1] M. Jayakumar m.jayakumar@curtin.edu.my.

been undoubtedly deteriorating the serviceability of pavement structure, particularly if the pavement is left unconcerned on upkeep and maintenance. Nowadays, critical signals are emerging for the road authorities and road users as it could be seen that pavement failures have been widespread occurring in many places. Common types of flexible pavement failures that could be observed in Malaysia consist of corrugations, depression, ravelling, cracking, potholes, patching etc. [2]. Rather than the vertical stress within all the pavement layers, horizontal stress is only effective on wearing and base course, which is the main cause of fatigue related pavement failures [3]. The crack spacing is strongly dependent on the surface layer thickness, tensile strength of the surface layer and elastic modulus of the surface layer [4]. Excess fines and very high liquid limit values combined with a very low CBR value are instigating the failure of the pavement [5]. Fuzzy regression method is better accounting the uncertainties of traditional method to assess the pavement performances [6]. The combined Overall Pavement Condition Index (OPCI) also is a good indicative of pavement condition and performance [7]. An efficient approach for reliability-based mechanistic-empirical pavement design considering fatigue and rutting failures requires much less computational effort to determine probability of failure and easy to be adopted in the engineering practice [8].
Recycled concrete aggregate and reclaimed asphalt pavement can be used as replacement for natural aggregates in road constructions [9]. Warm–mix asphalt (WMA) and high percentages of reclaimed asphalt pavement (RAP) mixtures expressed better fatigue resistance than HMA-high RAP [10]. As a solution to pavement failure, cement stabilized fly ash class F may be utilized with or without aggregate as a pavement layer [11]. Understanding on the types of pavement failures and investigating the relevant causes are the primary approaches for obtaining the effective solutions to reduce the deterioration and to provide a longer pavement lifespan.

2. Pavement condition assessment

Field condition assessment was conducted in Sarawak and Sabah states of Malaysia at ten predominant routes having heavy commercial vehicle traffic and various subgrade soil strata. In this phase, the total length of each road was taken as 4km where 40 segments had been taken into investigation. The results of the assessment were summarized for every 100 meter stretch in each lane. The study mainly focused in observing and recording the types of flexible pavement failures along each road, in which four major types of pavement distresses were classified, known as surface deformation, surface defects, cracking, potholes and patching. In order to further describe the condition of visible distresses recorded, scoring method as detailed in Table-1 was put into practice by rating the pavement distresses using different level of scores based on the severity and extent of occurrence observed during the assessment.

Table 1. Degree of pavement distress

Degree	**Severity**	**Description**
0	-	No distress Visible
1	Slight	Slight-(No attention required immediately)
2	Between slight and warning	Between slight and warning
3	warning	Warning-distress is distinct. Start of secondary defect
4	Between warning and severe	Between warning and severe
5	Severe	Severe – Distress is extreme. Secondary defects are well developed. Urgent attention required

Other significant aspects that had been taken into inspection during the assessment are road shoulder conditions for its maintained cross fall and adequacy of drainage system for its effectiveness because both of the elements play an integral role in maintaining the behaviour of pavement in terms of serviceability and structural strength. These were observed and judged based on the capability of the pavement structure to drain water away from the asphaltic-wearing surface to prevent any water accumulation on the surface as well as on the shoulder edges. The existing drainage was taken into inspection to identify whether it is normal expected water flow level is adequately lowered from the pavement level to avoid the return moisture to pavement structure.

3. Existing pavement strength

Field tests and laboratory tests on collected soil samples were conducted to assess the existing strength of pavement structure at outer wheel path zones. The conducted field tests were proving ring penetrometer test and vane shear test, whereas California Bearing Ratio (CBR) test was carried out at laboratory.

Proving Ring Penetrometer test and Vane Shear Test were used to determine the penetration resistance and shear strength of the subgrade soil respectively. Both of these tests were carried out at outer wheel path zone nearer to the shoulder of pavement. For the former test, a penetrometer was used to test subgrade soil in three different locations and an average result was taken from the three readings as the penetration resistance of that particular soil. A similar procedure was applied in vane shear test where a vane inspection tester was used to test subgrade soil in three different locations at outer wheel path zones. Apart from the top surface, additional two readings were taken in each location at depth of 0.5m and 1.0m from the top surface.

The most important subgrade strength parameter is CBR result, which was obtained from laboratory CBR test. It gives an indication of bearing capacity of the subgrade soil. Bulk sample of subgrade soil for each road were collected at depth of 500mm from the top of road base. The test was carried out in accordance with specifications in Australian Standard AS 1289.6.1.1-1998. For unsoaked-CBR test, load penetration was conducted on a particular soil sample after the sample was oven-dried for 24 hours and compacted according to the specifications. For soaked-CBR test, the sample was soaked in water for 4 days before taken into load penetration test. The focus in this test is the effect of moisture content on the variation of CBR strength of the soil sample.

4. Results and discussion

4.1. Pavement Condition Assessment

The distress types and its distribution for five roads in Miri are summarized and presented in Figure 1 and Table 2. Common types of distress recorded along study area in Miri were cracking, potholes, patching, corrugation, depression, raveling and delamination. Among all the distresses, corrugation and cracking contribute the highest distribution percentage of occurrences, which are 35% and 30.5% respectively. The chart in Figure 1 indicates that Road 2 is suffering of the highest occurrences distresses whereas Road 1 is of satisfactory condition having the lowest percentages of failures. It is seen that the pavement surface of Road 4 is currently in undulating condition, which is known as corrugation, a type of pavement deformation.

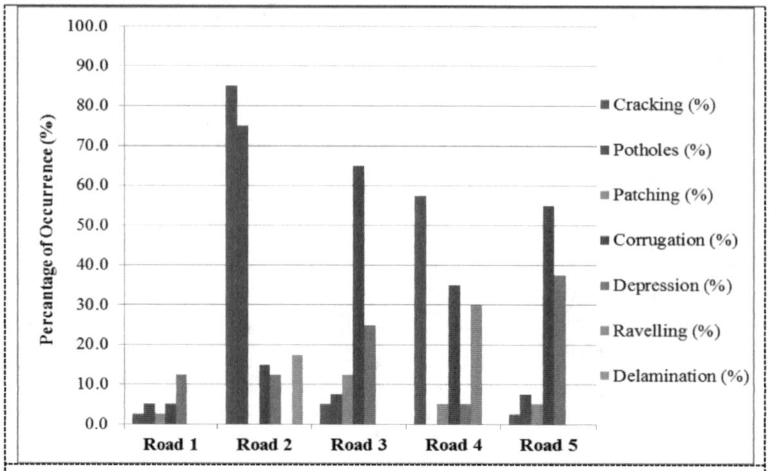

Figure 1. Summary of Distresses and its Distribution (Miri)

Table 2. Summary of distress type and its distribution (Miri)

Distress Type	Road 1	Road 2	Road 3	Road 4	Road 5	Total
Length (km)	4	4	4	4	4	20
Cracking	1	34	2	23	1	61
(%)	2.5	85.0	5.0	57.5	2.5	30.5
Potholes	2	30	3	0	3	38
(%)	5	75	7.5	0	7.5	19
Patching	1	0	5	2	2	10
(%)	2.5	0	12.5	5	5	5
Corrugation	2	6	26	14	22	70
(%)	5	15	65	35	55	35
Depression	5	5	10	2	15	37
(%)	12.5	12.5	25	5	37.5	18.5
Ravelling	0	0	0	12	0	12
(%)	0	0	0	30	0	6
Delamination	0	7	0	0	0	7
(%)	0	17.5	0	0	0	3.5
Total No. of Segments (100m stretch in each lane)	40	40	40	40	40	200

The distress types and its distribution for five roads in KK, No. 6-10 are summarized and presented in Figure 2 and Table 3. Among all the distresses, cracking and patching contribute the highest distribution percentage of occurrences, which are 54% and 40% respectively. Figure 2 shows that Road 8 is of poor quality of pavement condition due to high occurrences of failures. Conversely, the lowest occurrence of distresses was observed on Road 10.

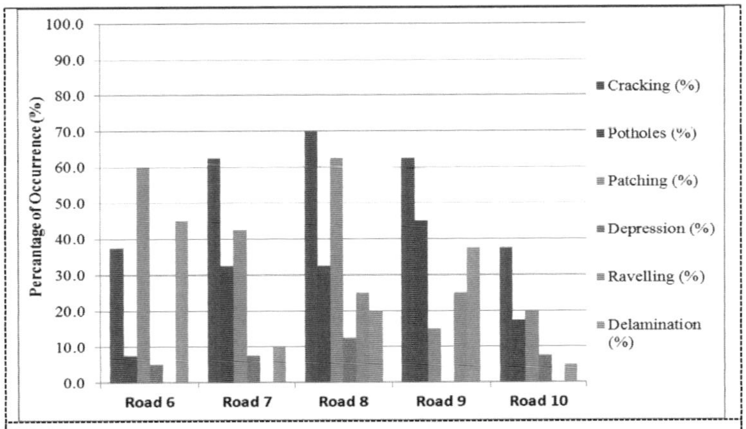

Figure 2. Summary of Distresses and its Distribution (KK)

Table 3. Summary of distress type and its distribution (kk)

Distress Type	Road 6	Road 7	Road 8	Road 9	Road 10	Total
Length (km)	4	4	4	4	4	20
Cracking	15	25	28	25	15	108
(%)	2.5	85.0	5.0	57.5	2.5	54
Potholes	3	13	13	18	7	54
(%)	7.5	32.5	32.5	45	17.5	27
Patching	24	17	25	6	8	80
(%)	60	42.5	62.5	15	20	40
Depression	2	3	5	0	3	13
(%)	5	7.5	12.5	0	7.5	6.5
Ravelling	0	0	10	10	0	20
(%)	0	0	25	25	0	10
Delamination	18	4	8	15	2	47
(%)	45	10	20	37.5	5	23.5
Total No. of Segments (100m stretch in each lane)	40	40	40	40	40	200

4.2. Proving Ring Penetrometer Test Results

The field test results for Road No. 1-6 are tabulated in Table 4. The obtained result shows that that Road 2 has the lowest penetration resistance of 394.66kPa corresponding to 0.255kN of penetration load. Conversely, the penetration load and penetration resistance are the highest in Road 4, which were recorded as 0.412kN and 1130.05kPa respectively. This shows that the subgrade soil of Road 4 is having relatively high bearing capacity in resisting the applied load. It was justified by the extent of penetration of penetrometer into the shoulder surface during the field test, the penetrometer could hardly penetrate into the surface. Large force was required to push the penetrometer into the ground and only 3.5cm of the cone was embedded into the soil.

Table 4. Summary of proving ring penetrometer test results

Road	Penetration Load (kN)	Penetration Resistance (kPa)
1	0.313	485.22
2	0.255	394.66
3	0.305	835.79
4	0.412	1130.05
5	0.591	915.36

4.3. Vane Shear Test

Shear strength results at various depth obtained from each road subgrade are presented in Table 5. It is shown that the shear strength increases when the vane instrument penetrates into ground and turned at deeper layer. However, there was a special case for Road 1, in which the shear strength decreases from depth of 0.5m to 1m. This could be caused by the presence of water layer at certain level between the range of 0.5m and 1m that weakens the soil and decreases the soil shear strength. In comparisons of all the roads, Road 2 soil has the lowest shear strength of 22kPa whereas Road 3 soil has the highest shear strength of 184kPa. For Road 2, the vane instrument was relatively easy to be pushed and rotated uniformly into the ground. Yet, the maximum shear strength is only 70kPa at 1.0m below the ground surface. For Road 3 and 5, the soil has very high shear strength that goes up to 260kPa and 240kPa respectively at 0.5m below the surface. During the field test, the instrument was hardly to be pushed and rotated in the soil beyond 0.5m depth.

Table 5. Summary of vane shear test results

	Shear Strength kPa		
Road	Surface	0.5m	1.0m
1	138	222	146
2	22	49	70
3	184	260	-
4	61	104	229
5	44	240	-

4.4. Sand Cone Replacement field density Test(FDT)

The FDT results are shown in Table 6. It shows that Road 2 clay subgrade has the highest bulk density of 14.15kN/m3 at moisture content of 20.47%, with dry density of 11.75kN/m3 and Road 5, silt subgrade, has the lowest moisture content 9.78%. The soil sample for Road 2 was collected during rainy day whereas for Road 5 soil sample was collected under sunny weather. The significant degradation in dry density may be due to infiltration of rain water into pavement.

Table 6. Summary of sand cone replacement test results

Road	Bulk Density (kN/m3)	Dry Density (kN/m3)	Moisture Content
1	12.82	11.28	13.65
2	14.15	11.75	20.47
3	13.75	12.42	10.76
4	11.89	10.44	13.91
5	13.08	11.92	9.78

4.5. Standard Proctor Compaction Test

The standard proctor compaction test results are summarized and shown in Table 7. Road 2 has the lowest optimum moisture content (OMC) of 9% and maximum dry density (MDD) of 16.35kN/m3 and Road 3 has the highest OMC of 12.4% and MDD of 16.87kN/m3. The degree of compaction was computed for each sample using the field dry density and maximum dry density from compaction curve. The degree of compaction for all the soils lies within the range of 70% which vary significantly from the general range of relative compaction of 90% -100%. This variation in the degree of compaction of all the road subgrades could be caused by intrusion of water into the soil that decreases the field dry density.

Table 7. Summary of standard proctor compaction test results

Road	Optimum Moisture Content, OMC (%)	Maximum Dry Density, MDD (kN/m3)	Degree of Compaction (%)
1	11.0	16.06	70.26
2	9.0	16.35	71.86
3	12.4	16.87	73.60
4	10.2	15.55	67.14
5	12.0	16.50	72.23

4.6. California Bearing Ratio (CBR)Test

The laboratory CBR test results are shown in Table 8 to 10. The unsoaked CBR test results shows that Road 2 clay subgrade has the highest CBR of 60.65% and dry density of 14.76kN/m3 at the lowest moisture content of 10.16%. As a result of soaking for 4 days, Road 2 soil shows considerable amount of reduction in CBR of 52.20% with only 2.82% increase in moisture content. In soaked-CBR test, it has the lowest CBR of 8.45% and highest dry density of 14.75kN/m3 at the moisture content of 12.98%. The result indicates that Road 2, clay subgrade is problematic in bearing capacity as a minor increase in water content leads to substantial amount of reduction in CBR.

On the contrary, Road 4, silt subgrade shows less water sensitiveness, the large increase in water content (9.95%) causes only a minor reduction in CBR (2%). The unsoaked and soaked CBR values for Road 4 are 32.88% and 30.88%. Figure 3, a comparison chart shows the variation between unsoaked CBR and soaked CBR for each road subgrade soil.

Table 8. Summary of unsoaked-CBR test results

Road	Moisture Content (%)	Dry Density (kN/m3)	CBR (%)
1	11.02	13.596	17.77
2	10.16	14.76	60.65
3	12.11	14.36	16.30
4	12.06	11.93	32.88
5	12.19	12.653	19.26

Table 9. Summary of soaked-CBR test results

Road	Moisture Content (%)	Dry Density (kN/m3)	CBR (%)
1	11.29	13.595	16.49
2	12.98	14.75	8.45
3	21.50	13.97	8.51
4	22.01	12.05	30.88
5	14.63	12.650	17.15

Table 10. Relationship between moisture content and CBR

Road	Increase in Moisture Content (%)	Loss of CBR (%)
1	0.27	1.28
2	2.82	52.20
3	9.39	7.79
4	9.95	2.00
5	2.44	2.11

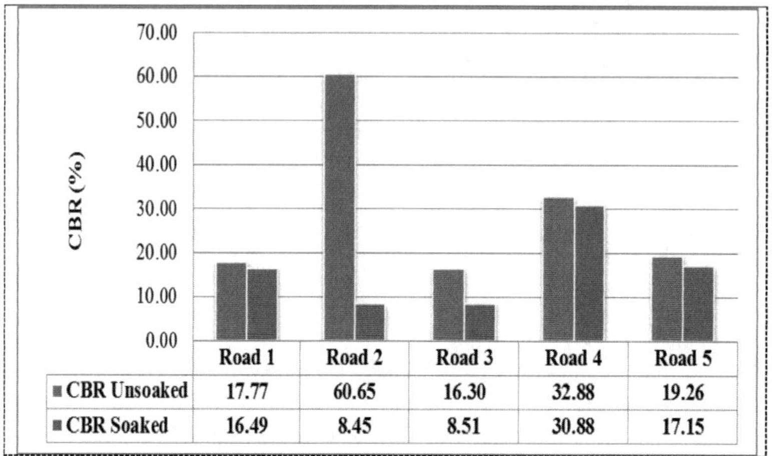

Figure 3. Comparison chart for Soaked and Unsoaked CBR of all of the Road subgrades

4.7. Resilient Modulus and Resistance Value

In addition to CBR, resilient modulus (Mr) and resistance value (R-Value) are the important parameters in evaluating the stiffness of subgrade soils and the capability of a soil structure in resisting lateral distribution of pressure from applied vertical loads. The Mr and R-Value for each of the road subgrade were computed by the equations [12] and summarized in Table 11.

$$Mr\ (psi) = 1500*CBR \quad (1)$$

$$R = (2555*CBR^{0.64} - 1555)/555 \quad (2)$$

Where Mr = Resilient Modulus in psi, or 0.00689MPa
CBR = Laboratory soaked CBR in %
R = Resistance Value

Road 2 clay subgrade has the lowest R-value of 15.94 whereas Road 4 silt subgrade has the highest value of 39.22. Based on the R-value for each subgrade, further classification of soil type can be identified. The R-values show that subgrade soil of Road 2 and Road 3 is sandy clay, whereas subgrade soil of Road 1, Road 4 and Road 5 is sandy silt. In overall, all of the road subgrade soils consist of sand constituents. The overall results of R-value indicate that sandy clay subgrade is having low resistance against vertical stress from traffic load. Furthermore, clay subgrade soils for Road 2 and Road 3 has the lowest Mr, 87.39MPa and 88.01MPa respectively.

Table 11. Correlation between CBR with resilient modulus and resistance-value

Road	CBR (%)	Resilient Modulus, Mr (MPa)	Resistance-Value	CBR (%)
1	16.49	170.55	25.56	Sandy Silts
2	8.45	87.39	15.94	Sandy Clay
3	8.51	88.01	16.02	Sandy Clay
4	30.88	319.38	39.22	Sandy Silts
5	17.15	177.37	26.27	Sandy Silts

5. Summary of Failures

While summarizing the results from all tests, nature of subgrade soil is considered as one of the factors driving to pavement failures. This is justified in Road 2, having the low soil strengths, including 49kPa of shear strength, 394.66kPa of penetration resistance and CBR of 8.45% which is less than CBR of 10%, deemed necessary to adequately support the heavy traffic load without causing severe deformation [13].

In addition to subgrade properties, inadequate and poorly maintained cross fall of shoulders and drainage systems are causing pavement failures. The pavement and shoulder condition observed at one of the junction in Road 5 during our survey is shown in Figure 4

The condition of drainage structure is also one of the decisive causes of pavement failures. The main issue concerned is blockage in drainage structure due to deposition of silts, presence of excess grasses and waste products. One of the blocked drains observed at Road 2 is shown in Figure 5.

Figure 4. Accumulation of water on turning junction of Road 5

Figure 5. Blockage in part of a longitudinal drain in Road 2

The effects of traffic loading could not be underrated along the entire life span of pavement. The heavy vehicle rolling loads are the critical causes for fatigue and rutting failures. Figure 6 shows the multi-axle heavy vehicle observed on Road 9 in an industrial area. Figure 7 shows the transportation of timber materials by heavy-loaded vehicles from Miri to Bintulu along Road 2. The pavement structures will be damaged significantly if they are not designed to carry the existing actual traffic loading capacity [14]. The soil sample collected in Road 2 showed large reduction of CBR in lab test for increasing very small amount of moisture content. The strength of the sensitive clay is highly susceptible to effect of moisture. The bearing capacity of sensitive clay would also suffer and has a constant reduction when the structure is subjected to cyclic loadings [15].

Figure 6. Multi-axle heavy vehicle on Road 9

Figure 7. Heavy loaded vehicles were transporting timbers along Road 2

6. Conclusion and recommendation

The study results show that subgrade strength, water penetration in pavement, defective condition of shoulder and drainage system, as well as under rated traffic loadings are the influential factors for failures. In tropical regions excessive infiltration of precipitated water into pavement layers and retaining for longer period due to inadequate drainage capacity is a critical cause for the pavement failures. Another main cause may be of providing inadequate pavement thickness by considering very conservative values of soil strengths in design stage by assuming the condition of optimum moisture content whereas actual soil strength during the life span are for behind due to prevailing equilibrium moisture content particularly at outer wheel path zones. Based on this study and our continuing research study on the performance of pavement structures and its life span, we suggested the following design and construction recommendations.

As a viable solutions for most of the deceive flexible pavement in Sarawak and Sabah, cut the existing asphalt surface along with base material for the minimum of 250 mm depth, recycle with required amount of stabilizing materials such as cement, lime or mixture of fly ash and lime and lay as a subcase. The open graded bitumen stabilized natural gravel or crushed rock by removing the fines up to 2mm size for better permeability, is recommended as a road base. Particularly in the tropical regions for having a consistent design drainage capacity of shoulders, always maintain its cross fall, avoid planting trees or bushes at the edge of the shoulder before table drains. The trees may be planted in the reserve after the table or storm water drains.

References

[1] Mohd Isa, Abdul Hamid Mohd, Dadang Mohamed Ma'some, and Tiek Hwa Law, "Pavement Performance Model For Federal Roads," Proceedings of the Eastern Asia Society for Transportation Studies 5: pp.428 – 440, 2005

[2] Kordi, Nurul Elma, Endut Intan Rohani, and Baharom Bahardin, "Types of Damages on Flexible Pavement for Malaysian Federal Road." Malaysian University Transportation Research Forum and Conferences (978-967-5770-08-1): pp.421-432, 2010.

[3] H. Akbulut and K. Aslantas, "Finite element analysis of stress distribution on bituminouspavement and failure mechanism", Materials and Design 26, pp. 383–387, 2005.
[4] Chang Xu, Yan Anzhi, Mingcheng Liao and Tang Chunan, "Analysis of periodic cracks in surface layer of pavement structures", Engineering Failure Analysis 18, pp. 411–420, 2011.
[5] G. Jegede, "Effect of soil properties on pavement failures along the F209 highway at Ado-Ekiti, south-western Nigeria", Construction and Building Materials 14, pp. 311-315, 2000.
[6] Nang-Fei Pa, Chien-Ho Ko, Ming-Der Yang and Kai-Chun Hsu, "Pavement performance prediction through fuzzy regression", Expert Systems with Applications 38, pp. 10010–10017, 2011.
[7] Yogesh U.Shaha, S.S. Jainb, Devesh Tiwaric and M.K. Jaind, "Development of Overall Pavement Condition Index for Urban Road Network", Procedia - Social and Behavioral Sciences 104, pp. 332 – 341, 2013.
[8] Zhe Luo, Feipeng Xiao and Radhey Sharma, "Efficient reliability-based approach for mechanistic-empirical asphalt pavement design", Construction and Building Materials 64, pp. 157–165, 2014.
[9] Harianto Rahardjo, Alfrendo Satyanaga, Eng-Choon Leong and Jing-Yuan Wang, "Unsaturated properties of recycled concrete aggregate and reclaimed asphalt pavement", Engineering Geology 161, pp. 44–54, 2013.
[10] Sheng Zhao, Baoshan Huang, Xiang Shu and Mark Woods, "Comparative evaluation of warm mix asphalt containing high percentages of reclaimed asphalt pavement", Construction and Building Materials 44, pp. 92–100, 2013.
[11] A. Hilmi Lav, M. Aysen Lav and A. Burak Goktepe, "Analysis and design of a stabilized fly ash as pavement base material", Fuel 85, pp. 2359–2370, 2006.
[12] Hashiro, S. Reyn,. "Correlation of Resistance Value (R-value) with California Bearing Ratio (CBR) For Use in the Design of Flexible Pavements." Master's thesis, University of Hawaii Library, 2005.
[13] M. Y., Ahmed, A. H. Nury, F. Islam, and M. J. B. Alam, "Evaluation of geotechnical properties and structural strength enhancing road pavement failure along Sylhet-Sunamganj highway, Bangladesh". Journal of Soil Science and Environmental Management 3(2141-2391): pp.110-117, 2012.doi: 10.5897/JSSEM12.024.
[14] Croney, P., and D. Croney, "The Design and Performance of Road Pavements". New York City: McGraw-Hill Professional, 1998.
[15] Javed, "Behavior of Sensitive Clay Subjected to Static And Cyclic Loading".PhD,Thesis, Concordia, Canada, 2012.

IOP Conf. Series: Materials Science and Engineering **78** (2015) 012003 doi:10.1088/1757-899X/78/1/012003

Review on Metallic and Plastic Flexible Dye Sensitized Solar Cell

A R Yugis, R F Mansa, C S Sipaut
Energy and Materials Research Group (EMRG), Faculty of Engineering, University Malaysia Sabah, UMS Road, 88400 Kota Kinabalu, Sabah, Malaysia

Email: AUGUSTUS_NEO@yahoo.com

Abstract. Dye sensitized solar cells (DSSCs) are a promising alternative for the development of a new generation of photovoltaic devices. DSSCs have promoted intense research due to their low cost and eco-friendly advantage over conventional silicon-based crystalline solar cells. In recent years, lightweight flexible types of DSSCs have attracted much intention because of drastic reduction in production cost and more extensive application. The substrate that used as electrode of the DSSCs has a dominant impact on the methods and materials that can be applied to the cell and consequently on the resulting performance of DSSCs. Furthermore, the substrates influence significantly the stability of the device. Although the power conversion efficiency still low compared to traditional glass based DSSCs, flexible DSSCs still have potential to be the most efficient and easily implemented technology.

1. Introduction of dye sensitized solar cell

The prosperity of human society largely relies on safe energy supply, and fossil fuel has been serving as the most reliable energy source. However, as a non-renewable energy source, the exhaustion of fossil fuel is inevitable and imminent in this century. To address this problem, renewable energy especially solar energy has attracted much attention, because it directly converts solar energy into electrical power leaving no environment affect. Among all the organic solar cells, dye-sensitized solar cells (DSSCs) are the most efficient and easily implemented technology [1].

Dye-sensitized solar cell (DSSC) has been under intense investigation since the first publication reported by O'Regan and Grätzel in 1991[2]. The possibility to produce these solar cells at low cost and their relatively high energy conversion efficiency, reaching 10.3% ± 0.34% (value for a 1cm^2 cell) [3] make such devices promising alternatives for cheap electricity compared to the classical silicon-based photovoltaic panels. Unlike silicon solar cells, electrons and holes in a DSSC are transported in two different phases, Titanium dioxide (TiO_2) and electrolyte respectively. Thus, the chances of recombination in the cell become low. In addition, DSSCs have been proved to perform better than conventional solar cells, without significant change in the power conversion efficiency, in diffused light and at moderate temperature – up to 50°C.Moreover, these cells can be made on flexible substrates and thus find applications on curved surfaces like clothes, bags, car tops, etc. Transparency of these cells and their performance under the diffused light make these cells more attractive for indoor applications and as panels on windows and doors.

In this review, we focus our study on operation principal and configuration of flexible DSSCs, cost overview of flexible DSSCs, recent achievement in flexible DSSCs and stability issues related to metal and plastic DSSCs.

2. Operation principle and structure of flexible dye sensitized solar cells

The DSSCs is a new type of photoelectrochemical solar cell composed of a dye-modified wide band semiconductor electrode, a counter electrode, and an electrolyte containing a redox couple (I^-/I_3^-).When a DSSC is illuminated by sun light, the dye molecules absorb light and become excited. The absorption of light by the dyes is followed by the injection of an electron from the excited state of the dye to the conduction band of the semiconductor and its subsequent transfer to the transparent oxide. Finally the electron flows through the external circuit. The oxidized dye is simultaneously reduced by a redox mediator in the electrolyte and returns to the ground state. The electrons that flow through the external circuit arrive at the counter electrode and have an effect on the reverse reaction of the redox couple. Radiant energy is converted to electricity by this process [4 - 6].

The DSSCs fabrication can be switched from traditional glass substrates to either flexible plastic or metal and can be fabricated in different combination. As indicated in Figure. 1, the flexible DSSCs can be prepared in three basic configurations. There are key challenges associated with each type of configuration. For instance, the whole cell assembly can be fabricated on transparent conducting oxide coated flexible plastic sheets like traditional glass based DSSC as Figure 1(a) if the materials could be deposited or coated using low temperature paste or inks. On the other hands, metals also can be used as either counter electrode or photo electrode as figure 1(b) and 1(c). As all know metal can stand high temperature, the high temperature deposition of paste or ink in not the issue, but configuration must be different from conventional DSSCs assembly which is the counter electrode on top and photo electrode on bottom of the DSSCs fabrication. However, this configuration needed careful optimization to achieve maximum output power since the cell is illuminated from the counter electrode side and the light is absorbed by semi-transparent catalyst and electrolyte. In case the metal used as counter electrode, low temperature paste or ink deposition needed to be coated on plastic photo electrode.

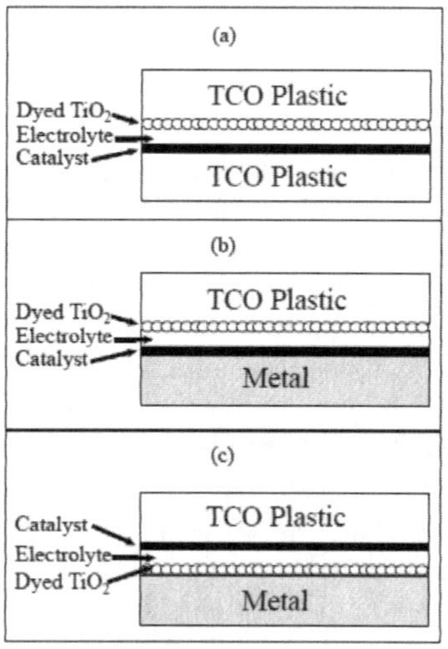

Figure 1. Different configuration of flexible DSSCs

3. Cost overview of flexible dye sensitized solar cells

A key issue in the commercialization of DSSCs is bringing down the production cost further while keeping up a decent efficiencies and lifetime. The total expense of volume DSSCs production principally relies on upon the materials cost and the fabrication methods. The rigid glass substrate is the most costly part in a DSSCs which is in excess of 60% of the aggregate expense of all parts in DSSCs[7].Also, overall cost can be decreased by replaced with cheap and flexible substrates such as plastic, metal or even paper[7;8]. Another advantage of flexible DSSCs is avoiding the cost for inactive materials such as metallic frames, strut and bolds and additional glass encapsulation used in the installation of silicon solar cells [9]. Another favorable aspect for the DSSCs is the cost of its manufacturing tools and equipment. For a 10 MW/year manufacturing capacity, the production costs for glass based DSSCs has been estimated in between $\$7.25/m^2$ - $\$ 11.60/m^2$[7] compared to $\$ 32/m^2$ [9] production cost for thin film solar cells. Hence, by fabricating DSSCs on light weight and flexible substrates and also implementing roll-to-roll processing, large cost reductions could be expected even be halved compared to glass based batch process DSSCs technology. In case of DSSCs, which is mostly still in a laboratory scale phase, scaling up to mass production should most likely bring down the cost further through different learning and scale effects in production. The easy and non-vacuum based fabrication equipment for DSSCs may potentially further reduce the overall cost of DSSCs production.

4. Flexible photo electrode

Flexible photo electrode can be fabricated by utilizing fast roll-to-roll fabrication devices. The metal based photo electrode offers an advantage for high temperature sintering processes which is a key requirement for high quality TiO_2 nano particle film. These high temperature TiO_2 paste have been employed in reverse illumination configuration of DSSCs and has achieved an efficiency of 8.6 % for stainless steel based substrate. However in the case of stainless steel, the stability is a critical problem as the cell can lose performance in couple of hour under sun illumination [22;23].So far, Titanium is among metal based photoelectrode that show good stability under sun[22]. The highest efficiencies been report on Titanium based flexible DSSCs is 9.2 % .On the other hand, the high temperature sintering process cannot be applied on plastic substrates such as polyethylene terephthalate (PET) or Polyethylene naphthalate (PEN) because deformation at 150°C. Transferring the high temperature sintered TiO_2 film has been demonstrated through 'lift off' process and mechanical compressing. For this technique, efficiency of cell is 7.6 % been achieved. The key issue related to low temperature paste or ink preparation is the weak inter particle bonding between individual TiO_2 particles which is basically obtained through high temperature sintering process. Due to that problem, the movement of electron within TiO_2 film experiences high resistance and cause lower current compared to high temperature sintering film. Several researchers have been demonstrated different low temperature methods such as chemical sintering, vacuum sealing and pressing, hydrothermal treatment, UV treatment, spray deposition methods. Efficiency level of 2 -7 % has been obtained using those method which is still lower compared to the glass based DSSCs.

Table 1. Performance of DSSC with different materials as photoelectrode

Substrate	Preparation method	Efficiency, η (%)	Reference
Ti[a]	Surface treatment	9.2	[24]
StSt[b]	Surface treatment and TiO underlayer	8.6	[10]
W[c]	ITO interlayer	3.6	[36]
Zn[d]	ZnO nanorod	2.4	[37]
ITO-PEN[e]	Mechanical pressing and UV treatment	7.6	[11]
ITO-PEN[e]	Vacuum seal and pressing	6.3	[12]
ITO-PEN[e]	Hydrothermal and mechanical pressing	5.8	[11]
ITO-PET[f]	Electrophoretic deposition and CVD/UV	3.8	[25]
ITO-PET[f]	Spray deposition and UV treatment	3.3	[27]
ITO-PET[f]	Hydrothermal crystallization and autoclaving	2.5	[26]

[a]Ti: Titanium. [b]StSt: stainless steel. [c]W: tungsten. [d]Zn: Zinc. [e]ITO-PEN:Polyethylene Naphthalate doped Indium Titanium Oxide. [f] ITO-PET: Polyethylene Terephthalate doped Indium Titanium Oxide

5. Flexible counter electrode

Different catalyst layer has been tested by several researchers such as Platinum (Pt), Carbon composites and conducting polymer such as poly(3,4-ethylenedioxythiophene)(PEDOT) and polystyrene sulfonate (PSS). Among all the materials, Pt can be deposited through various low temperature processes such as chemical reduction, electrodepostion and mechanical pressing which is suitable for plastic substrate. For plastic substrate, efficiency level of 5 - 8 % has been obtained using those methods. In the case of metal based counter electrode, using low temperature methods such as sputtering in preparation of the Pt has been a much better option compared to high temperature treatment [19;20]. This has been thought to be caused by the increased oxidation of the metal surface in high temperatures which would hinder the charge transfer. However, the selection of viable deposition methods for a catalyst in the case of metals is larger than with plastic. For metal substrate such as stainless steel, the best efficiency is 5.3 % been achieved with Pt as catalyst, which is relatively poor compared to glass substrate DSSCs. Furthermore, the DSSC fabricated using stainless based counter electrode with thermal deposition corroded [28]. Ti exhibits good chemical stability with electrolyte compared to stainless steel. The best conversion efficiency of DSSC obtained with Ti based counter electrode is 7.7%. However, to reach lower cost fabrication of DSSC, alternative materials have been investigated such as conducting polymer. Conducting polymers such PEDOT-PSS have been tested as alternative catalyst via spin coating and screen printing on polymer. The best efficiency for these polymers on plastic substrates is about 4.4%. For metal based substrate, with Ti as counter electrode with PEDOT as catalyst, 6.3 % efficiency has been reached. Stainless steel based counter electrode also work well with carbon by applied composite film made of carbon aerogel and poly(tetrafluoroethyelene) and in such case give efficiencies as high as 9.2 %. As corrosion is the main challenge for stainless steel based counter electrode, the next step was to move toward more corrosion resistant metals.

Table 2. Performance of DSSC with different materials as counter electrode and different catalyst

Substrate	Catalyst	Preparation method	Efficiency, η (%)	Reference
StSt[a]	Carbon	High temperature sintering	9.2	[14]
Ti[b]	Pt[c]	High temperature sintering	7.7	[29]
Ti[b]	PEDOT[d]	Electrodeposition	6.3	[21]
StSt[a]	Pt[c]	Sputtering	5.3	[30]
Ni[e]	Pt[c]	Sputtering	5.1	[33]
Ni-PET[f]	Pt[c]	Electrodeposition and film transferring	7.9	[15]
ITO-PET[g]	Pt[c]	Screen print and pressing	7.2	[16]
ITO-PEN[h]	Pt[c]	Electrodeposition	6.5	[18]
ITO-PEN[h]	Pt[c]	Chemical reduction	5.4	[17]
ITO-PEN[h]	PEDOT-PSS[i]	Screen printing	4.4	[31]

[a]StSt: stainless steel. [b]Ti:titanium. [c]Pt: platinum. [d]PEDOT: Zinc. [e]Ni:nickel. [f] Ni: Polyethylene Terephthalate doped Nickel. [g] ITO-PET: Polyethylene Terephthalate doped Indium Titanium Oxide. [h]ITO-PEN:Polyethylene Naphthalate doped Indium. [i]PEDOT-PSS:

6. Stability of flexible dye sensitized solar cells

DSSCs manufactured on glass substrates have exhibited high efficiencies along with excellent stabilities under 60°C and 80°C accelerated aging test and also in outdoor natural environmental condition. This is mainly due to the excellent resistance of glass against the moisture intrusion as well as the temperature. The case for flexible substrates such as plastic and metallic sheet is totally different. The plastic sheets are permeable for intrusion of water or moisture within the cells. In this case, the selection and suitability of material that could resist moisture is crucial because some component in DSSCs such as N719 dye is known to be hydrophilic which is capable of absorbing water [32]. The plastic substrates also sensitive to UV light [34] as well as harsh volatile solvents such as acetonitrile [35]. Whereas, flexible metallic sheet such as stainless steel are facing corrosion problem from redox couple [28;31].

Long term stability is vital for the success of flexible dye sensitized solar cells. These cells need to stand high humidity and extreme temperature. The UV sensitivity of plastic can also be severe problem for flexible DSSCs compared to glass. In this regard, the integrations of flexible DSSCs need to be reviewed according to its application in different operating conditions. The good news about the DSSCs is its higher relative performance in low light condition compared to Silicon based solar cell. This can broaden the utilization of flexible DSSCs for daily uscs such as battery charging of mobile phones, traveling bags, and clothes where both bright light and extreme weather condition are not necessary required for the optimum performance of DSSCs

7. Conclusion

Overall, DSSCs have been recognized as highly potential solar energy conversion devices. Toward the low cost and roll-to-roll processing of flexible DSSCs, alternative substrate needed to replace conventional glass substrate mainly on plastic and metal substrates. However, stability is the main problem in development of flexible DSSCs for practical applications. Hence, further research needs to be carried put in order to commercialize flexible DSSCs.

8. Acknowledgment

Authors wishing to acknowledge the funding assistance of the Ministry of Education of Malaysia (MOE) under Fundamental Research Grant Schemes (FRGS), grant no. FRG0308-TK-1/2012 and FRG0351-TK-2/2013.

9. References

[1] Gong J, Liang J & Sumathy K 2012 Review on dye-sensitized solar cells (DSSCs): fundamental concepts and novel materials *Renewable and Sustainable Energy Reviewsv* **16** 5848-60

[2] O'regan B & Gratzel M 1991 A low-cost, high-efficiency solar cell based on dye-sensitized colloidal TiO2 films *Nature* **353** 737-40

[3] Green M A, Emery K, Hishikawa Y & Warta W 2011 Solar cell efficiency tables (version 37). *Progress in Photovoltaics: Research and* Applications **19** 84-92

[4] Kalyanasundaram K and Grazel M 1998 *Chem. Rev.* **77** 347–414

[5] Shen Y, Deng H, Fang Y and Lu Z 2000 *Colloids and Surfaces A: Physicochem. Eng. Aspects* **175** 135–140

[6] Smestad G, Bignozzi C and Argazzi R 1994 *Sol. Energy Mater. Sol. Cells* **32** 259–72

[7] Kalowekamo J and Baker E 2009 Estimating the manufacturing cost of purely organic solar cell *Solar Energy* **83** 1224-31

[8] Wang B and Kerr L L 2011 Dye sensitized solar cells on paper substrates *Solar Energy Materials & Solar Cells.* **95** 2531-35

[9] Zweibel K 2000 Thin film PV manufacturing: materials costs and their optimization *Solar Energy Materials & Solar Cells.* **63** 375-86

[10] Park J H, Jun Y , Yun H G, Lee S Y and Kang M G 2008 Fabrication of an efficient dye-sensitized solar cell with stainless steel substrate *J. Electrochem. Soc.* **155** 145-49

[11] Yamaguchi T, Tobe N, Matsumoto D, Nagai T and Arakawa H 2010 Highly efficient plastic-substrate dye-sensitized solar cells with validated conversion efficiency of 7.6% *Solar Energy Materials & Solar Cells.* **94** 812-16

[12] Weerasinghe H C, Sirimanne P M, Simon G P and Cheng Y-B 2010 Cold isostatic pressing technique for producing highly effiecint flexible dye-sensitised solar cells on plastic substrates *Progress in Photovoltaic Research and Application.* **20** 321-32

[13] Ito S,Ha N C, Rothenberger G, Liska P, Comte P and Zakeeruddin S M 2006 High efficiency (7.2%) flexible dye-sensitized solar cells with Ti-metal substrates for nanocrystalline-TiO_2 photoanode *Chem Commun* **38** 4004-06

[14] Murakami T N and Grätzel M 2008 Counter electrodes for DSC : application of functional materials as catalysts *Inorganica Chimica Act* **362** 572-80

[15] Su H, Zhang M, Chang Y-H, Zhai P, Hau N Y, Huang Y-T, Liu C, Soh A K and Feng S-P 2014 Highly conductive and low cost Ni-PET flexible substrate for efficient dye-sensitized solar cell *Applied Material & Interfaces* **6** 5577-84

[16] Gong Y, Li C, Huang X,Luo Y, Li D, Meng Q and Iverson B B 2013 Simple method for manufacturing Pt counter electrode on conductive plastic substrates for dye-sensitized solar cells *Applied Materials & Interfaces.* **5** 795-800
[17] Chen L L, Tan W W, Zhang J B, Zhou X W, Zhang X L and Lin Y 2010 Fabrication of high performance Pt counter electrodes on conductive electrodes on conductive plastic substares for flexible dye-sensitizd solar cells *Electrochim Acta* **55** 3721-26
[18] Adriana I, Hisayoshi M, Andreas B and Benedetto B 2009 Nucleation and growth of thin nickel layers under the influence of a magnetic field *J Electroanal Chem.* **626** 174-82
[19] Ma T, Fang X, Akiyama M, Inoue K,Noma H and Abe E 2004 Properties of several types of novel counter electrodes for dye-sensitized solar cel *J. Electroanal Chem* **574** 77-83
[20] Miettunen K, Asghar M I, Ruan X, Saukkonen S, Halme J and Lund P 2011 Stabilizing metal counter electrodes for dye sensitized solar cells *J. Electroanal. Chem.* **653** 93-99
[21] Xiao Y, Wu J, Yue G, Lin J, Huang M, Lan Z and Fan L 2012 Electrodeposition of high performance PEDOT/TI counter electrodes on Ti meshes for large-area flexible dye-sensitized solar cells *Electrochimica Acta.* **85** 432-37
[22] Miettunen K, Ruan X, Saukkoken T,Halme J,Toivola M and Guangsheng H 2009 Stability of dye solar cells with photoelectrode on metal substrates *J. Electrochem. Soc.* **157(6)** 814-19
[23] Miettunen K, Halme J Lund P 2009 Segmented cell design for improved factoring of aging effects in dye solar cell *J. Phys. Chem C.* **113(3)** 10297-302
[24] Yun H G, Bae B S and Kang M G 2011 A simple and highly efficient method for surface treatment of Ti substrates for use in dye-sensitized solar cells *Adv.Energy Mater.* **1** 1-6
[25] Mukarami T N, Kijitori Y, Kawashima N and Miyasaka T 2003 UV light-assited chemical vapor deposition of TiO_2 for efficiency development at dye-sensitized mesoporous layers on plastic film electrodes *Chem. Lett.* **32(11)** 1076-77
[26] Zhang D, Yoshida T, Furuta K and Minoura H 2004 Hydrothermal preparation of porous nanocrystalline TiO_2 electrodes for flexible solar cells *J. Photochem. Photobiol. A.* **164(1-3)** 159-66
[27] Pan H, Ko S H, Misra N and Grigoropoulos C P 2009 Laser anneled composite titanium dioxide electrodes for dye sensitized solar cells on glass and plastic *Appl. Phys. Lett.* **94(7)** 1-3
[28] Miettunen K, Halme J, Saukkonen T, Toivola M and Lund P D 2009 Performance degradation of dye sensitized solar cells on flexible stainless steel sbstrates *Proc. of the 24th Europen photovoltaic solar energy Conf.* (Hamburg, Germany, 21-25 September 2009)
[29] Wang G and Lin Y 2007 Novel counter elctrodes based on NiP-plated glass and Ti plate substrate for dye-sensitized solar cells *.J. Mater. Sci*, **42(13)** 5281-85
[30] Ma T, Fang X, Akiyama M, Inoue K, Noma H and Abe E 2004 Properties of severals types of novel counter electrodes for dye-sensitizd solar cells *J. Electroanal. Chem.* **574(1)** 77-83
[31] Muto T, Ikegami M, Kobayashi K and Miyasaka T 2007 Conductive polymer-based mesoscopic counter electrodes for plastic dye-sensitized solar cell *Chem. Lett.* **36(6)** 804-05
[32] Kitamura T, Okada K, Matsui H and Tanabe N 2010 Durability of dye-sensitized solar cells and modules *J.Sol. energy. Eng.* **132**
[33] Ma T, Fang X, Akiyama M, Inoue K, Noma H and Abe E 2004 Properties of several types of novel counter electrodes for dye-sensitized solar cells *J. Electroanal. Chem* **574** 77-83
[34] Scheirs J and Gardette J L 1997 Photo-oxidation of photolysis of poly(ethylene napthalate) *Polym. Degrad. Stab.* **56** 339-50
[35] Sommeling P M, Späth M, Smith H J P, Bakker N J and Kroon JM 2004 Long-term stability testing of dye sensitized solar cells *J. Phtoochem Photobiol A.* **164** 137-44
[36] Jun Y and Kang M G 2007 The characterization of nanocrystalline dye-sensitized solar cells with flexible metal substrates by electrochemical impedance spectroscopy *J. Electrochm.Soc* **154** B68

[37] Gao R, Tian J, Zhang Q, Wang L and Cao G 2013 Nanorod-nanosheet hierarchirally sturctured ZnOcrystals on zinc foil as flexible photoanodes for dye-sensitized solar cells *Nanoscale* **5** 1894-901

Effect of Varying Acid Hydrolysis Condition in *Gracilaria Sp.* Fermentation Using *Sasad*

H Mansuit[1], M D C Samsuri[1], C S Sipaut[1], C F Yee[3], S M Yasir[2], R Mansa[1]
[1] Energy and Materials Research Group, Faculty of Engineering, Universiti Malaysia Sabah, Jalan UMS, 88400 Kota Kinabalu Sabah
[2] Seaweed Research Unit, Faculty of Science and Natural Resources, Universiti Malaysia Sabah, Jalan UMS, 88400 Kota Kinabalu Sabah
[3] Faculty of Food Science and Nutrition, Universiti Malaysia Sabah, Universiti Malaysia Sabah, Jalan UMS, 88400 Kota Kinabalu Sabah

E-mail: elvyeva@yahoo.com, rmansa@gmail.com

Abstract. Macroalgae or seaweed is being considered as promising feedstock for bioalcohol production due to high polysaccharides content. Polysaccharides can be converted into fermentable sugar through acid hydrolysis pre-treatment. In this study, the potential of using carbohydrate-rich macroalgae, *Gracilaria sp.* as feedstock for bioalcohol production via various acid hydrolysis conditions prior to the fermentation process was investigated and evaluated. The seaweed used in this research was from the red algae group, using species of *Gracilaria sp.* which was collected from Sg. Petani Kedah, Malaysia. Pre-treatment of substrate was done using H_2SO_4 and HCl with molarity ranging from 0.2M to 0.8M. The pre-treatment time were varied in the range of 15 to 30 minutes. Fermentation was conducted using *Sasad*, a local Sabahan fermentation agent as a starter culture. Alcohol extraction was done using a distillation unit. Reducing sugar analysis was done by Benedict test method. Alcohol content analysis was done using specific gravity test. After hydrolysis, it was found out that acid hydrolysis at 0.2M H_2SO_4 and pre-treated for 20 minutes at 121°C has shown the highest reducing sugar content which has yield (10.06 mg/g) of reducing sugar. It was followed by other samples hydrolysis using 0.4M HCl with 30 minutes pre-treatment and 0.2M H_2SO_4, 15 minutes pre-treatment with yield of 8.06 mg/g and 5.75 mg/g reducing sugar content respectively. In conclusion, acid hydrolysis of *Gracilaria sp.* can produce higher reducing sugar yield and thus it can further enhance the bioalcohol production yield. Hence, acid hydrolysis of *Gracilaria sp* should be studied more as it is an important step in the bioalcohol production and upscaling process.

1. Introduction

Marine biomass comprise of macro and microalgae, and both have been proposed as potential biofuel feedstock. Microalgae are potential sources of bio-oils whilst macroalgae are potential sources of carbohydrates for fermentation or thermo-chemical based conversion. In fact, previous research on the wide range of macroalgae species was thought to be a promising next generation biofuel feedstock. Seaweeds consist of carbohydrates which can be used as main feedstock for alcohol production.

However, the conversion of those carbohydrates to fermentable sugar is the key issue for the utilization of seaweeds for the alcohol production. One major challenge in using algae biomass is its native recalcitrant structure due to their natural physicochemical barriers, which provides its a significant resistance to hydrolysis for further conversion by anaerobic fermentative bacteria. Most of the algae species present a rigid cellulose-based wall and accumulate various complex polysaccharides conferring a resistance to bacterial attack and limiting their degradation during the anaerobic fermentation process [1].

To overcome these natural barriers, several types of pre-treatment technologies have been used for bioalcohol production from algal biomass. Among them, thermal and thermo-chemical pre-treatments have gained consideration in contributing solubilisation of carbohydrate polymers into soluble sugars such as glucose, xylose, arabinose and galactose. Thus, thermo-chemical pre-treatments had been investigated to overcome the natural physicochemical barriers of such biomass and to enhance biofuel production from marine biomass (macroalgae). This particular pre-treatment used diluted acid pre-treatment by manipulating several variables such as type and concentration of acid as well as duration of thermal process.

Even through pre-treatments are often efficient in increasing the accessibility of biodegradable compounds to microorganism by weakening the physicochemical barriers of the algal biomass, they release also soluble sugar-derived by-products such as furfural, 5 HMF (hydroxylmethyl furfural), or lignin-derived by-products such as vanillin, syringaldehyde and other phenolic compounds [9]. Due to their strong inhibitor effects on productivity and end products formation, these by-products may constitute a limiting factor in the feasibility of using algal biomass as substrate for biofuel conversion.

To avoid the negative effect of such by-products on alcohol, several processes for hydrolyzate detoxification, including evaporation, adsorption on active charcoal, and adsorption on ion exchangers, solvent extraction, alkaline treatment or enzymatic treatment has been proposed [2]. However, detoxification methods increase significantly the overall costs due not only to capital and chemical costs, but also to the loss of sugars from primary material [2].

Until recently, there are numerous studies focuses on production of biodiesel from microalgae but still lack of research on production of bioalcohol from seaweeds. Here, it is attempted to use *Gracilaria sp* , the third largest genus of class Rhodophyta or red algae for the bioalcohol production considering the pre-treatment process could contribute significantly to whole research. Due to the formation of fermentation inhibitors during the acid hydrolysis reaction, this will decrease the alcohol production from the biomass fermentation. This situation must be avoided as the fermenting agent used, *Sasad* is quite novel in the alcohol fermentation application which the formation of particular inhibitors could led to the deactivation of *Sasad* activity.

The diluted acid hydrolysis is very complex, mainly because the substrate is in a solid phase and the catalyst is in a liquid phase. One of the disadvantages is the formation of undesired by products that may affect the cell growth and alcohol fermentation. Eventually, the high temperature and acidic condition during the hydrolysis process will led to the formation of chemical inhibitors such as hydroxymethylfurfural (HMF) and levulinic acid which considered as undesired by-products [10]. Some by products such as HMF and levulinic acid might inhabit the cell growth and decrease the alcohol production. Hence, it is important to maximize the monosaccharide conversion yield and minimize the formation of by-products from this pre-treatment process.

In fact, the application of *Sasad* as fermenting agent in this bioalcohol production could also lead to several undesired implication during this research. Currently, *Sasad* application is quite well known in Sabah for its rice fermentation producing the traditional alcoholic beverages [5]. Thus, practically it's been assumed that the seaweed could substitute the rice as a substrate for bioalcohol production. Since the content of *Gracilaria sp.* that has been used as substrate might slightly different from the rice, the end product probably will contain several by-products which perhaps significantly affected by the fermenting agent, *Sasad*.

The main objective of this research is to develop bioalcohol production technique from *Gracilaria sp.* through acid hydrolysis, neutralization and fermentation using *Sasad*.

IOP Conf. Series: Materials Science and Engineering **78** (2015) 012004 doi:10.1088/1757-899X/78/1/012004

2. Materials and Methods

2.1. Raw materials

Samples of *Gracilaria sp.* were purchased from Sg. Petani Kedah, Malaysia. The seaweed samples were soaked and washed with tap water to remove any foreign materials, salt and debris. Samples were then oven dried at 50°C until constant weight. Dried *Gracilaria sp.* was then kept in air tight container for further usage. Materials used in this research are HCl, H_2SO_4, NaOH, *Sasad*, local Borneo fermentation agent, and Benedict Reagent.

2.2. Substrate preparation

600g of dried *Gracilaria sp.* was cooked at 100°C in distilled water with volume ratio of 1:4 for 2 hours to ease the polysaccharide extraction process. Then, the substrate was size reduced using blender for 15 minutes, producing finely shredded slurry. 16 samples were prepared in 500ml flasks containing 200g of substrates. Aliquot of 2ml from substrate prepared was taken and filtered for reducing sugar concentration analysis.

2.3. Acid hydrolysis

H_2SO_4 and HCl were used in this experiment. Acid stocks solution was prepared in the range of 0.2M to 0.8M. Hydrolysis times were varied in the range of 15 to 30 minutes and it was conducted using an autoclave machine at 121°C. After hydrolysis process was done, the residues were separated by filtration. The liquid was then further analysed for reducing sugar content. Hydrolysis were done as in experimental design in table 1.

Table 1. Experimental Design for Acid Hydrolysis

Experiment	Acid Type	Molarity (M)	Hydrolysis Time (min)
1	H_2SO_4	0.2	30
2	H_2SO_4	0.4	30
3	H_2SO_4	0.6	30
4	H_2SO_4	0.8	30
5	HCl	0.2	30
6	HCl	0.4	30
7	HCl	0.6	30
8	HCl	0.8	30
9	H_2SO_4	0.2	15
10	H_2SO_4	0.2	20
11	H_2SO_4	0.2	25
12	H_2SO_4	0.2	30
13	HCl	0.2	15
14	HCl	0.2	20
15	HCl	0.2	25
16	HCl	0.2	30

2.4. Reducing sugar analysis

Reducing sugar concentration was analyzed using the Benedict reagent method. Glucose is used as the standard for this analysis. The ratio of Benedict reagent to standard/sample solution is 1:10. Then, the standard solution and a sample containing Benedict reagent will need to be heated using water bath at 80°C about 30 minutes. After cooled down to room temperature, the reducing sugar concentration was read at 550 nm using Jasco UV-Vis Spectrophotometer 650.

2.5. Hydrolyzate neutralization

After acid hydrolysis of Gracilaria sp, hydrolyzates were neutralized with 5M NaOH. Sodium hydroxide was added until the pH of hydrolyzate is in the range of pH5.0 – 6.0.

2.6. Fermentation agent preparation

Sasad, was primarily size reduced using a manual mortar until it turns into powder form and stored in airtight container for further used. 10 g of *Sasad* was added into all samples. The anaerobic fermentative processes are allowed to take place as the conical flask will be sealed using the wool and parafilm for 10 days.

2.7. pH analysis

pH analysis is crucial in order to determine the presence of microbial activity during the fermentation process. pH was obtained using pH meter.

2.8. Distillation

The fermented product was distilled using Distillation Unit Buchi K-350. 50 ml of sample was measured at each time of distillation. The final volume of distillate was measured and recorded. It was stored in 5-8°C prior to estimation of percent alcohol in distillate.

2.9. Estimation of alcohol percentage in distillate

The determination of percent alcohol is estimated from the relationship between specific gravity and percent alcohol. The data was derived from Amerine and Ough [11].

3. Results and Discussions

3.1. pH Analysis

Table 2.0 shows pH profile for all samples in 3 different conditions:

i. pH before substrate undergo pretreatment process.
ii. pH after substrate undergo thermal-chemical pretreatment.
iii. pH after hydrolysate neutralization (before adding *Sasad*).

Table 3.0 shows the pH profile for all samples which was obtained during fermentation. pH data was recorded once for every two days throughout the 10 days of fermentation.

It can be seen from table 2.0 and table 3.0 that most of pH samples decreased from around pH 5 to around pH 3. These results are quit significant with the pH of inoculated rice with Sasad which was rapidly drop from pH 6.6 to pH 3.4 [3]. Based on the pH analysis, we can conclude that the pH profile is acidic during fermentation. In fact, the presence of some by products (formic and levulinic acid) after been autoclaved also contributes to the acidic condition [9].

It was also believed that the correlation between acidity and pH was due to the connection with both yeasts and lactic acid bacteria (LAB). LAB were well recognized as acids producer in fermentation as well as some yeasts were also producing acid which may results in the variation of products flavor [4].

Table 2. pH profile for three different conditions

Experiment	pH		
	1	**2**	**3**
1	7.35	2.11	5.86
2	7.13	1.45	5.49
3	8.70	1.33	5.38
4	8.36	1.33	5.87
5	8.11	1.85	5.42
6	9.27	1.63	5.75

7	8.63	1.35	5.23
8	7.35	2.11	5.86
9	7.97	1.52	5.75
10	7.95	2.12	5.27
11	7.86	1.97	5.63
12	8.25	1.62	5.74
13	8.69	1.81	5.52
14	8.14	1.83	5.28
15	8.70	1.87	5.50
16	7.97	1.52	5.75

Table 3. pH profile for samples during fermentation

	Fermentation Days				
Experiment	**2**	**4**	**6**	**8**	**10**
			pH		
1	3.43	3.54	3.23	2.52	2.58
2	3.85	3.71	3.63	2.30	2.23
3	3.84	3.79	3.62	3.63	3.63
4	3.68	3.76	3.51	3.21	3.24
5	3.40	3.38	3.24	3.23	3.00
6	3.35	3.27	3.11	2.75	2.50
7	3.35	3.27	3.11	2.75	2.50
8	3.52	3.23	3.60	3.45	3.36
9	3.84	3.74	3.55	3.29	3.22
10	3.90	3.78	3.64	3.54	3.41
11	3.84	3.72	3.63	3.62	3.36
12	3.65	3.76	3.54	3.37	3.18
13	3.23	3.24	3.07	3.04	3.11
14	3.35	3.36	2.94	3.23	3.03
15	3.41	3.42	3.35	2.76	3.21
16	3.84	3.74	3.55	3.29	3.22

3.2. Effect of acid concentration and hydrolysis time on reducing sugar yield

Based on figure 1, experiment 6 (0.4M HCl, 30min) shows the highest reducing sugar content (8.06mg/g) compared to others for the fixed duration of hydrolyzate in autoclave. As we can see from the graph, the reducing sugar contents are increasing proportionally with acid molarity. However, there are decreasing in reducing sugar content which can be noticed at 0.8M concentration for both type of acids.

Figure 2 shows that sample with sulphuric acid show an increasing amount of reducing sugar from 15min to 20min hydrolysis. However, the amount was decreasing at longer hydrolysis time. Compared to sample hydrolyzed with HCl, the reducing sugar content was increasing over the hydrolysis time, but the overall amount is lower than samples treated with H_2SO_4. Longer hydrolysis time results in the decreasing of reducing sugar concentration might be due to sugar decomposition [10].

From figure 2, the highest content of reducing sugar was produced from sample treated with 0.2M H_2SO_4 in 20min. This is comparable to the previous study, whereby high reducing sugar concentration can be achieved with low reaction temperature at 120°C at shorter time [6]. It can be concluded that extreme condition during the thermal-chemical pretreatment of *Gracilaria sp* was not suitable for this project as the reducing sugar might be further converted into some by- products such as furfural, 5-HMF, formic acid and levulinic acid [7].

This present work proved that mild condition of acid hydrolysis at 0.2M H_2SO_4, 20 min duration in autoclave machine (10.06 mg/g) produced the most reducing sugar. It was followed by the hydrolysis of 0.4M HCl, 30 min, (8.06 mg/g) and 0.2M H_2SO_4, 15 min (5.75 mg/g).

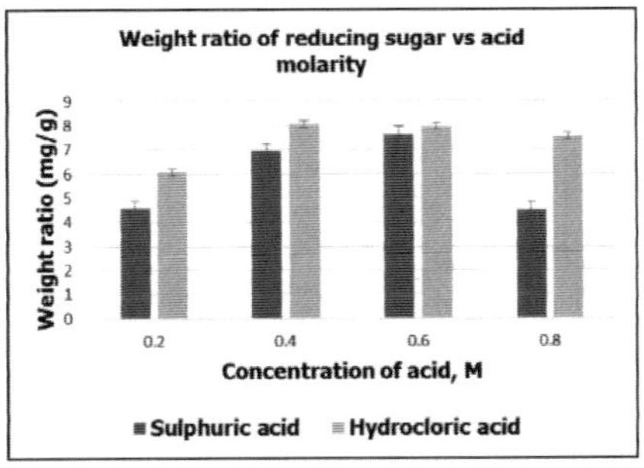

Figure 1. Reducing sugar concentration for different acid concentration

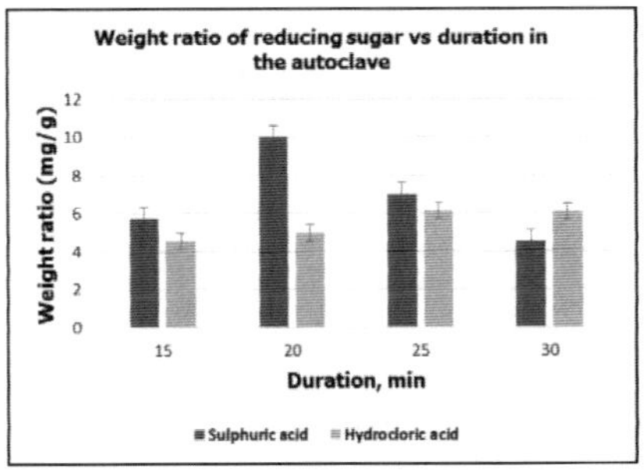

Figure 2. Reducing sugar concentration for different pre-treatment time

3.3. Effect of acid hydrolysis on alcohol yield
The concentration of alcohol in each fermentation sample after various hydrolysis conditions on *Gracilaria sp* can be seen in figure 3 and 4. Hydrolysis at 0.4M HCl, 121^{0}C, 30min clearly produced the most alcohol (15.22 % v/v), followed by sample hydrolysed with 0.2M HCl, 121^{0}C, 30min (12.52 % v/v); 0.2M HCl,121^{0}C, 20min (7.21 % v/v); 0.2M H_2SO_4, 121^{0}C, 20min (6.98 % v/v) and 0.8M H_2SO_4, 121^{0}C, 30min (6.72 % v/v) with distilled products (figure 5).

Interestingly, the alcohol yield is not depending on the amount of reducing sugar produced after being hydrolyzate. For example, sample treated with 0.4M HCl, 121^{0}C, 30min producing maximum alcohol although its reducing sugar content is lower than sample treated at 0.2M H_2SO_4, 20 min prior to fermentation process. Yet, experiment 10 still produced the fourth highest amount of alcohol which was 6.98% v/v. The reason which leads to this result is probably because the suitability of acid hydrochloric compared to sulphuric acid which can provide a mild conditions by minimizing the degradation of reducing sugar into inhibitor.

From figure 3, it shows that the alcohol production is fluctuating as the acid molarity during pre-treatment was increased. Hence, as the duration in the autoclave were fixed into 30 minutes, hydrolysis using 0.4M HCl producing the higher amount of alcohol. The effect of temperature was not significant as compared to molarity change.

Elevated temperature will produce poor fermentability hydrolysate. Sugars can be degraded to furfural which is formed from pentoses and 5-hydroxymethylfurfural (5-HMF) which is formed from hexoses. 5-HMF can be further degraded, forming levulinic acid and formic acid [8].

In addition, formic acid can be formed from furfural under acidic conditions at elevated temperatures. Acetate is liberated from hemicellulose during hydrolysis. Since the application of *Sasad* as the alcohologenic bacteria is quite novel for the bioalcohol production from macroalgaee biomass, the solubilisation of inhibitory compound could lead to the significant effect which disrupting the multicellular microbial activity.

Overall, the highest production of alcohol was produced from the hydrolyzate which by using hydrochloric acid during the thermal-chemical pre-treatment. Experiment 5, 6 and 14 produced 15.21 % v/v, 12.52% v/v, 7.21% v/v of alcohol respectively. It proves that the HCl could contribute significant yield for the mild acid hydrolysis condition which was more conducive for the microbes from *Sasad.*

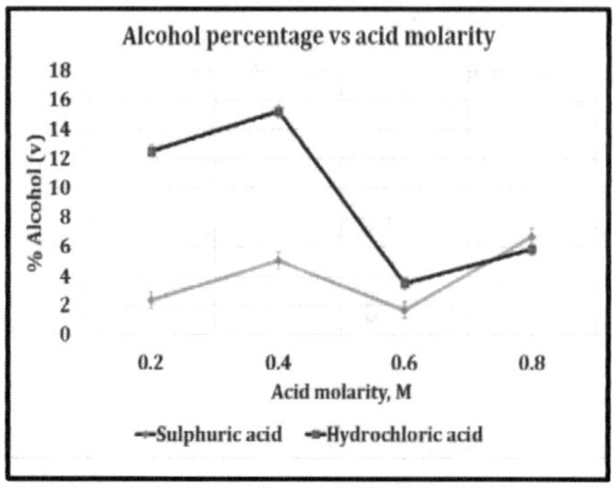

Figure 3. Alcohol concentration based on different acid molarity

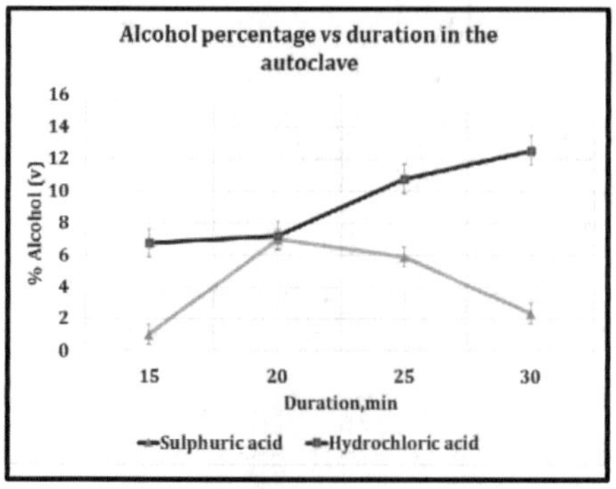

Figure 4. Alcohol concentration based on different hydrolysis time

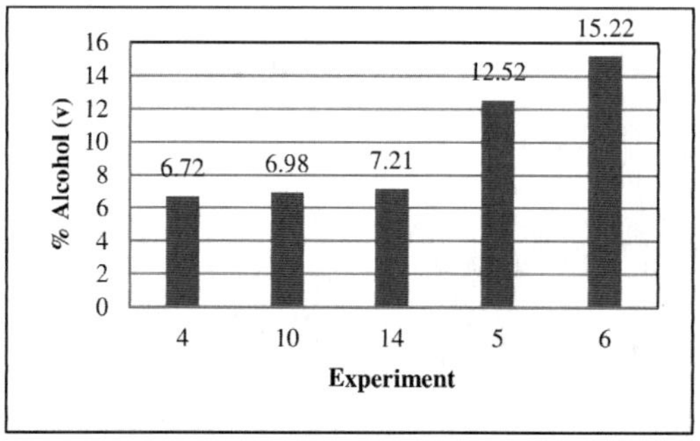

Figure 5. Alcohol percentage for five significant samples

4. Conclusion
This present's preliminary work has shown that fermentation of *Gracilaria sp.* with *Sasad* and via acid hydrolysis pre-treatment has been done. The data has shown that the highest estimated alcohol yield obtained was 15.22% (v/v) from the *Gracilaria sp* treated with 0.4M HCl, and 30min pre-treated at 121°C. It was followed by *Gracilaria sp.* treated with 0.2M HCl, and 30min pre-treated at 121°C which yield 12.52% (v/v) alcohol. Whereas, acid hydrolysis of *Gracilaria sp.* using 0.8M H_2SO_4 and autoclaved at 121°C for 30min yield 6.72% (v/v) alcohol.

The reducing sugar analysis shows highest amount from the hydrolysis of *Gracilaria sp.* with 0.2M H_2SO_4 and pre-treated for 20 minutes at 121°C. It was followed by other samples hydrolysis using 0.4M HCl, 30 minutes pre-treatment and 0.2M H_2SO_4, 15 minutes pre-treatment yield (8.06 mg/g) and (5.75 mg/g) reducing sugar content respectively.

However, further work in analyzing the alcohol yield would be needed to obtain an accurate result besides to detect the components in the fermentation products.

Acknowledgments
The authors would like to extend their appreciation and gratitude for Ministry of Agriculture for funding Seaweed Research Unit Grant GPRL013 (Biofuel from Seaweed).

6. References

[1] Adams J M M, Toop T a, Donnison I S and Gallagher J a 2011 Seasonal variation in *Laminaria digitata* and its impact on biochemical conversion routes to biofuels *Bioresour. Technol.* **102 (21)** 9976–84 (doi:10.1016/j.biortech.2011.08.032)

[2] Almeida J R M, Bertilsson M, Gorwa-Grauslund M F, Gorsich S and Lidén G 2009 Metabolic effects of furaldehydes and impacts on biotechnological processes *Appl. Microbiol. Biotechnol.*, **82 (4)** 625–38 (doi:10.1007/s00253-009-1875-1)

[3] Chiang Y, Chye F and Ismail A M 2006 Microbial diversity and proximate composition of *Tapai*, A Sabah's fermented beverage *Malays. J. Microbiol.* **2 (1)** 1–6 Retrieved from http://www.usm.my/mjm/issues/vol2no1/research1.pdf

[4] Fleet G H 2003 Yeast interactions and wine flavour *Int. J. of Food Micr.* **86** 11-22 (doi:10.1016/S0168-1605(03)00245-9)

[5] Bakar H J H A 2000 Fermented Rice - *Tapai* 16 – 28 Retrieved from http://fos.ubd.edu.bn/sites/default/files/2000-Paper3.pdf

[6] Jang S, Shirai Y, Uchida M and Wakisaka M 2012 Production of mono sugar from acid hydrolysis of seaweed. *African J. Biotechnol.* **11** (8) 1953–1963. (doi:10.5897/AJB10.1681)

[7] Klinke H B, Thomsen a B and Ahring B K 2004 Inhibition of ethanol-producing yeast and bacteria by degradation products produced during pre-treatment of biomass *Appl. Microbiol. Biotechnol.* **66 (1)** 10–26 (doi:10.1007/s00253-004-1642-2)

[8] Larsson S and Palmqvist E 1999 The generation of fermentation inhibitors during dilute acid hydrolysis of softwood *Enzym. Microb.* **0229 (98)** 151–159 Retrieved from http://www.sciencedirect.com/science/article/pii/S014102299800101X

[9] Monlau F, Sambusiti C, Barakat a, Quéméneur M, Trably E, Steyer J-P and Carrère H 2014. Do furanic and phenolic compounds of lignocellulosic and algae biomass hydrolyzate inhibit anaerobic mixed cultures? A comprehensive review *Biotechnol. Adv.* (doi:10.1016/j.biotechadv.2014.04.007)

[10] Setyaningsih D, Windarwati S, Khayati I, Muna N and Hernowo P 2012. Acid hydrolysis technique and yeast adaptation to increase red macroalgae bioethanol production *Int. J. Environ. Bioener.* **3 (2)**, 98–110

[11] Amerine M A and Ough C S 1974 *Wine and Must Analysis* (New York: John Wiley and Sons)

Sol – Gel synthesis and characterization of magnesium peroxide nanoparticles

J Jaison[1, 2], C Ashok raja[2], S Balakumar[2], Y S Chan [*1]

[1]Department of Chemical and Petroleum Engineering, School of Engineering and Science, Curtin University CDT 250, 98009 Miri Sarawak Malaysia
[2]National Center for Nanoscience and Nanotechnology, University of Madras, Chennai, 600025, India.

E-mail: chanyensan@curtin.edu.my

Abstract. Magnesium peroxide is an excellent source of oxygen in agriculture applications, for instance it is used in waste management as a material for soil bioremediation to remove contaminants from polluted underground water, biological wastes treatment to break down hydrocarbon, etc. In the present study, sol-gel synthesis of magnesium peroxide (MgO_2) nanoparticles is reported. Magnesium peroxide is odourless; fine peroxide which releases oxygen when reacts with water. During the sol-gel synthesis, the magnesium malonate intermediate is formed which was then calcinated to obtain MgO_2 nanoparticles. The synthesized nanoparticles were characterized using Thermo gravimetric –Differential Thermal Analysis (TG- DTA), X-Ray Diffraction studies (XRD) and High Resolution Transmission Electron Microscope (HRTEM). Our study provides a clear insight that the formation of magnesium malonate during the synthesis was due to the reaction between magnesium acetate, oxalic acid and ethanol. In our study, we can conclude that the calcination temperature has a strong influence on particle size, morphology, monodispersity and the chemistry of the particles.

1. Introduction

Magnesium oxide has extensive applications in the detection and remediation of chemical waste and warfare agents [1]. It is also used as an additive in heavy fuel oils [2]; reflecting and anti-reflecting coatings [3] in optical applications; and is used as the substrate in superconducting and ferroelectric thin films [4-6]. Besides, magnesium oxide has also shown a promising application in catalysis applications of many organic reactions [7-12]. The properties of novel MgO are further magnified when its size is reduced to nanoscale. Many synthesis routes like sol-gel, hydrothermal, flame spray pyrolysis, laser vaporization, chemical gas phase deposition, combustion aerosol synthesis, aqueous wet chemical, and surfactant methods have been studied for nano-size synthesis of MgO [13-28]. But other forms of magnesium nanoparticles like MgOH and MgO_2 were given less importance because their applications are still unknown in many fields [29].

For a long time, MgO was believed to be among the least polymorphic solids[30]. Experimental simulation also predicted that MgO remains non-metallic up to extremely high pressure (20.7 TPa) [30, 31], making it to our knowledge the most difficult mineral to metalize. Thermodynamic equilibriain the Mg-O system at 0.1 MPa shows that only MgO is a stable compound synthesized, though metastable compounds (MgO_2, MgO_4), which can be prepared at very high oxygen fugacities [32-34]. In the experiment performed by Qiang Zhu *et al*, two new stable compounds (MgO_2 and Mg_3O_2) has been found to exhibit interesting crystal structures with unusual chemical bonding and insulating and semiconducting electronic structures, respectively [35].

As magnesium peroxide is a stable oxygen releasing compound than Mg_3O_2, it can be used in agriculture to improve soil quality and in environmental industries as a bioremediation agent that converts contaminated soil into a fertile soil [36, 37]. In certain circumstances MgO_2 has also been shown to inhibit growth of bacteria. For example, the growth of sulfate –reducing bacteria can be inhibited in an environment containing MgO_2[38]. The nano form of MgO_2 can be used to enhance the oxygen releasing ability of the compound [39] and thereby increase its potentiality to be used in enormous applications. In our study, we had synthesized magnesium peroxide nanoparticles (MgO_2) by simple sol-gel method using magnesium acetate as precursor.

2. Experimental methods

2.1 Chemicals used

The commercially available magnesium acetate (98% pure) was obtained from Alfa Aesar, oxalic acid (99.5% pure) and ethanol (99.9%) from Merck was used in this reaction without further purification.

2.2. Synthesis of MgO_2 nanoparticles.

AnEquimolar ratio of magnesium acetate and oxalic acid were prepared separately by dissolving the salts in ethanol before they were mixed at 300rpm. The formation of gel starts immediately after mixing. Due to the addition of gelating agent, the sol of precursor dissolved in solvent forms a three dimensional network of magnesium through three major steps that were hydrolysis, polycondensation and thereafter gelation. After this, the gel was allowed for aging for about 12 hours in order to increase the thickness of the gel. Then the above aged gel was dried in hot air oven for about 24 hours at 100°C in order to remove impurities through evaporation [13, 40]. The obtained powder was then calcinated at 600°C for 2 hours to obtain MgO_2 nanoparticles. The calcination was done in a box furnace (Sentrotech ST 1200C-7812) at a heating rate of 5°C/min. The calcination temperature was fixed by Thermo Gravimetric-Differential Thermal Analysis and the calcinated MgO_2 nanoparticles were characterized by X-Ray Diffraction spectroscopy for phase confirmation and High Resolution Transmission Electron Microscope for morphological analysis. Figure 1 shows the flow chart of magnesium peroxide nanoparticles synthesis by sol-gel method.

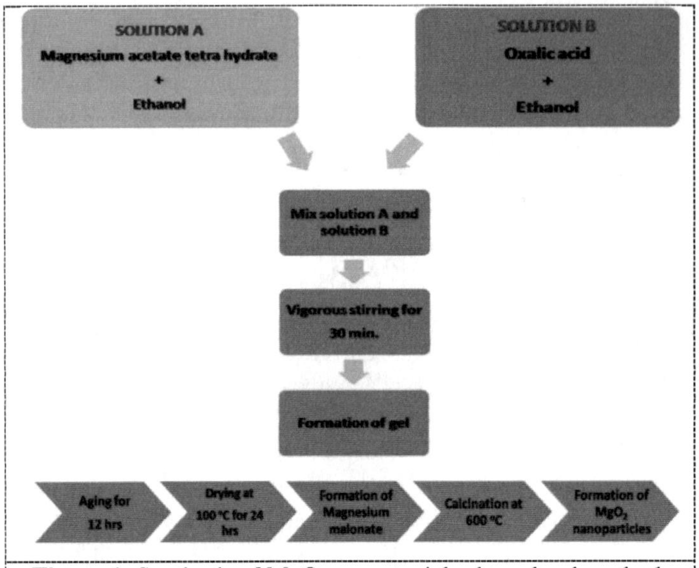

Figure 1: Synthesis of MgO_2 nanoparticles by sol-gel method.

2.3. Characterization of MgO_2 nanoparticles

The thermal analysis was carried out with a Thermo Gravimetry Analysis (TGA, ***S II TG/DTA 6300***) which will heat the sample up to 1200°C to find out thermal degradation of the sample. The TG-DTA was conducted at a heating rate of 5°C/min in nitrogen atmosphere. The phase purity and crystallinity of the MgO_2 nanoparticles were determined by X-ray diffraction (XRD) recorded on an X-ray Diffractometer system (***RIGAKU X ray diffractometer***). High resolution transmission electron microscopy (HR-TEM) measurement and the SAED pattern were studied using a HR-TEM, ***Tecnai T30 G^2 S-Twin, FEI Company*** which has a point resolution 0.24nm and its magnification ranges from 25x – 1030kx.

3. Results and Discussions

Sol–Gel method is a wet chemical technique widely used in the fields of materials science to synthesize ceramics with an intermediate stage. It is the precursor that, by its chemistry, leads the reaction towards the formation of either colloidal particles or polymeric gels [40]. In our experiment, the magnesium acetate precursor reacts with oxalic acid and ethanol to give magnesium malonate and the excess volatile substances in the reaction gets evaporated during aging and drying process [Equation 1]. Due to high temperature calcination at 600°C, the magnesium malonate gets transformed into magnesium peroxide [equation 2].

$$Mg\,(CH_3COO)_2 + C_2H_2O_4 + C_2H_5OH \longrightarrow Mg.\,[CH_2\,(COO)_2] + C_2H_2O_4\uparrow + C_2H_5OH\uparrow \quad (1)$$

$$Mg.\,[CH_2\,(COO)_2] \xrightarrow[\Delta]{600^{o}C} MgO_2 \text{ nanoparticles} \quad (2)$$

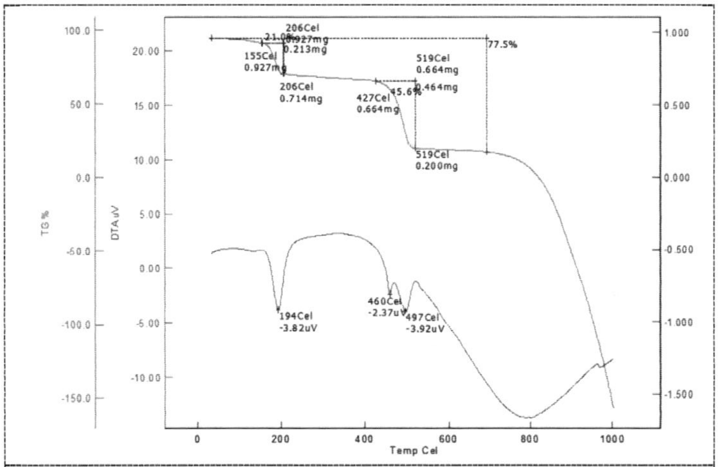

Figure 2: TG-DTA of pre calcinated magnesium malonate.

The TG curve of the compound shows the thermal decomposition in three stages. The water molecules get evaporated completely during drying of pre calcinated powder after aging in sol- gel process as shown in Fig 2. Dehydration occurred in the first stage (155-206 °C) in which the carboxyl molecules from the malonate group were lost due to decarboxylation (equation 3) [41]. After the first degradation, the end product contains unstable Mg. (CH_2) and it was degraded at 427 – 519 °C, in the second stage (equation 4). Though complete weight loss observed at 1000°C, we were able to study the effect of calcination at lower temperature (< 1000°C) with the intention of controlling the particle size and morphology [42, 43]. But at low temperature, calcination ended in the formation of unstable oxide (equation 5). MgO_2which was formed were confirmed by X-ray diffraction studies and displayed in figure 3. The thermal decomposition scheme is exemplified below

$$Mg.[CH_2(COO)_2] \xrightarrow{155\text{-}206\,^{\circ}C} Mg.(CH_2) + 2CO_2 \quad \text{(First decomposition)} \tag{3}$$

$$Mg.(CH_2) \xrightarrow{427\text{-}519^{\circ}C} Mg + CH_2 \quad \text{(Second decomposition)} \tag{4}$$

$$Mg + O_2 \xrightarrow{600\,^{\circ}C} MgO_2 \quad \text{(Third decomposition)} \tag{5}$$

In the experiment performed by Davoodnia *et al*, the first decomposition of pre calcinated sample was found to be 359°C. The reason for this high temperature for the first decomposition was due to the usage of magnesium nitrate as a precursor in the sol-gel method. As the nitrate bond is very strong, only high temperature can break that bond to form stable oxides of magnesium [12]. But in our experiment, we can able to get first decomposition of the compound at 155°C due to the presence of magnesium acetate as the precursor in the sol-gel reaction. As the first decomposition is at 155°C, all the subsequent decomposition also occurs at low temperature compared to the literature. Then the sample is subjected to X-ray diffraction studies in order to confirm the crystalline phases of the pre calcinated and the calcinated sample.

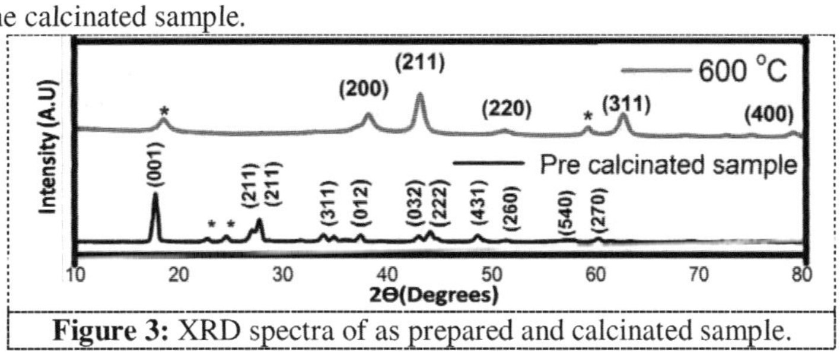

Figure 3: XRD spectra of as prepared and calcinated sample.

The * symbol in certain peaks indicates that those peaks are due to impurity in reaction atmosphere or due to unreacted components in the reaction mixture.

Figure 3 shows the XRD spectra for the formation of magnesium malonate in pre calcinated samples and the formation of MgO_2 nanoparticles in the calcinated sample at 600°C. The formation of magnesium malonate (PDF 24-1794) in pre calcinated sample is due to the reaction between the solvent (ethanol) and the gelating agent used (oxalic acid). The XRD peak of magnesium malonate was suppressed in the calcinated sample, which indicates the formation of magnesium peroxide. The XRD peak of the calcinated sample at 600 °C was matching with the MgO_2 (PDF 75-1585) crystal structure with some peak shifts. The peak shifts were due to the presence of certain impurities in the samples.
In the experiments performed by Stengl *et al*, Davoodnia *et al* and Athar *et al*, magnesium hydroxide [PDF 45-0946] is formed as the pre calcinated sample which was subjected to high temperature treatment up to 1000°C for several hours to get magnesium per oxide and magnesium oxide [12, 44, 45]. But in our experiment, the hkl (Miller indices) planes of magnesium malonate and magnesium peroxide is perfectly matching with the JCPDS (Joint Committee on Powder Diffraction Standards) files of the compound and the final product is obtained at low temperature (600°C) than compared to the above literatures.
The sample is then subjected to HRTEM for structural analysis. Figure 4 (a) shows the HRTEM micrograph of pre calcinated sample. It shows that the magnesium malonate formed are porous due to the evaporation of volatile compounds during drying process in sol-gel method at 100 °C. The SAED pattern [fig. 4(b)] reveals that the magnesium malonate is polycrystalline in nature as there are only two rings without proper spots for crystal pattern which is consistent with the result reported by Stengl *et al* and Gedanken *et al*[44, 46].The HRTEM micrograph of calcinated MgO_2 nanoparticles [fig. (c)] at 600°C shows a spherical morphology, which is agglomerated with each other due to high temperature. At 20nm scale, we observe that there is no porous structure in the calcinated samples. The calcination of pre calcinated magnesium malonate powder breaks the micro particles into pieces to form nanoparticles of magnesium peroxide [45].

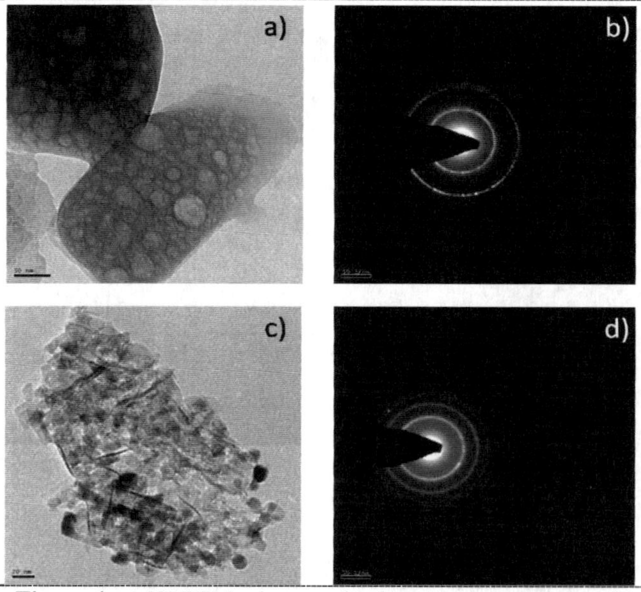

Figure 4: a) HRTEM micrograph of magnesium malonate b) SAED pattern of magnesium malonate c) HRTEM micrograph of magnesium peroxide d) SAED pattern of magnesium peroxide.

IOP Conf. Series: Materials Science and Engineering **78** (2015) 012005 doi:10.1088/1757-899X/78/1/012005

Further calcination of pre calcinated sample at higher temperature will lead to the formation of highly stable MgO nanoparticles[13]. We recommend that increasing the calcination time and calcinating the sample in ambient atmosphere can help to obtain monodispersed MgO_2 nanoparticles and to avoid agglomeration [47]. By obtaining monodispersed nanoparticles, it can be used in bioremediation applications and in improving soil quality on par with bulk magnesium peroxides.

4. Conclusion

In this work, MgO_2 nanoparticles have been synthesized by simple sol – gel method using inorganic precursor such as magnesium acetate, oxalic acid as gelating agent and ethanol as solvent. Initially, magnesium malonate was formed which was due to the inorganic precursor and shows that precursor plays a major role in the formation of nanoparticles. Further, the as prepared samples were calcinated at 600°C in order to obtain MgO_2 nanoparticles, which show that calcination temperature plays a pivotal role in the formation of the end product as well as the morphology of the nanoparticles. Thus the synthesized MgO_2 nanoparticles can be used as an excellent oxygen source in agriculture and in environment application.

5. References

[1] Wagner G W, Bartram P W, Koper O and Klabunde K J 1999 Reactions of VX, GD, and HD with nanosize MgO *The Journal of Physical Chemistry B***103** 3225-8

[2] Davini P and Tartarelli R 1985 Emissions from the combustion of heavy oils: effect of granulometry and surface area of MgO additive *Fuel***64** 380-3

[3] Duyar O and Durusoy H Z 2004 Design and preparation of antireflection and reflection optical coatings *Turk. J. Phys***28** 139-44

[4] Phillips J M 1996 Substrate selection for high-temperature superconducting thin films *Journal of Applied Physics***79** 1829-48

[5] Hamet J, Mercey B, Hervieu M and Raveau B 1992 < i> a</i>-Axis oriented YBa< sub> 2</sub> Cu< sub> 3</sub> O< sub> 7</sub> superconducting thin film grown on MgO substrate *Physica C: Superconductivity***193** 465-70

[6] Liu S, Weaver J, Yuan Z, Donner W, Chen C, Jiang J, Meletis E, Chang W, Kirchoefer S and Horwitz J 2005 Ferroelectric (Pb, Sr) TiO 3 epitaxial thin films on (001) MgO for room temperature high-frequency tunable microwave elements *Applied Physics Letters***87** 142905--3

[7] Hattori H 1995 Heterogeneous basic catalysis *Chemical Reviews***95** 537-58

[8] Mishakov I V, Bedilo A F, Richards R M, Chesnokov V V, Volodin A M, Zaikovskii V I, Buyanov R A and Klabunde K J 2002 Nanocrystalline MgO as a dehydrohalogenation catalyst *Journal of Catalysis***206** 40-8

[9] Chesnokov V V, Bedilo A F, Heroux D S, Mishakov I V and Klabunde K J 2003 Oxidative dehydrogenation of butane over nanocrystalline MgO, Al< sub> 2</sub> O< sub> 3</sub>, and VO< sub> x</sub>/MgO catalysts in the presence of small amounts of iodine *Journal of Catalysis***218** 438-46

[10] Kumar D, Reddy V B, Mishra B G, Rana R, Nadagouda M N and Varma R S 2007 Nanosized magnesium oxide as catalyst for the rapid and green synthesis of substituted 2-amino-2-chromenes *Tetrahedron***63** 3093-7

[11] Climent M J, Corma A, Iborra S and Mifsud M 2007 MgO nanoparticle-based multifunctional catalysts in the cascade reaction allows the green synthesis of anti-inflammatory agents *Journal of Catalysis***247** 223-30

[12] Mirzaei H and Davoodnia A 2012 Microwave assisted sol-gel synthesis of MgO nanoparticles and their catalytic activity in the synthesis of hantzsch 1, 4-dihydropyridines *Chinese Journal of Catalysis* **33** 1502-7
[13] Kumar A and Kumar J 2008 On the synthesis and optical absorption studies of nano-size magnesium oxide powder *Journal of Physics and Chemistry of Solids* **69** 2764-72
[14] Jiu J, Kurumada K-i, Tanigaki M, Adachi M and Yoshikawa S 2004 Preparation of nanoporous MgO using gel as structure-direct template *Materials Letters* **58** 44-7
[15] Chadwick A V, Poplett I J, Maitland D T and Smith M E 1998 Oxygen speciation in nanophase MgO from solid-state 17O NMR *Chemistry of materials* **10** 864-70
[16] Bokhimi X, Morales A, Portilla M and García-Ruiz A 1999 Hydroxides as precursors of nanocrystalline oxides *Nanostructured Materials* **12** 589-92
[17] Subramania A, Kumar G V, Priya A S and Vasudevan T 2007 Polyol-mediated thermolysis process for the synthesis of MgO nanoparticles and nanowires *Nanotechnology* **18** 225601
[18] López T, Gómez R, Navarrete J and López-Salinas E 1998 Evidence for Lewis and Brønsted Acid Sites on MgO Obtained by Sol-Gel *Journal of sol-gel science and technology* **13** 1043-7
[19] Utamapanya S, Klabunde K J and Schlup J R 1991 Nanoscale metal oxide particles/clusters as chemical reagents. Synthesis and properties of ultrahigh surface area magnesium hydroxide and magnesium oxide *Chemistry of Materials* **3** 175-81
[20] Kim J Y, Jung H S and Hong K S 2005 Effects of Acetic Acid on the Crystallization Temperature of Sol–Gel-Derived MgO Nano-Powders and Thin Films *Journal of the American Ceramic Society* **88** 784-7
[21] Wang J, Novaro O, Bokhimi X, Lopez T, Gomez R, Navarrete J, Llanos M and Lopez-Salinas E 1998 Characterizations of the thermal decomposition of brucite prepared by sol–gel technique for synthesis of nanocrystalline MgO *Materials letters* **35** 317-23
[22] Wang J, Bokhimi X, Novaro O, Lopez T and Gomez R 1999 Effects of the surface structure and experimental parameters on the isopropanol decomposition catalyzed with sol–gel MgO *Journal of Molecular Catalysis A: Chemical* **145** 291-300
[23] Ding Y, Zhang G, Wu H, Hai B, Wang L and Qian Y 2001 Nanoscale magnesium hydroxide and magnesium oxide powders: control over size, shape, and structure via hydrothermal synthesis *Chemistry of Materials* **13** 435-40
[24] El-Shall M S, Slack W, Vann W, Kane D and Hanley D 1994 Synthesis of nanoscale metal oxide particles using laser vaporization/condensation in a diffusion cloud chamber *The Journal of Physical Chemistry* **98** 3067-70
[25] Matthews J S, Just O, Obi-Johnson B and Rees W 2000 CVD of MgO from a Mg (β-ketoiminate) 2: Preparation, Characterization, and Utilization of an Intramolecularly Stabilized, Highly Volatile, Thermally Robust Precursor *Chemical Vapor Deposition* **6** 129-32
[26] Helble J J 1998 Combustion aerosol synthesis of nanoscale ceramic powders *Journal of Aerosol Science* **29** 721-36
[27] Bhargava A, Alarco J A, Mackinnon I D, Page D and Ilyushechkin A 1998 Synthesis and characterisation of nanoscale magnesium oxide powders and their application in thick films of Bi< sub> 2</sub> Sr< sub> 2</sub> CaCu< sub> 2</sub> O< sub> 8</sub>*Materials Letters* **34** 133-42
[28] Kordas G 2000 Sol–gel preparation of MgO fibers *Journal of Materials Chemistry* **10** 1157-60
[29] Zhu Q, Oganov A R and Lyakhov A O 2013 Novel stable compounds in the Mg–O system under high pressure *Physical Chemistry Chemical Physics* **15** 7696-700
[30] Oganov A R, Gillan M J and Price G D 2003 Ab initio lattice dynamics and structural stability of MgO *The Journal of chemical physics* **118** 10174-82
[31] Umemoto K, Wentzcovitch R M and Allen P B 2006 Dissociation of MgSiO3 in the cores of gas giants and terrestrial exoplanets *Science* **311** 983-6
[32] Wriedt H 1987 The Mg– O (Magnesium-Oxygen) system *Bulletin of Alloy Phase Diagrams* **8** 227-33

[33] Recio J and Pandey R 1993 Ab initio study of neutral and ionized microclusters of MgO *Physical Review A***47** 2075
[34] Wang Z, Bentley J, Kenik E, Horton L and McKee R 1992 In-situ formation of MgO< sub> 2</sub> thin films on MgO single-crystal surfaces at high temperatures *Surface science***273** 88-108
[35] Znaidi L, Chhor K and Pommier C 1996 Batch and semi-continuous synthesis of magnesium oxide powders from hydrolysis and supercritical treatment of Mg (OCH< sub> 3</sub>)< sub> 2</sub>*Materials Research Bulletin***31** 1527-35
[36] Vidali M 2001 Bioremediation. An overview *Pure and Applied Chemistry***73** 1163-72
[37] Rockne K J and Reddy K R 2003 Bioremediation of contaminated sites. In: *Invited Theme Paper, International e-Conference on Modern Trends in Foundation Engineering: Geotechnical Challenges and Solutions, Indian Institute of Technology, Madras, India,*
[38] Chang Y-J, Chang Y-T and Hung C-H 2008 The use of magnesium peroxide for the inhibition of sulfate-reducing bacteria under anoxic conditions *Journal of industrial microbiology & biotechnology***35** 1481-91
[39] Tabrizi A, Ayhan F and Ayhan H 2009 Gold nanoparticle synthesis and characterisation *Journal of biology and chemistry Hacettepe J. Biol. & Chem***37** 217-26
[40] Hench L L and West J K 1990 The sol-gel process *Chemical Reviews***90** 33-72
[41] Doreswamy B, Mahendra M, Sridhar M, Shashidhara Prasad J, Varughese P, George J and Varghese G 2005 A novel three-dimensional polymeric structure of crystalline neodymium malonate hydrate *Materials Letters***59** 1206-13
[42] Tang Z-X and Shi L-E 2008 Preparation of nano-MgO using ultrasonic method and its characteristics *Eclética Química***33** 15-20
[43] Zhang A, Ma Q, Wang K, Liu X, Shuler P and Tang Y 2006 Naphthenic acid removal from crude oil through catalytic decarboxylation on magnesium oxide *Applied Catalysis A: General***303** 103-9
[44] Štengl V, Bakardjieva S, Marıková M, Bezdicka P and Šubrt J 2003 Magnesium oxide nanoparticles prepared by ultrasound enhanced hydrolysis of Mg-alkoxides *Materials Letters***57** 3998-4003
[45] Athar T, Hakeem A and Ahmed W 2012 Synthesis of MgO Nanopowder via Non Aqueous Sol–Gel Method *Advanced Science Letters***7** 27-9
[46] Makhluf S, Dror R, Nitzan Y, Abramovich Y, Jelinek R and Gedanken A 2005 Microwave-Assisted Synthesis of Nanocrystalline MgO and Its Use as a Bacteriocide *Advanced Functional Materials***15** 1708-15
[47] Sundrarajan M, Suresh J and Gandhi R R 2012 A comparative study on antibacterial properties of MgO nanoparticles prepared under different calcination temperature *Digest Journal of Nanomaterials and Biostructures***7** 983-9

Acknowledgement
The authors would like to thank Department of Chemical and Petroleum Engineering, School of Engineering and Science, Curtin University and National Centre for Nanoscience and Nanotechnology (NCNSNT), University of Madras for providing support to carry out this project.

Optimization and performance evaluation for nutrient removal from palm oil mill effluent wastewater using microalgae

Raheek I Ibrahim[1,3], **Z H Wong**[2] **and A W Mohammad**[2]
[1] Chemical engineering department, University of Technology, Baghdad, Iraq
[2] Chemical and process engineering department, Universiti Kebangsan, Malaysia, Selangor, Malaysia

E-mail: doctorraheek@yahoo.com

Abstract. Palm oil mill effluent (POME) wastewater was produced in huge amounts in Malaysia, and if it discharged into the environment, it causes a serious problem regarding its high content of nutrients. This study was devoted to POME wastewater treatment with microalgae.The main objective was to find the optimum conditions (retention time, and pH) in the microalgae treatment of POME wastewater considering retention time as a most important parameter in algae treatment, since after the optimum conditions there is a diverse effect of time and pH and so, the process becomes costly. According to our knowledge, there is no existing study optimized the retention time and pH with % removal of nutrients (ammonia nitrogen NH_3-N, and orthophosphorous PO_4^{3-}) for microalgae treatment of POME wastewater. In order to achieve with optimization, a central composite rotatable design with a second order polynomial model was used, regression coefficients and goodness of fit results in removal percentages of nutrients (NH_3-N, and PO_4^{3-}) were estimated.WinQSB technique was used to optimize the surface response objective functionfor the developed model. Also experiments were done to validate the model results.The optimum conditions were found to be 18 day retention time for ammonia nitrogen, and pH of 9.22, while for orthophosphorous, 15 days were indicated as the optimum retention time with a pH value of 9.2.

1. Introduction

Raw palm oil mill effluent (POME) wastewater is a colloidal suspension containing 95-96% water, 0.6-0.7% oil, and 4-5% total solids, including 2-4% suspended solids, also POME contains high concentrations of proteins, carbohydrate, nitrogenous compounds, lipids and minerals that may be converted into useful materials using microbial processes. If this untreated POME was thrown out to the environment, it has a negative effect on it, therefore it must be a critical needful solution savings the equilibrium between environmental safety and sustainable energy that can be produced from nutrient materials existing in the POME [1]. Bio-treatment of POME with microalgae is the solution because the algae grows in photosynthesis method that has the ability to use solar energy and transforming it to a profitable biomass take advantages from nitrogen and phosphorous compounds presented already in POME. Microalgae culture becomes an economizing solution to the problem treats the wastewater by consuming nutrients by microalgae growth and transform it to a biomass

[3] To whom any correspondence should be addressed.

which is in turn a major source of biofuel [2].Colak et al (1988) [3], used the microalgae for industrial wastewater treatment. Mohammed (1994) [4], used algae to treat wastewater with high rates ofnitrogenous compounds.The usual methods of experimentation in any process have been neglecting the interaction effects between the operating parameters rather than it's considered as time consuming methods. Thus, the efficiency of the optimization process is negatively affected by these restrictions, and as a result, the response surface methodology (RSM) statistical experimental design is detected. Some researchers optimized the biological treatment using different microalgae strains to treat the wastewater of various sources [5-10]. This study is devoted to the treatment of POME wastewater using biological method, consisting of microalgae treatment of the wastewater, then optimization of the parameters affecting the biological treatment process using RSM statistical experimental design. Because there is no specific literature existed on how the retention time and pH of algal solution would affect along with the treatment of POME in range of applicable experimental variables. Thus the main goal of the present study was to investigate the effect of retention time (0-23 days) and pH (7.48-9.26) on the efficiency of algae treatment of POME, and optimize these operating parameters to find the optimum conditions to reach the highly efficient algae treatment with highest removal percentages of ammonia nitrogen (AN), and orthophosphorus (OP). Also, this process is considered as a completely green process without using any chemicals that produces environmental injured byproducts.

2. Materials and methods

2.1 Palm oil mill effluent (POME)

Palm oil mill effluent (POME) samples were taken directly after the last anaerobic disgestor from East Mill Sime Darby Plantation, Carey Island, Selangor, Malaysia, the samples were kept in a cool room at a temperature around 4°C before using, they were being centrifuged at 8000 rpm for 5 minutes to obtain the supernatant of the samples.The characteristics after centrifugation are shown in Table 1.

Table 1. Characteristic of POME after centrifugation.

Parameter	Value
pH (-)	7.48
Chemical Oxygen Demand (COD), (mg/L)	4045.0
Ammonia Nitrogen (NH_3-), (mg/L)	334.0
Orthophosphorus (PO_4^{-3}), (mg/L)	465.0
Total Suspended Solids (TSS), (mg/L)	607.0
Turbidity (NTU)	395.5

2.2 Microalgae

In this research, the microalgae strain selected was *Botryococcus Braunii* which was obtained from the Department of Chemical and Process Engineering, Faculty of Engineering and Built Environment in the National University of Malaysia. *Botryococcus braunii* in the concentration of 0.3 g/l was cultivated in Bold Basal Medium (BBM) for 10 days before transferring it into a conical flask containing two litters of AnPOME. The medium was aerated using a peristaltic pump with aeration of 1 l/hr placed on the bottom. The flask was placed under two continuous fluorescent lamps from both sides with an average light intensity of 1000 lux and a constant temperature of 21 °C.

2.3 Analytical method

Ammonia nitrogen (NH_3-N) was measured using Nessler method while orthophosphorus (PO_4^{3-}) was measured using Molybdovanadate method. NH_3-N, and PO_4^{3-} were analysed with the help of HACH DR/3900 spectrophotometer. For pH, a pH meter was used.

2.4 Experimental Procedure

After the microalgae strain of *B. braunii* was cultivated for 10 days, 200 ml was inoculated into 2 liters of POME that was pretreated with centrifugation for removal of solids. The reactor consists of a 2 liter conical flask with providing aeration from the bottom using a peristaltic pump. The top of the conical flask was covered with double layer aluminum foils for prevention of contamination with other microalgae strains. Two continuous fluorescent lamps were placed from both sides above the reactor with a constant temperature of 21 °C. The samples were taken with an interval of 3 days, from day 0 to day 15 retention time (RT) and the sample taken was on day 23 retention time. Throughout the entire treatment, the pH was not adjusted by any chemicals.As the samples were collected, another round of centrifugation was done to obtain the supernatant of all the samples. Then analysis of NH_3-H, PO_4^{-3}, and pH were done to observe the performance of microalgae treatment with different retention time.

3. Optimization technique

In order to achieve with the RSM of experimental design, we must firstly calculate the number of experiments required as $2^K + 2K + 1$,where, K is the number of variables. It means that for two operating variables, we need at least 9 experiments, plus two experiments represent the probability of experimental errors. According to RSM of experimental design, the operating variables that need to be optimized such as retention time (X_1), and pH (X_2), were correlated in a non-linear second order polynomial model equation, which is more sufficient proposed model to show the interaction effects between variables. The coefficients of model equation were estimated using a STATISTICA program (version 8.0). The proposed model was:

$$Y = a_0 + a_1X_1 + a_2X_2 + a_3X_1X_2 + a_4X_1^2 + a_5X_2^2 \tag{1}$$

Where: x1 and x2 are operating variables representing retention time (day), and pH, while, Y represents the objective function as percentage removal of nutrients (%). The optimization process was done using a winQSB technique (winQSB software version 1.0) in order to find the optimum values of operating conditions for the purpose of enhancing the efficiency of the POME treatment process using the microalgae.

4. Results and discussion

4.1 Multiple regression analysis results

In order to achieve with the objective, a polynomial model regression coefficients and goodness of fit results in removal percentages of nutrients were estimated as:

4.1.1 Ammonia nitrogen (AN)

For ammonia nitrogen (AN), the second order polynomial is:

$$Y = 3389.118 + 124.819X_1 - 856.201X_2 - 13.4514X_1X_2 + 0.007139X_1^2 + 53.8917X_2^2$$

With goodness of fit analysis as:Correlation coefficient (R)=0.99955, the variance explained= 99.91%, mean square error=0.84943, and mean absolute error=0.7115.

4.1.2 Orthophosphorous (OP)

For orthophosphorous (OP), the second order polynomial is:

$$Y = -6494.96 - 105.026X_1 + 1522.328X_2 + 11.9975X_1X_2 - 0.19423X_1^2 - 88.5063X_2^2$$

With goodness of fit analysis as:Correlation coefficient (R)=0.9992, the variance explained= 99.84%, mean square error=0.51285, and mean absolute error= 0.5535. Table 2, shows the results of experimental design.

Table 2. Results of RSM experimental design.

Coded values		Real values		OP removal (%)		AN removal (%)	
Retention time (day)	pH (-)	Retention time (day)	pH (-)	predicate	observed	predicate	observed
-1	-1	0	7.48	0.00223	0.000	-0.00289	0.000
-1	0	0	9.14	25.362	25.255	65.473	65.320
-1	+0	0	9.26	42.127	43.101	81.720	80.980
0	-1	12	7.50	44.532	44.962	83.209	83.090
0	0	12	9.26	48.833	47.995	85.857	85.160
0	+0	12	9.14	52.948	52.600	89.050	88.600
+1	-1	23	7.50	47.899	47.560	90.642	89.940
+1	0	23	9.14	48.552	48.521	91.012	90.970
+1	+0	23	9.26	49.490	49.400	91.517	91.400
0	0	12	9.14	52.948	52.600	89.050	88.600
0	0	12	9.14	52.948	52.600	89.050	88.600

4.2 Optimum condition results and effect of operating parameters

The optimum conditions that give optimum objective functions are explained as follows: The optimum conditions for % removal of AN are:Time = 18 days, and pH = 9.22 gives maximum removal of 92.857%. Figure 1, shows the optimum retention time, and figure 4, shows the effect of time and pH on % removal of AN as a response surface.It was clearly obvious that 18 days is the best time for maximum removal of AN, since before day 18, the microalgae still grown by feeding with nitrogen and phosphorous compounds in the POME wastewater, and after a day (18), the values of removal percentages of AN were lowered because that microalgae capacity to eliminate AN was reduced and the microalgae became older. Also, when the microalgae removed AN, it was simultaneously transferred it to the biomass as a result of concentration equilibrium and mass conservation, that's clarified why the removal percentage was decreased after optimum retention time. But at the optimum it has a maximum removal and maximum conversion of AN into biomass, which is more useful in biofuel production. The optimum conditions for % removal of OP are: Time = 15 days, and pH = 9.2 gives maximum removal of 56.5%. Figure 2, shows the optimum retention time. The retention time (15) day represents the optimum retention time for OP removal. It is the same reason for AN removal, dealing with equilibrium and mass conservation of OP in the algal solution, figure 5, shows the effect of time and pH on % removal of OP as a response surface. Figure 3, shows the effect of pH values on AN and OP concentrations, it can be clearly distinguished that at optimum value of pH, found a minimum concentration of nutrients. In addition to that the mirage consumption of AN is more significant and has a sharper decreasing than OP within the experimental range of operating variables, this can be attributed to chemical composition and initial concentration of each compound in the raw POME wastewater.

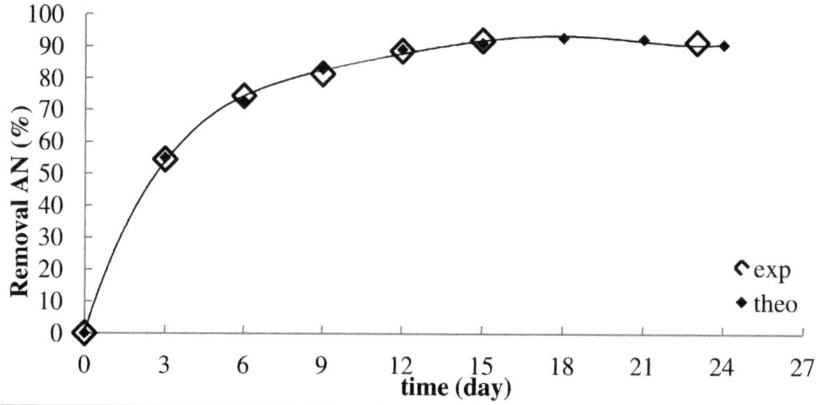

Figure 1. Effect of retention time on % removal of ammonia nitrogen.

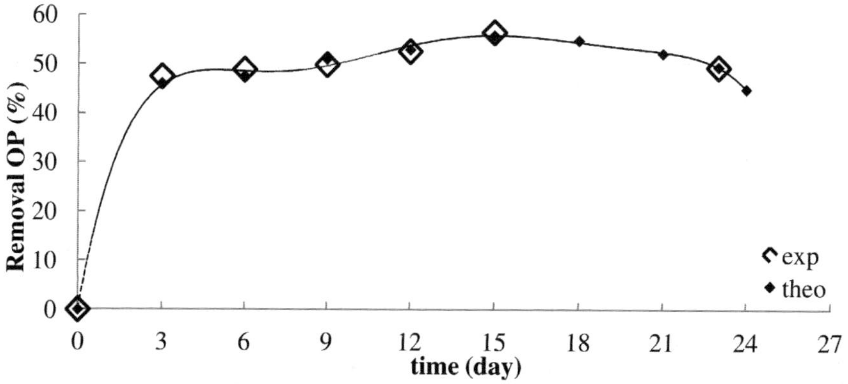

Figure 2. Effect of retention time on % removal of orthophosphorous.

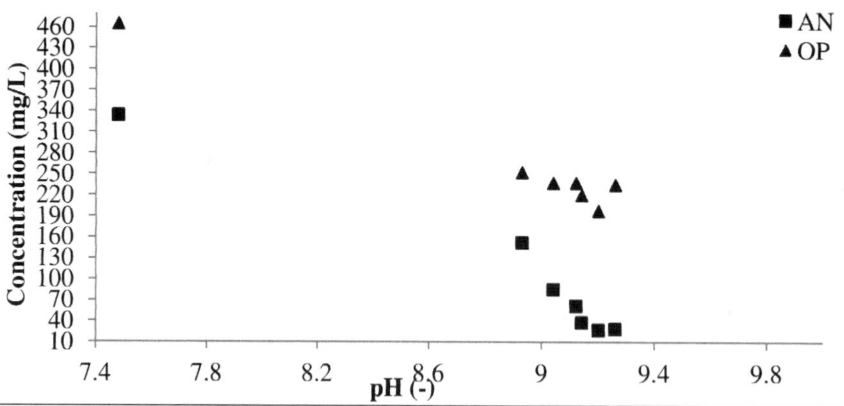

Figure 3. Effect of pH on ammonia nitrogen and orthophosphorous concentrations.

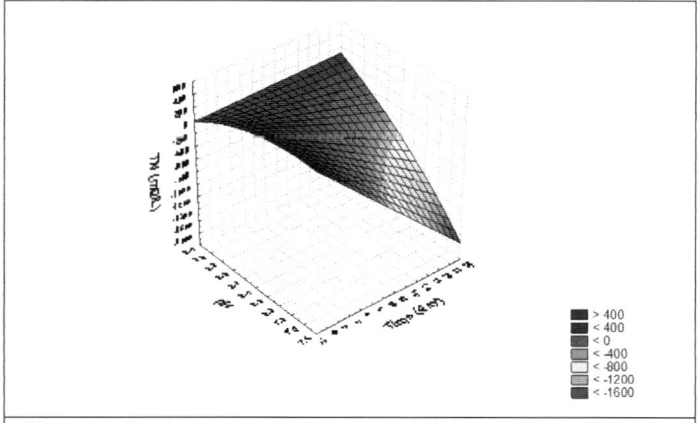

Figure 4. Surface response of ammonia nitrogen concentrations.

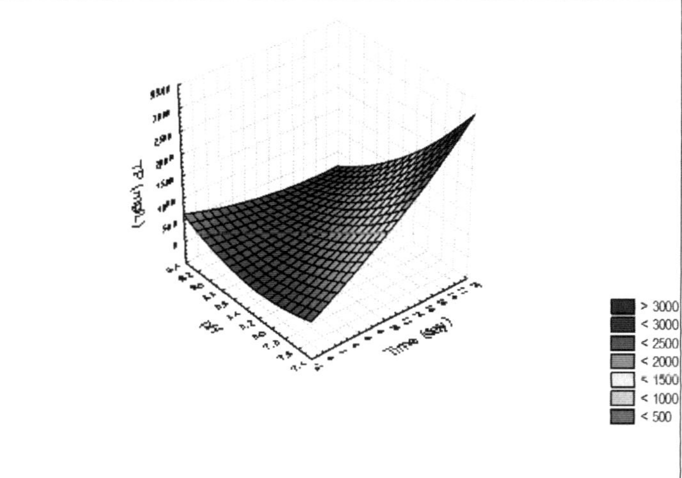

Figure 5. Surface response of orthophosphorous concentrations.

4.3 Evaluation of the process efficiency

From the point of view of researcher, the process is succeed and cost effective, since it used only biological source in the treatment and no chemicals were present, therefore it considered as a totally green treatment process and environmentally friendly. In addition, its useful biomass could be use in a recently highlydeveloped biofuel production.

5. Conclusions

From this experimental work and optimization, we can conclude that: biotreatment using microalgae is considered as a promising method for POME wastewater treatment in addition to biofuel production, distinguished as totally green process without any chemicals used. From optimization done on operating variables of microalgae treatment, it was shown that retention time is a very important parameter needs to be estimated carefully in order to keep the treatment process useful and cost effective. According to optimization process using experimental design technique, the optimum conditions for % removal of ammonia nitrogen was 18 days retention time, and pH value of 9.22, while for % removal of orthophosphorous, it was15 days retention time, and 9.2 for pH value.

6. References

[1] Wu Ty , Mohammad AW, Jahim J Md, and, Anuar N 2007 Palm oil mill effluent (POME) treatment and bioresources recovery using ultrafiltration membrane: Effect of pressure on membrane fouling *Biochemical Eng. J.* **35** 309-317.

[2] Abdel-Raouf N , Al-Humaidan AA , and Ibrahim IBM 2012 Review: Microalgae and wastewater treatment *Saudi J.Bio. Sci* **19** 257-275.

[3] Colak O, and Z Kaya 1988 A study on the possibilities of biological wastewater treatment using algae *Doga Biyoliji Serisi* **12**(1) 18-29.

[4] Mohammed NA 1994 Application of algal ponds for wastewater treatment and algal production *MSc Thesis* (Cairo Univ.).

[5] Somong OT , Prasertsan P , Dhamwichukorn S, and Birkeland N 2008 Optimization of simultaneous thermophilic fermentative hydrogen production and COD reduction from palm oil effluent by Thermoanaerobacterium- rich sludge *Int. J. Hydrogen Energy* **33** 1221-1231.

[6] Poonsuk P, Thong SO and Birkeland NK 2009 Optimization and microbial community analysis for production of biohydrogen from palm oil mill effluent by thermophilic fermentative process *Int. J. Hydrogen Energy* **34** 7448-7459.

[7] Shaikh AR, Hossain MM, Lucky RA and Bassi AS 2013 Intigrated CO2 capture, wastewater treatment and biofuel production by microalgae culturing- a review *Renew. & Sus. Energ. Rev.* **27** 622-653.

[8] Giorgas M and Nerantzis E 2013 Microalgae for high- value compounds and biofuels production: A review with focus on cultivation under stress conditions *Biotech. Advances.* **31**1532-1542.

[9] Yasser RA and Olama ZA 2002 L-asparaginase production by pseudomonas aeruginosa in solid- state culture: evaluation and optimization of culture conditions using factorial designs *Process Biotech.* **38** 115-122.

[10] Li C, Bai J, Cai Z and Ouyang F 2002 Optimization of cultured medium for bacteriocin production by Lactococcus lactis using response surface methodogy *J. Biotech.* **93** 27-34.

Acknowledgments

The authors would like to thank the staff and support of Faculty of Engineering and Build Environment in Universiti Kebangsaan Malaysia/ UKM for their financial support in doing the research.

Experimental Investigation on Sandstone Rock Permeability of Pakistan Gas Fields

Arshad Raza[1], **Chua Han Bing**[1], **Ramasamy Nagarajan**[1] **and Mohamed Ali Hamid** [1]

[1]School of Engineering and Science, Curtin University, Sarawak, Malaysia

Email: arshad.raza@postgrad.curtin.edu.my

Abstract. Permeability is the ability of formation to produce hydrocarbon which is affected by compaction, pore size, sorting, cementation, layering and clay swelling. The effect of texture on permeability in term of grain size, sorting, sphericity, degree of cementing has been reported in literature. Also, the effect of permeability on capillary pressure, irreducible water saturation, displacement pressure and pore geometry constant has been studied separately. This preliminary study presents the experimental results of eight samples to understand the effect of similar factors of texture on permeability. With the knowledge of the results, it can be said that the effect of grain size, cementation, texture material, sphericity, and porosity can't be observed on permeability except sorting when less than ten samples are considered from different depositional environment. The results also show the impact of permeability on capillary pressure, irreducible water saturation, and displacement pressure and pore geometry index as similar as published in the literature.

1. Introduction

Permeability is a key petrophysical parameter that has the capability to conduit the fluid flow, which has been used as a vital factor in the development and management of petroleum reservoir [1, 2]. As a matter of fact, permeability is indicative of reservoir system quality. It generally ranges from 0.1 to 1000 mD or more [3]. Permeability depends on effective porosity of rock, whereas, it is also controlled by composition, texture and structure of rocks [4]. Texture of sandstone is different due to variation in depositional environment, and it consists of grain size, sphericity, sorting, porosity, texture material and textural maturity of formation. A number of researchers [5-7] reported grain size ranged from 0.125 mm to 2 mm. Furthermore, sandstone exists in a variety of colors ranging from white to black due to extensive fluctuation in depositional environment [8].

It has been shown by previous researchers that grain size, and sorting are fundamental variables which are controlling permeability in unconsolidated rocks. As better sorting steadily increases porosity as well as permeability. Other researchers [9, 10] also investigated the effect of grain size on permeability. Their common conclusion was that the increase in gravel and coarse grain size types of particle in sandstone causes an increase in permeability whereas reducing the porosity. Many relationships of permeability with the porosity; irreducible water saturation and sorting are presented in literature [10, 11, 12].

Also, as a result of the extensive research conducted to date on the subject of permeability, it has proved the effect of permeability on capillary pressure, displacement pressure, irreducible water saturation and pore geometry index. All these parameters except pore geometry index (λ) dramatically increase with the decrease in permeability due to squeezing of pore throat radius [13]. Whereas, pore geometry index (λ) which is a measure of the degree of pore sorting shows a direct relationship with permeability [14].

Recently, a comparative study of porous plate method, oil displacement method and centrifugation method have been presented to show that irreducible water saturation decreases with the increasing permeability [15]. Additionally, the impact of petrophysical control on irreducible saturation, displacement pressure and pore throat size distribution have also been investigated by centrifuge experimental displacement technique [16].

The objective of this experimental study is to provide a source review of key factors (grain size, cementation, texture material, sphericity, sorting and porosity) influence on permeability.

2. Materials and Methods

Eight driller's samples of different formations were obtained for this study as mentioned in table 1. Core fittings of 1.5-inch in the width and 3-inch in length were prepared to carry out necessary experiments (porosimeter, permeameter, forced injection and capillary pressure measurement) by following the standard procedure of apparatus [17, 18].

By means of Manual Dynoptic Polarizing Microscope (Model L.M. 1966) [19], rock texture of representative core samples was directly predicted visually on prepared thin sections as can be seen in figure 1.

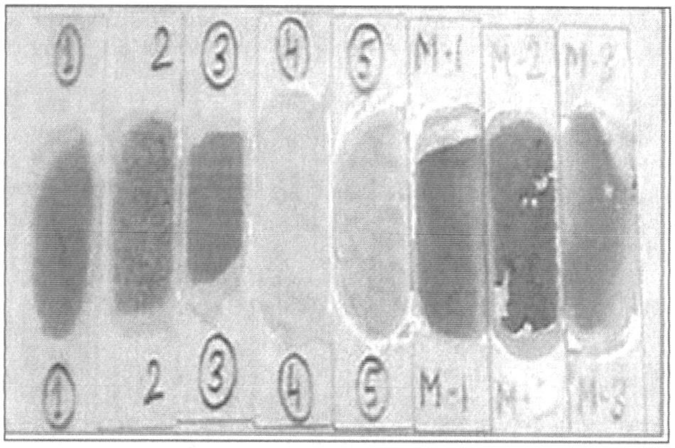

Figure 1. Thin section films.

Direct effective pore volume of the same cores was measured by using a Helium Porosimeter, TPI-219 Cortest system [20]. In order to calculate the porosity, bulk volume was measured gravimetrically. Permeability was analyzed by Ruska Gas Permeameter-1011-801 [21]. In third place, core samples were 100% saturated using Manual Saturator-535 Coretest System [22] with prepared saline water (density of 1.01g/cc). Gravimetric capillary pressure system, TGC - 764 based on porous plate method [23] was used to determine primary drainage capillary pressure curve and the irreducible water saturation. The equipment used for air-brine porous plate method consisted of a pressure control system, a balance and a pressure vessel with a porous ceramic plate of 15 bar (threshold pressure). In addition, stable saturation was gravimetrically determined at different desaturation pressures to obtain capillary pressure curves.

Grain sizes of all grains on thin sections were measured under attached microscopic scale with the lense during the examination and average grain size of samples were gotten by arithmetic mean as accounted for in table 1. Sorting and sphericity were visually examined and matched with the Ehlers & Blatt's sorting and sphericity classification [24]. Table 1 lists the results of texture obtained during microscopic analysis.

All the studied clastic rocks are classified as sandstone based on their textural parameters such as grain size, sorting & sphericity by using Ehlers & Blatt's [20] classification and the results are shown in table 1. As evident from the results that rock samples of different locality in this study have widely diversified rock texture.

Table 1. Summary of sandstones texture material description.

Sample ID	1	2	3	4	5	M-1	M-2	M-3
Formation	Mari	Mari	Mari	Dhandhot	Warcha	XYZ	XYZ	XYZ
Grain size (mm)	0.244	0.374	0.08	0.159	0.53	0.214	0.08	0.063
Texture material	Fine Sand	Med. Sand	Very Fine Sand	Fine Sand	Coarse Sand	Fine Sand	Very Fine Sand	Very Fine Sand
Cementing	No	Poor	Poor	Very poor	Highly	Poor	Very poor	Poor
Sphericity	Angular	Angular	Rounded	Sub Angular	Rounded	Very Angular	Rounded	Sub Rounded
Sorting	Poor	Very Poor	Well	Very Well	Well	Poor	Well	Well

Listed in the table 2 are the experimental outcome of parameters such as effective porosity, absolute permeability, irreducible water saturation and displacement pressure. The pore geometry index was determined by matching the measured capillary pressure data with the Brooks & Corey [25] calculated data.

Table 2. Summary of measured properties.

Sample ID	Porosity (%)	Permeability (mD)	Displacement Pressure (psi)	Pore Geometry Index	Irreducible Water Saturation (%)
1	13.4	4	14	1.64	59
2	9.3	0.4	19	1.5	84
3	21.6	26	9	1.824	47
4	17	60	6	2.32	36
5	9.3	30	9	1.94	40
M-1	12.8	2	18	1.63	75
M-2	18.9	25	9	1.8	47
M-3	20	20	9	1.79	50

3. Results Analysis

3.1. Factors Control the Permeability

All the rock texture factors were compared to know their influence on permeability. Texture material mainly varies from very fine grained (n=3) to coarse sand (n=1) and cementing ranges from none or very poor to high. The sphericity ranges between sub angular to rounded. Sample No# 3, 5, M-2 are textural mature. Almost all the samples are well sorted except Sample No# 1, 2 & M-1 as presented in table 1. Sandstone samples under study have porosity ranging from 9.3% to 21.6% and permeability ranging from 0.4 mD to 60 mD as given in table 2.

Figure 2 presents the variation in grain size as reported in table 1 which are plotted versus permeability as can be seen in figure 2. It was found that there is no influence exists on permeability. As described earlier [9] that grain size is a fundamental variable that control permeability in unconsolidated rocks. This contrast may be either due to less number of the samples or almost all the samples except S#1 are consolidated.

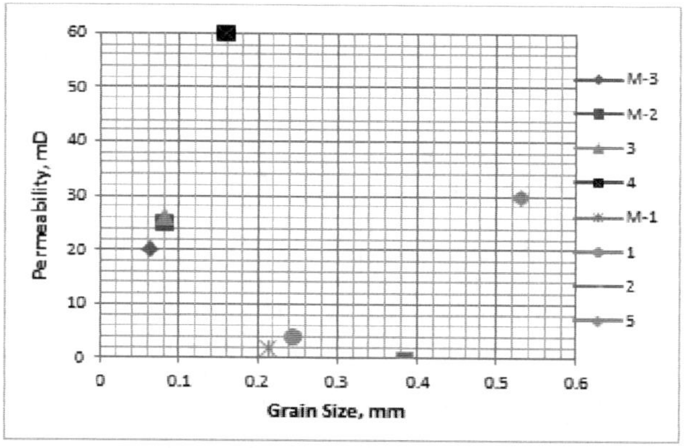

Figure 2. Plot of grain size versus permeability.

It can be seen in figure 3 that there is no positive relationship of the degree of cementing with the permeability if effect of all other parameters are assumed negligible. Albeit, there is no positive correlation exists, however, it can't be denied based on this study that there is no effect of cementing on permeability as it has been indicated earlier about the effect of cementation on permeability [3]. This unique behavior of cementing in sandstone sample can be due to type of texture material and the degree of sphericity in the samples.

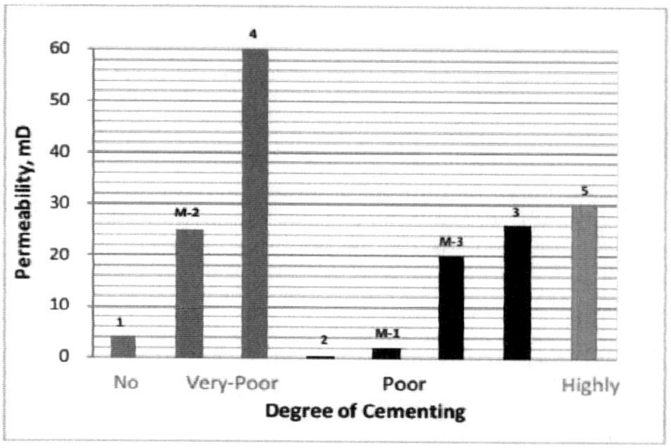

Figure 3. Plot of degree of cementing versus permeability.

Similar to previous described trends, the particle type (very fine sand to coarse sand) doesn't show any remarkable relationship with permeability individually as seen in figure 4.

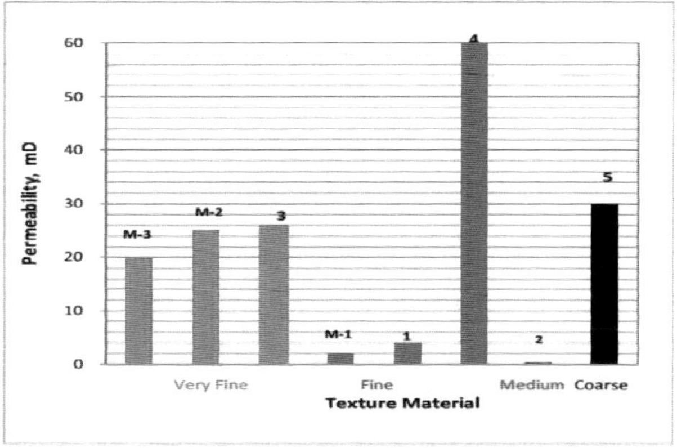

Figure 4. Plot of texture material versus permeability.

Also, with the increase in degree of roundness, permeability doesn't show a positive trend to call it a relationship that can be seen in figure 5. Clearly, permeability is increasing linearly with the degree of sphericity from sub rounded to round grains. Particularly, samples with more angularity showing very less permeability than rounded grain samples.

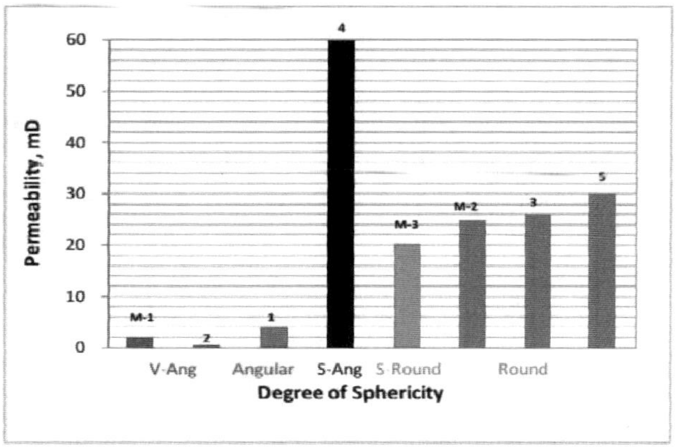

Figure 5. Plot of degree of sphericity versus permeability.

A clear trend can be observed between permeability and level of sorting such that permeability decreases with the decrease in degree of sorting as shown in figure 6. Hence, it can be concluded without any hesitation that individual parameters of texture except sorting don't have any positive impact on the permeability in this study. In fact, this study is not denying the secondary integrated effect of size, shape of grains and the degree of cementing on permeability but it is worth noting that sorting is a major controlling factor in comparison to other factors in consolidated sandstone which even shows positive relation independent of number and locality of the samples. Since, sedimentation process control sorting, therefore, sorting has an influence on the permeability as well as in relationship with it [10].

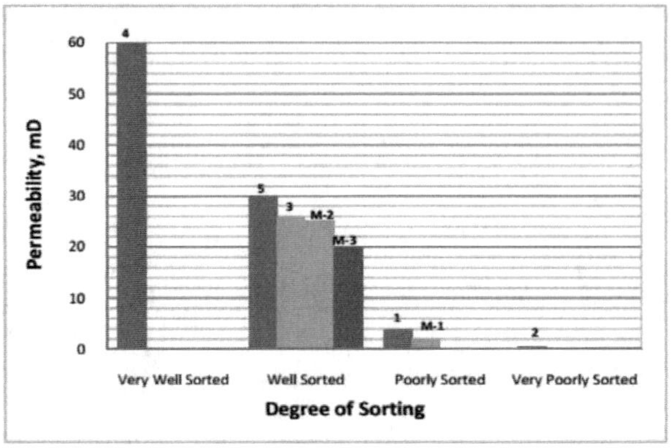

Figure 6. Plot of degree of sorting versus permeability.

As evident from figure 7 that porosity cannot manifest any relationship with permeability in this study. It can be because of a wide variety of fabrics and less number of samples.

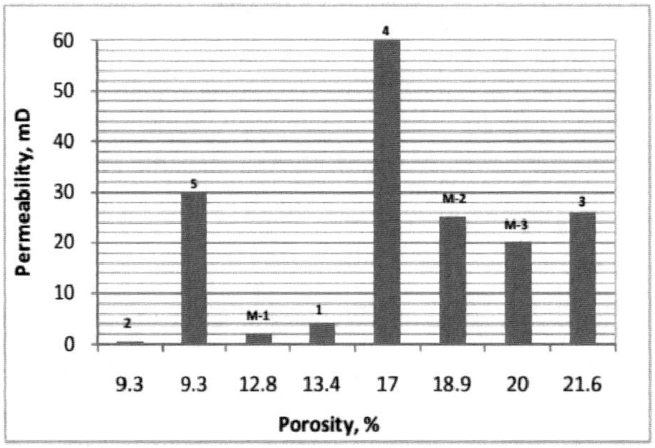

Figure 7. Plot of permeability versus porosity.

3.2. Effect of Permeability on Capillary Pressure, Irreducible Water Saturation, Pore Geometry Index and Displacement Pressure

Although, permeability effects a number of parameters, but capillary pressure, irreducible water saturation, pore geometry index and displacement pressure are considered in this study.

A trend can be observed between permeability and capillary pressure as presented in figure 8. It was found that with a decrease in permeability, the capillary pressure curve is shifting towards right implying increase in irreducible water saturation (S_{wir}); and implying an upward increase in displacement pressure (p_d) and its curvature is decreasing implying broader pore size distribution. The net effect is that as permeability is decreasing, capillary pressure became higher at a given value of saturation. Thus, the study confirms the established behavior of capillary pressure curve changes with the permeability as currently published [13]. A change in capillary pressure curve due to a change in sorting is analogous to changes in permeability because permeability itself is related to sorting as described earlier [10].

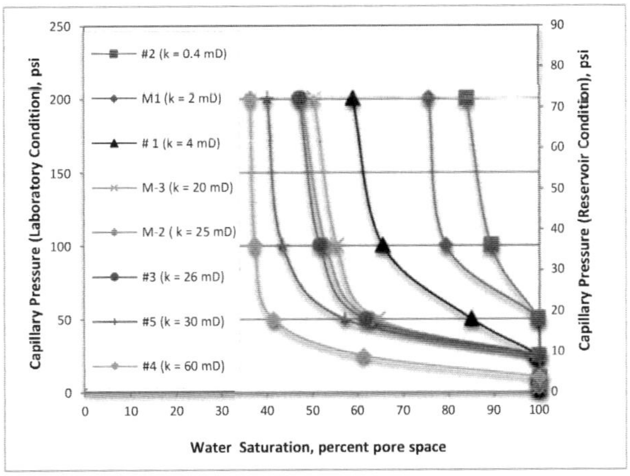

Figure 8. Plot between water saturation versus capillary pressure.

Permeability is plotted versus irreducible water saturation in figure 9. It shows that with the increase in permeability resulting in the decrease in irreducible water saturation. Since, permeability is related to capillary pressure, the irreducible water saturation also relates with capillary pressure. Hence, this trend confirms the concept as published in literature [13, 15, 16].

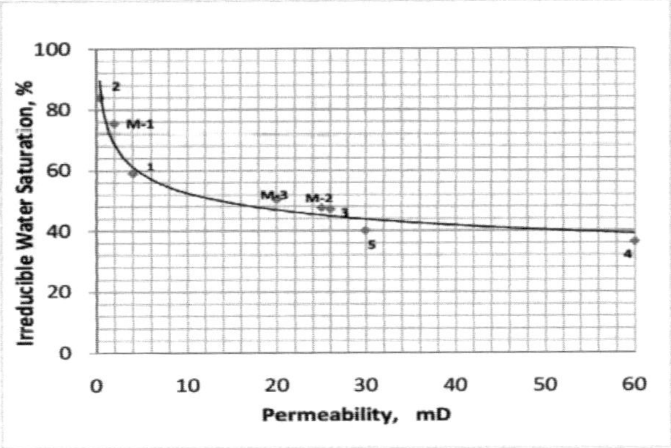

Figure 9. Plot of permeability versus irreducible water saturation.

Figure 10 is a plot of permeability against the pore geometry index. The pore geometry Index (pore sorting) is directly related to the arrangement of grains. It is endorsed that the increase in permeability represents well connected pore geometry as presented earlier [14].

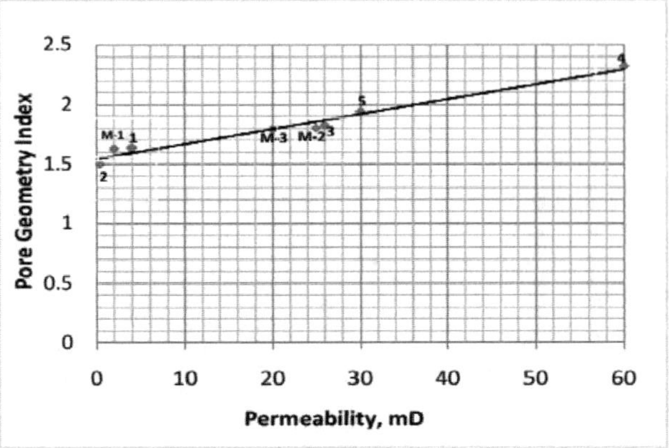

Figure 10. Plot of permeability versus pore geometry index.

Displacement pressure is directly linked to large pore throat radius which can be seen in figure 11. The trend shows that an increase in permeability causes a decrease in displacement pressure due to availability of large pore throats so verified researcher's findings on displacement pressure [13, 16].

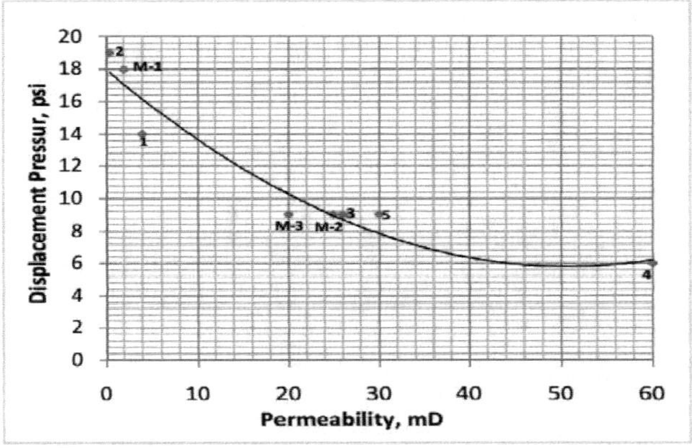

Figure 11. Plot of permeability versus displacement pressure.

4. Conclusion

Sorting is the only parameter of rock texture that is in positive correlation with the permeability. With the increase in degree of sorting from very poor sorted grains to very well sorted grains results increase in sample permeability. Parameters such as grain size, degree of cementing, texture, degree of sphericity and porosity have integrated influence on variation of permeability but any particular relationship with permeability can't be observed with the few numbers of samples. Variation in permeability has great influence on capillary pressure, irreducible water saturation, pore geometry index, displacement pressure.

References

[1]. Hansgeorg P, Clauser C and Joachim I 1998 Permeability Prediction for Reservoir Sandstones and Basement Rocks Based on Fractal Pore Space Geometry, *SEG Expanded abstracts.*

[2]. Ahmed T 2001 *Reservoir Engineering Handbook, 2nd Edition (Texas: Gulf Publishing Company)* p **183-251**.

[3]. Djebbar T and Erle C 2004 *Petrophysics, Theory and Practice of Measuring Reservoir Rock and Fluid Transport Properties 2nd Edition, (USA: Elsevier)* p **81-120**.

[4]. Al-laboun A et al 2014 Reservoir Characterization of the Burqan Formation sandstone from Midyan Basin, northwestern Saudi Arabia *Turkish Journal of Earth Sciences.*

[5]. Nelson S A 2012 Occurrence, Mineralogy, Textures, and Structures of Sedimentary Rocks.??

[6]. Weimer R J and Tillman R W 1982 Sandstone Reservoirs *Society of Petroleum Engineers* (SPE).

[7]. Tucker M E 2011 Sedimentary Rocks in the Field *A Practical Guide, Fourth Edition, John Wiley & Sons, Ltd.*

[8]. Boggs S 2009 *Petrology of Sedimentary Rocks 2nd Edition, (USA: Cambridge University)*

[9]. Russell G 1989 Correlations of Permeability and Grain Size **Vol. 27, No.5**.

[10]. Philip H N 1994 Permeability-Porosity Relationships in Sedimentary Rocks *U.S. Geological Survey, Denver, Colorado.*

[11]. Timur A 1968 An Investigation of Permeability, Porosity, and Residual Water Saturation Relationships *SPWLA Ninth Annual Logging Symposium*, California.

[12]. Torskaya T, Jin G and Verdin C T 2007 Pore-Level Analysis of the Relationship between Porosity, Irreducible Water Saturation, and Permeability of Clastic Rocks *Annual Technical Conference, California.*

[13]. Ashtiani A, Behzad S, Hassanizadeh M, and M.A. Celia 2002 Effects of heterogeneities on capillary pressure–saturation–relative permeability relationships *Journal of contaminant hydrology.*

[14]. Amir Maher S L 2013 Using Capillary Pressure Derived Parameters for Improving Permeability Prediction International *Journal of Geosciences.*

[15]. Zubo Z, Manli L and Weifeng C 2014 An Experimental Study of Irreducible Water Saturation Establishment, *International Symposium of Core Analysts, Avignon, France.*

[16]. Jing X D and Van Wunnik J N M 1998 A capillary pressure functions for interpretation of core-scale displacement experiments In *SCA* , Vol. 9807, p14-16.

[17]. Recommended Practices for core analysis, **rp 40**, *American Petroleum Institute.*

[18]. Petroleum Rock and Fluid Properties Manual, Middle East Technical University, 2003 *Department of Petroleum and Natural Gas Engineering, Ankara.*

[19]. B & L Dynoptic Polarizing Microscope Instructions, 1966. Web. 2 Oct. 2014. Helium Porosimeter Operating Manual TPI-219, *Core Test Systems*, Inc.

[20]. Ruska Permeameter Operating Manual 1011-801, *Ruska Instrumentation Corporation, Houston, TX.*

[21]. Manual Saturator Operating Manual MS-535, *Core Test Systems*, 2005.

[22]. Gravimetric Capillary Pressure System Manual User Manual, TGC-764, *Core Test Systems*, 2006.

[23]. Ernest E and Harvey B 1982 Petrology: Igneous, Sedimentary, and Metamorphic *(San Francisco: W.H. Freeman and Company)* **p 247-359**.

[24]. Brooks R H and Corey A T 1966 Properties of porous media affecting fluid flow *Journal of Irrigation and Drainage Division, Proceedings of the American Society of Civil Engineers.*

Acknowledgment

The University of Engineering and Technology, Lahore, Pakistan is greatly acknowledged for providing the research facilities.

Comparison of Buffer Effect of Different Acids During Sandstone Acidizing

Mian Umer Shafiq[1], **Hisham Khaled Ben Mahmud**[1] **and Mohamed Ali Hamid**[1]

[1]Petroleum Engineering Department, Curtin University, Sarawak, Malaysia

E-mail: umer.engr@hotmail.com

Abstract. The most important concern of sandstone matrix acidizing is to increase the formation permeability by removing the silica particles. To accomplish this, the mud acid (HF: HCl) has been utilized successfully for many years to stimulate the sandstone formations, but still it has many complexities. This paper presents the results of laboratory investigations of different acid combinations (HF: HCl, HF: H_3PO_4 and HF: HCOOH). Hydrofluoric acid and fluoboric acid are used to dissolve clays and feldspar. Phosphoric and formic acids are added as a buffer to maintain the pH of the solution; also it allows the maximum penetration of acid into the core sample. Different tests have been performed on the core samples before and after the acidizing to do the comparative study on the buffer effect of these acids. The analysis consists of permeability, porosity, color change and pH value tests. There is more increase in permeability and porosity while less change in pH when phosphoric and formic acids were used compared to mud acid. From these results it has been found that the buffer effect of phosphoric acid and formic acid is better than hydrochloric acid.

1. Introduction

The matrix acidizing is a stimulation technique commonly used to increase the formation permeability. This can be done by injecting different acids into the reservoir; it can dissolve soluble minerals that can be present in the rock. Sandstone acidizing is proved to be an extremely challenging due to various stages of acid injection and reactions of acids with different minerals [1]. These reactions are potentially dangerous and can cause precipitation reactions, which should be avoided because they can cause damage to the formation. Over the years, researchers have developed many acidizing systems for sandstone formation [2-7] to obtain desired results by minimizing the precipitation problems. The three main acid systems developed are: Mineral, Organic (slower reacting – less corrosive) and Powdered (acid sticks).

Retarded acid systems had also been developed in the past to solve precipitation problems related to mud acid but the problems remain the same and some precipitates formed, which are not formed with standard acids. In sandstone acidizing, the role of hydrofluoric acid (HF) is to dissolve aluminosilicates while the role of accompanying acid is to keep the reaction products soluble in the spent acid and to keep the change of pH value at minimum [7]. HF is the only acid, which can dissolve siliceous minerals [8]. Therefore, all the acid combinations that might be used in sandstone matrix acidizing must contain HF acid or its precursors. The ability of these acids to dissolve siliceous particles is due to the presence of fluoride ion. It is the only ion, which can react with the quartz [9-11] but the fast reaction of this ion leads to precipitation reaction. The most probable of precipitation reactions that may lead to the formation damage and reduction in the permeability and porosity are as following:

- Precipitation of potassium and sodium silicates
- Precipitation of calcium fluoride
- Precipitation of hydrated silica

Therefore, the sandstone acidizing is carried out in three stages to minimize the precipitation reactions as result of different ions present in the formation [12]: Preflush stage, main acid and after flush.

The reaction of HF acid with sandstone formation is very complex and involves several interactions. This is due to the diverse nature of the sandstone formation. The most important and dominant minerals in sandstone are quartz, clays, aluminosilicates and carbonate. The reaction of hydrofluoric acid is fast with clays and feldspar, but slower with quartz. The silica reacts with HF to produce silica tetra-fluoride (SiF_4), which further reacts with HF acid to produce fluosilicic acid (H_2SiF_6). Reactions can be written as follows:

$$4HF + SiO_2 \longrightarrow SiF_4 + 2H_2O \quad (1)$$

$$SiF_4 + 6HF \longrightarrow H_2SiF_6 \quad (2)$$

Thus, at a low concentration of acid, the reaction reverses itself to regenerate HF and precipitate silica [3]:

$$H_2SiF_6 + 4H_2O \longrightarrow Si(OH)_4 + 6HF \quad (3)$$

Therefore, one of the main purpose of the sandstone acidizing acids is to control the formation of silica during the reaction. This can be done by adding some organic or a mineral acid with essential HF acid in the main acid stage. Therefore, significant progress has been made in this field and numerous research papers have been published. This paper presents the laboratory study of different acid combinations used in the main acid stage of sandstone acidizing. The main objective of this research is to investigate the buffer effect of different acids used in sandstone acidizing. The buffer effect is responsible for permeability change and deep penetration of acid into the sample. There will be less change in the pH value, more change in permeability and more penetration of acid into the core sample if buffer effect is good.

2. Experimental Work

The cores were saturated with acid using desiccators by inducing vacuum conditions using a vacuum pump. In a vacuum condition, the acid can penetrate deeper in the samples. Saturation time of 4 hours was allowed to ensure maximum volume of acid to penetrate into the core sample. The core was dried before and after the acidizing process for 24 hours at 80 °C. Total seven core samples were used during experiments mentioned in table 1 (A, B… G). The size of the core is 7.62 cm in length and 3.81 cm in diameter. The total volume used is 175 ml, which includes both acid and distilled water. The volume should be enough to fully immerse the core sample in it [13]. Different acid combinations were prepared as shown in table 1. The chemical composition of each sample was same, i.e., 93% silica, 4% alumina and small quantities of calcium oxide and ferric oxide.

2.1. Saturation of the Sample

- Desiccators and vacuum pump were prepared.
- The sample was immersed in 175 ml of acid solution.
- The lid was placed on the desiccators before the tap was opened.
- Tap was connected to the vacuum pump and opened slowly to evacuate the desiccators.

- Tap was closed after achieving a vacuum in desiccators
- The vacuum supply was disconnected and waited for 4 hours to ensure reaction between the core and acid.
- The entire step was repeated with all the acid combinations. Different ratios of acids were selected to find the effect of concentration of each acid. A low concentration may not produce best results while a high concentration can cause damage to the formation.

Table 1. Acid Combinations

Sr. No	Acid Combination	Mixing %
A	HF: HCl	3:12
B	HF: H_3PO_4	3:12
C	HF: H_3PO_4	3:9
D	HF: H_3PO_4	1.5:9
E	HF: HCOOH	3:12
F	HF: HCOOH	3:9
G	HF: HCOOH	1.5:9

3. Results and Discussion

During acidizing, core samples were allowed to react with different acid combinations mentioned in table 1. After the reaction, different properties of these core samples were measured and analyzed as follows:

3.1. Porosity Measurement

After completing the core-acid reaction, the change of porosity was measured and analyzed using Poro-perm apparatus. The results of porosity (initial and final porosity) are reported in table 2.

- HF: H_3PO_4 is a new combination, which was never used before in the petroleum industry. It is observed an increased in the porosity using acid combinations B, C and D was 44.78%, 75.68% and 55.14% respectively. The results of later two combinations were better than standard mud acid (A), which increased the porosity by 47% only (figure 1).
- HF: HCOOH is known as organic mud acid in petroleum industry. It is observed that acid combinations E, F and G lead to increase the porosity by 40.65%, 50.27% and 53.71% respectively. The results of later two combinations were better than standard mud acid.

Table 2. Porosity Values

Sr. No	Acid Combination	Initial Porosity (%)	Final Porosity (%)
A	3%HF: 12%HCl	10.28	15.12
B	3%HF: 12%H_3PO_4	9.89	14.32
C	3%HF: 9% H_3PO_4	9.17	16.11
D	1.5%HF: 9% H_3PO_4	10.01	15.53
E	3%HF: 12%HCOOH	10.11	14.22
F	3%HF: 9%HCOOH	10.98	16.50
G	1.5%HF: 9%HCOOH	10.52	16.17

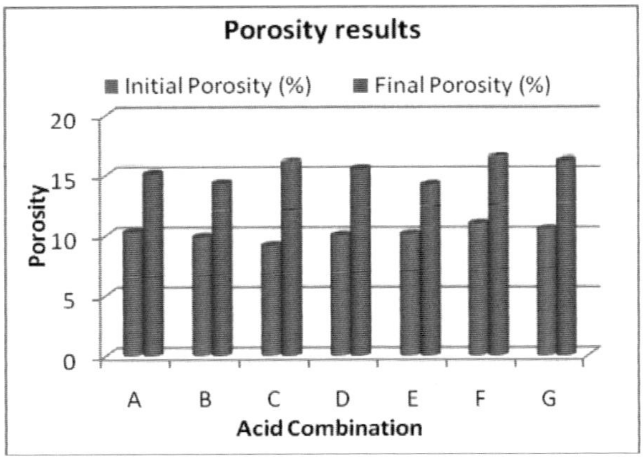

Figure 1. Porosity Comparison.

3.2. Permeability Measurement

Permeability is one of the most important parameters, which controls the flow rate of the oil from the reservoir. It is also measured and analyzed using Poro-perm apparatus. Table 3 shows the results of permeability including (initial and final permeability).

- HF: H_3PO_4 is a new combination, which showed a significant result. It is observed an increased in the permeability using acid combinations B, C and D was 68.35%, 110.20% and 46.62% respectively. The result of C is better than standard mud acid, which increased the permeability by 76% (figure 2).
- HF: HCOOH also showed significant results. It is observed that acid combinations E, F and G lead to increase permeability by 56.21%, 95.37% and 57.06% respectively. The result of C is better than standard mud acid, which increased the permeability by 76%.

Table 3. Permeability Values

Sr No	Acid Combination	Initial Permeability (md)	Final Permeability (md)
A	3%HF: 12%HCl	70.26	123.876
B	3%HF: 12%H_3PO_4	71.46	120.305
C	3%HF: 9% H_3PO_4	70.5	148.223
D	1.5%HF: 9% H_3PO_4	72.89	106.868
E	3%HF: 12%HCOOH	71.58	112.511
F	3%HF: 9%HCOOH	74.36	145.278
G	1.5%HF: 9%HCOOH	72.25	113.474

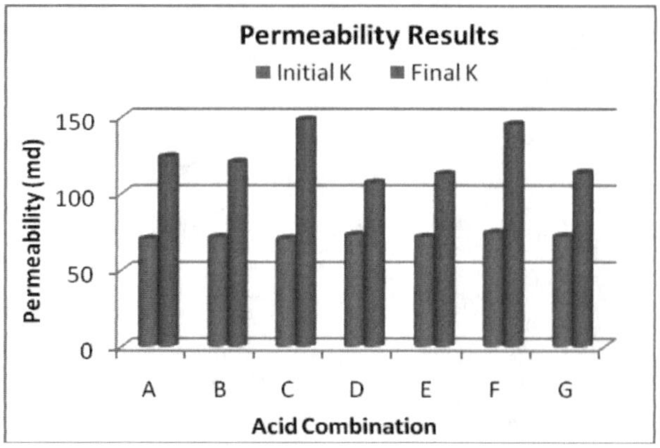

Figure 2. Permeability Comparison.

3.3. Color Change Test

The original color of the sample was light gray, but the change in the color was observed when the samples were dried in the oven after acidizing. This change in color was occurring because different reaction products were formed during the reaction.

- *After reaction with HF: HCl:* The change in color of the sample which was reacted with HF: HCl shown in figure 3(a). It indicates that the reaction is very fast and most of the acid reacted with and penetrated into the core at the early stages of reaction. It is clear from the figure 3(a) that some portions of the core remain un-reacted. Therefore, most of the acid used at the start due to very fast reaction of mud acid with the core. This fast reaction is due to the presence of HCl [3].

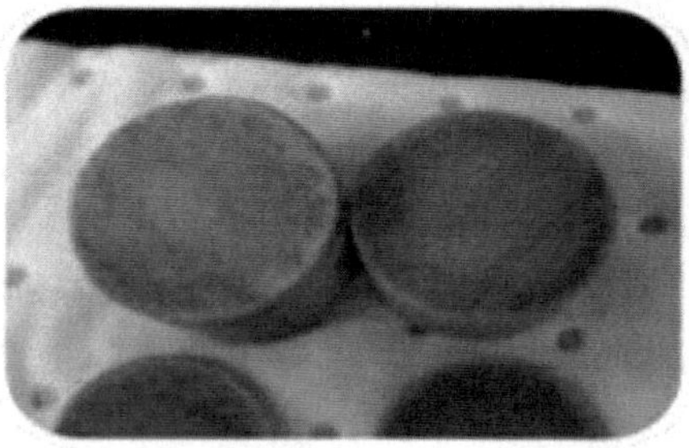

Figure 3(a). Change in color after reaction with HF: HCl.

- *After reaction with HF: H_3PO_4:* The core reacted with HF: H_3PO_4 is shown in figure 3(b). The main difference between the results of this combination and previous one is the iron content (dark brown color) which is formed by iron precipitate or corrosion products. These components can easily be observed in other reacted core samples. Therefore, it indicates that this acid may not corrode the tubing and other things. Figure 3(b) also shows that the penetration of this acid is at its maximum rate and which leads to a color change throughout the core sample.

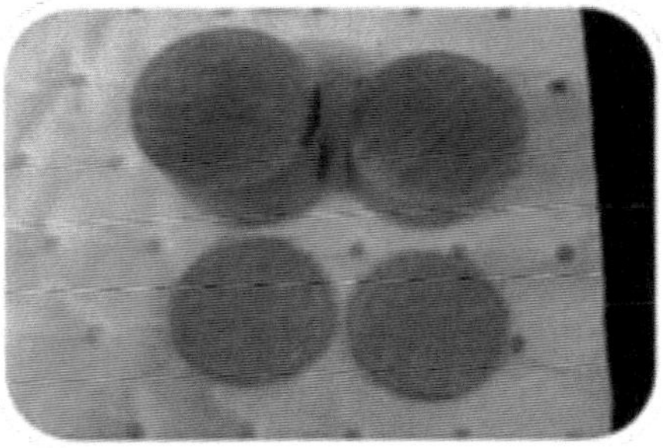

Figure 3(b). Change in color after reaction with HF: H_3PO_4.

- *After reaction with HF: HCOOH:* The sample reacted with HF: HCOOH can be seen in (figure 3(c)).It shows that the change in color also darker at the outer edges and some parts of the inner core reacts with the acid. This indicates that the reaction is slow compared to mud acid. This slow reaction is due to presence of organic acid [7].

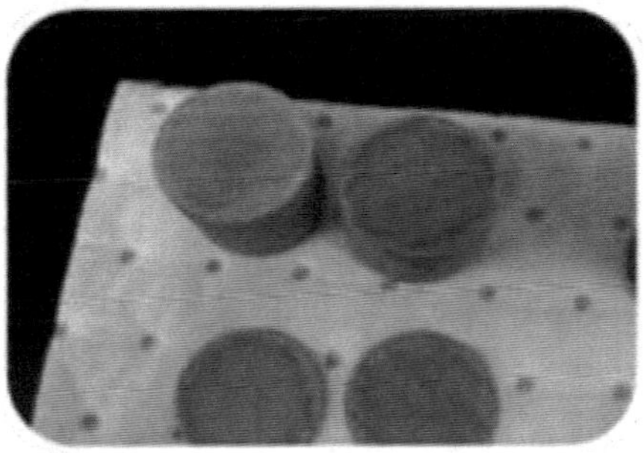

Figure 3(c). Change in color after reaction with HF: HCOOH.

3.4. pH Value Test

The change of acid solution pH value during acidizing process is very important to determine the extent of acid penetration in the formation. An increase in the permeability and porosity of the core sample also depends on the change in the pH value of acid solution. Moreover, the change in the pH value of also gives information about the buffer action of the acid The HCl contains one hydrogen ion; HCOOH contains two hydrogen ions, while H_3PO_4 contains three hydrogen ions. However, frequent changes in the pH value refer to the fast reaction rate of the acid solution with the core and less penetration. The pH value change was obviously noted when HF: HCl reacted with the core sample reported in table 4. It indicates that most of the acid has been spent due to the fast rate of reaction of acid with the core. Whereas a change in pH value remained less when HF: H_3PO_4 and HF: HCOOH were used which is the indication of the low reaction rate. Figure 4 represents the change in the pH value of each sample. The maximum change (1200%) was observed when mud acid was used, while minimum change (866%) was noted when phosphoric acid has been added. This effect shows that the buffer effect of phosphoric acid was better than HCl acid.

Table 4. pH Values

Sr No	Acid Combination	Initial pH	Final pH	% Change
A	3%HF: 12%HCl	0.023	0.32	1291.30
B	3%HF: 12%H_3PO_4	0.033	0.39	1081.81
C	3%HF: 9% H_3PO_4	0.036	0.37	927.777
D	1.5%HF: 9% H_3PO_4	0.040	0.44	1000
E	3%HF: 12%HCOOH	0.029	0.34	1072.41
F	3%HF: 9%HCOOH	0.030	0.29	866.667
G	1.5%HF: 9%HCOOH	0.036	0.37	927.8

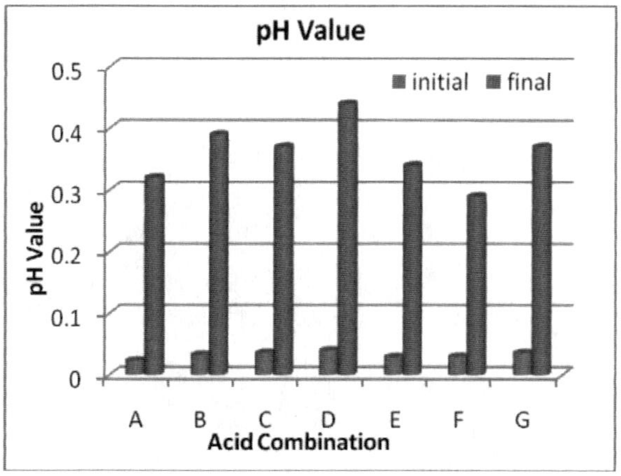

Figure 4. Porosity Comparison.

4. Critical Analysis

The color change test of HF: HCl shows that most of the acid was reacting at the edges and the penetration is not deep because when the reaction is fast; pH value of the solution increases and acidity becomes less. That is why; the permeability and porosity change was also less with mud acid.

The permeability and porosity change by HF: HCOOH and HF: H_3PO_4 was better than mud acid mentioned in Tables 2 & 3 and also the pH value change is less as reported in Table 4. Therefore, maximum penetration can be achieved with these combinations. These results are also supported by the color change test. Therefore, it is concluded that the buffer action of phosphoric (H_3PO_4) and formic acid (HCOOH) is better than hydrochloric acid (HCl).

5. Conclusion

Deep penetration into the formation can be achieved using phosphoric acid in the main stage as compared to standard hydrochloric acid, is good for acidizing perspective. This effect is due to less change in pH value during the acidizing process. The combination of (Hydrofluoric and Phosphoric acid) can be used as a main acid in sandstone acidizing, the most successful combinations are (3% HF: 9% H_3PO_4). The buffer action of phosphoric (H_3PO_4) and formic acid (HCOOH) perform better than hydrochloric acid (HCl).

IOP Conf. Series: Materials Science and Engineering **78** (2015) 012008 doi:10.1088/1757-899X/78/1/012008

References

[1] Muecke T W, Principles of Acid Stimulation, 1982, *Society of Petroleum Engineers.*

[2] Al-Dahlan M N, Nasr-El-Din and Al-Qahtani A A, Evaluation of Retarded HF Acid Systems, 2001, *Society of Petroleum Engineers.*

[3] Al-Harbi B G, Al-Khaldi M H and AlDossary K A, Interactions of Organic-HF Systems with Aluminosilicates: Lab Testing and Field Recommendations, 2011, *Society of Petroleum Engineers.*

[4] Ali A H A et al, Chelating Agent-Based Fluids for Optimal Stimulation of High-Temperature Wells, 2002, *Society of Petroleum Engineers.*

[5] Da Motta E P and Dos Santos, New Fluosilicic Acid System Removes Deep Clay Damage, 1999, *Society of Petroleum Engineers.*

[6] Mian Umer Shafiq, Aung Kyaw and Muahnnad Talib Shuker A Comprehensive Research to Find Suitable Acid for Sandstone Acidizing, *Advanced Materials Research,* 2013. **787**: p. 274-280.

[7] Yang F, Nasr-El-Din, and Al-Harbi, Acidizing Sandstone Reservoirs Using HF and Formic Acids, 2012, *Society of Petroleum Engineers.*

[8] Smith C F and Hendrickson A R, Hydrofluoric Acid Stimulation of Sandstone Reservoirs, 1965, *Socity of Petroleum Engineers.*

[9] Mahmoud M A et al, Sandstone Acidizing Using A New Class of Chelating Agents, 2011, *Society of Petroleum Engineers.*

[10] Ali S A et al, Stimulation of High-Temperature Sandstone Formations From West Africa With Chelating Agent-Based Fluids, *SPE,* 2008.

[11] Frenier W et al, Hot Oil and Gas Wells Can Be Stimulated Without Acids, *SPE*, 2004.

[12] McLeod H O, Ledlow L B and Till M V, The Planning, Execution, and Evaluation of Acid Treatments in Sandstone Formations, 1983, *Society of Petroleum Engineers.*

[13] Recommended Practices for core analysis, **rp40**. *American Petroleum Institute*

Cuttings-liquid frictional pressure loss model for horizontal narrow annular flow with rotating drillpipe

T N Ofei[1]*, S Irawan[1] and W Pao[2]

[1] Petroleum Engineering Department, Universiti Teknologi PETRONAS, Bandar Seri Iskandar, 31750 Tronoh, Malaysia.
[2] Mechanical Engineering Department, Universiti Teknologi PETRONAS, Bandar Seri Iskandar, 31750 Tronoh, Malaysia.

*Email: titus.ofei@petronas.com.my; titusofei@hotmail.com

Abstract. During oil and gas drilling operations, frictional pressure loss is experienced as the drilling fluid transports the drilled cuttings from the bottom-hole, through the annulus, to the surface. Estimation of these pressure losses is critical when designing the drilling hydraulic program. Two-phase frictional pressure loss in the annulus is very difficult to predict, and even more complex when there is drillpipe rotation. Accurate prediction will ensure that the correct equivalent circulating density (ECD) is applied in the wellbore to prevent formation fracture, especially in formations with narrow window between the pore pressure and fracture gradient. Few researchers have attempted to propose cuttings-liquid frictional pressure loss models, nevertheless, these models fail when they are applied to narrow wellbores such as in casing-while-drilling and slimhole applications. This study proposes improved cuttings-liquid frictional pressure loss models for narrow horizontal annuli with drillpipe rotation using Dimensional Analysis. Both Newtonian and non-Newtonian fluids were considered. The proposed model constants were fitted by generated data from a full-scale simulation study using ANSYS-CFX. The models showed improvement over existing cuttings-liquid pressure loss correlations in literature.

1. Introduction

The prediction of annular frictional pressure losses is an important parameter during drilling operations. This is vital when designing drilling hydraulic program, particularly, the equivalent circulating density (ECD) required for the efficient circulation of drilled cuttings from the bottom of the drill bit, through the annulus, to the surface without any formation damage. Erroneous predictions may result in wellbore catastrophes such as burst, collapse, mud influxes, or induced lost circulation events, especially, in formations where there is a narrow window between the pore pressure and fracture gradient.

Previous field, laboratory and simulation studies [1-15] have been conducted to estimate frictional pressure loss for single-phase Newtonian and non-Newtonian fluids with drillpipe rotation in annular geometries of diameter ratio ranging from 0.37-0.92. Till date however, few experimental studies [16-20] on two-phase cuttings-liquid flow in annulus have been carried out to predict the annular frictional pressure losses. These studies have attempted to propose empirical pressure loss correlations for cuttings-liquid flow in concentric and eccentric annular geometries. These pressure loss correlations are however, only applicable within a diameter ratio of 0.57-0.64, hence, will under-predict for cases where the diameter ratio exceeds 0.70.

Recently, other researchers [21, 22] have implemented computational fluid dynamics (CFD) methods to simulate two-phase cuttings-liquid flow in annuli, where the effects of several drilling parameters such as fluid velocity, drillpipe rotation, diameter ratio, and fluid type were analysed. Validation of CFD model with experimental data showed a very good match. In current drilling operations, such as, casing-while-drilling and slimhole drilling, the annular geometry is very narrow (i.e. diameter ratio above 0.70). This results in extreme pressure losses as the drilling fluid and cuttings travel in the narrow annular region from the bottom-hole to the surface. Current two-phase cuttings-liquid pressure loss correlations are only limited to annular geometries of diameter ratio of 0.64, hence, may under-predict pressure losses in narrow annular geometries.

The present study proposes an improved frictional pressure loss model for narrow horizontal wellbores with drillpipe rotation and having diameter ratios from 0.64 to 0.90 using dimensional analysis method. The model is compared to CFD simulation pressure loss data generated using ANSYS-CFX 14.0. Furthermore, the accuracy of the proposed model is tested against other existing models in literature.

2. Materials and methods

2.1. *Dimensionless analysis*

There are several drilling parameters that affect frictional pressure loss estimation of cuttings-liquid flow in annular geometry. These include: annular geometry(D_i, D_o), cuttings concentration (C_c), drillpipe rotation (Ω), axial fluid velocity (v_a), tangential fluid velocity (v_Ω), fluid density (ρ_f) and viscosity (μ_f), cuttings density (ρ_c) and size (d_c), and wellbore inclination (θ). The pressure gradient is mathematically related to these parameters in Eq. (1) as:

$$\frac{\Delta P}{\Delta L} = f\left(g, \Omega, D_i, D_o, C_c, \rho_f, \rho_c, d_c, \mu_f, \theta, v_a, v_\Omega\right) \quad (1)$$

To develop a comprehensive model which is valid for a wide range of drilling parameters, as mentioned above, dimensionless groups of these parameters are defined. Assuming a homogeneous mixture of the cuttings-liquid flow, mixture density and viscosity [23] will be defined respectively as:

$$\rho_m = C_c\rho_c + (1 - C_c)\rho_f \quad (2)$$

$$\mu_m = \left(1 + 2.5C_c + 10.05C_c^2 + 0.00273\exp(16.6C_c)\right)\mu_f \quad (3)$$

where the cuttings feed concentration is given as:

$$C_c = \frac{(ROP)A_{bit}}{R_T Q} \quad (4)$$

For a shear-thinning Power Law fluid where the flow behaviour index $n < 1$, the fluid apparent viscosity is defined in SI units as:

$$\mu_f = K\left(\frac{D_o - D_i}{v_a}\right)^{1-n}\left(\frac{2n+1}{n}\right)\left(\frac{2^{2n}}{12}\right) \quad (5)$$

Where K is the fluid's consistency index in $Pa\ s^n$

The rotation of the drillpipe generates tangential velocity in addition to the axial velocity. With the knowledge of the fluid's annular flow rate (Q), the axial velocity can be defined as:

$$v_a = \frac{4Q}{\pi(D_o - D_i)} \quad (6)$$

According to Slattery [24], couette flow model is considered in determining the tangential velocity component as a function of radius (r), rotational speed (Ω), and diameter ratio ($\kappa = D_i/D_o$). An average tangential velocity in the annulus can be computed as:

$$v_\Omega = \frac{\int_{\frac{D_i}{2}}^{\frac{D_o}{2}} \int_0^{2\pi} \left\{ \frac{1}{2}\kappa^2 D_o \Omega \left(\frac{\frac{2r}{D_o} - \frac{D_o}{2r}}{\kappa^2 - 1} \right) \right\} r d\xi dr}{\int_{\frac{D_i}{2}}^{\frac{D_o}{2}} \int_0^{2\pi} r d\xi dr} \quad (7)$$

where $\frac{D_i}{2} \leq r \leq \frac{D_o}{2}$

By solving the Eq. (7) analytically, the tangential velocity is expressed as:

$$v_\Omega = \frac{\kappa^2 D_o \Omega}{(\kappa^2 - 1)\left(D_o^2 - D_i^2\right)} \times \left\{ \left(\frac{D_o^3 - D_i^3}{3D_o} \right) - \left(D_o^2 - D_o D_i\right) \right\} \quad (8)$$

The pressure gradient equation in Eq. 1 can be summarised by considering the mixture density and viscosity in Eq. 2 and 3 respectively as:

$$\frac{\Delta P}{\Delta L} = f(g, \Omega, D_i, D_o, C_c, \rho_m, d_c, \mu_m, \theta, v_a, v_\Omega) \quad (9)$$

Since there are eleven (11) independent variables in Eq. (9), i.e., gravity (g), drillpipe rotation (Ω), outer diameter of drillpipe (D_i), diameter of hole (D_o), cuttings feed concentration (C_c), mixture density (ρ_m), cuttings size (d_c), mixture viscosity (μ_m), wellbore inclination (θ), axial fluid velocity (v_a), and tangential fluid velocity (v_Ω), and three (3) dimensions (M, L, T), thus, eight (8) dimensionless groups will be defined. Applying the Buckingham-π theorem, the dimensionless groups are defined as follows:

$$\Pi_1 = \frac{D_i}{D_o} \quad (10)$$

$$\Pi_2 = C_c \quad (11)$$

$$\Pi_3 = \frac{\rho_m v_a (D_o - D_i)}{\mu_m} \quad (12)$$

$$\Pi_4 = \frac{\rho_m v_\Omega (D_o - D_i)}{\mu_m} \quad (13)$$

$$\Pi_5 = \frac{v_a^2}{g(D_o - D_i)} \quad (14)$$

$$\Pi_6 = \frac{\Omega (D_o - D_i)}{v_a} \quad (15)$$

$$\Pi_7 = \theta \quad (16)$$

$$\Pi_8 = \frac{\rho_m v_a d_c}{\mu_m} \quad (17)$$

The above dimensionless groups represent the following: $\prod_1$ is the diameter ratio (ratio of drillpipe diameter to hole diameter), $\prod_2$ is the amount of cuttings generated by the drill bit, $\prod_3$ is the ratio of inertial to viscous forces in the axial direction, $\prod_4$ is the ratio of inertial to viscous forces in the tangential direction, $\prod_5$ is the ratio of inertial to gravitational forces, $\prod_6$ is a function of drillpipe rotation, $\prod_7$ is the wellbore inclination angle, and $\prod_8$ is a function of cuttings size. The dimensionless groups defined in Eq. (10) to (17) are combined to formulate the frictional pressure loss model for cuttings-liquid flow in narrow horizontal wellbore.

2.1.1. CFD simulation study. Three-dimensional simulations for cuttings-liquid flowing in annular geometries were carried out to generate annular frictional pressure losses data using full-scale experimental setup data. Four annular geometries with diameter ratios $\kappa = 0.64, 0.70, 0.80, \text{and } 0.90$ were modelled and meshed using ANSYS Workbench 14.0. The continuity and momentum equations for both fluid and solids were solved using CFX solver [25] which is based on finite volume technique. The Eulerian-Eulerian model in ANSYS CFX, which regards both continuous (fluid) and dispersed (solid) phases as continuous media was adopted due to its ability to account for particle-particle interaction as well as handling high solid volume fractions greater than 20%. The simulation solution is assumed to be converged when the root mean square (RMS) of the normalised residual error reached 10^{-4}. The boundary conditions applied to the annular geometries are as follows: a mixture mass flow rate was specified at the inlet while a zero gauge pressure was specified at the outlet. No slip condition was also imposed on both inner and outer pipe walls for both fluids and solids. Both Newtonian (water) and Power Law fluid models were used in this study.

2.1.2. Meshing and model validation. The annular meshed geometries are shown in Figure 1. In this study, the geometries were meshed into unstructured tetrahedral elements of approximately $0.66\text{x}10^6$ to $2.15\text{x}10^6$ depending on the diameter ratio, with about 20% inflation layers near the wall regions to accurately capture the flow effect. Mesh independence study was carried out to optimise the number of elements until results were independent of the number of elements. The optimum number of elements used for the annular geometries of diameter ratios $\kappa = 0.64$, $\kappa = 0.70$, $\kappa = 0.80$, and $\kappa = 0.90$ are respectively, $0.66\text{x}10^6$, $0.72\text{x}10^6$, $1.13\text{x}10^6$ and $2.15\text{x}10^6$. The operating parameters, fluid and cuttings properties, and wellbore parameters are listed in Table 1. Full-scale experimental frictional pressure loss data from literature [22, 26] are used to validate the simulation model. Excellent agreement is observed between the measured and predicted data for (a) cuttings-water flow and (b) cuttings-power law flow as shown in Figure 2a and 2b respectively.

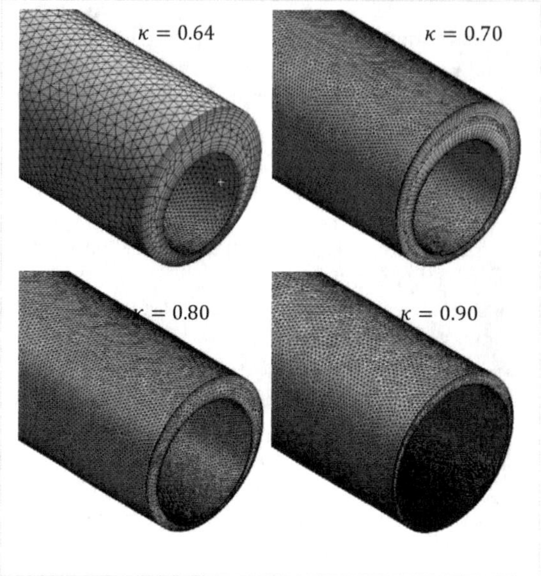

Figure 1. 3D section of meshed annular geometry.

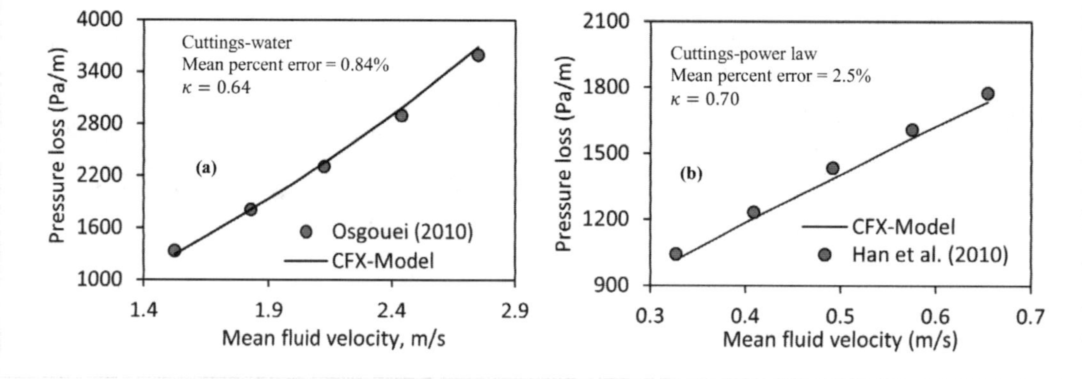

Figure 2. Experimental and simulation pressure loss data comparison (a) cuttings-water flow (b) cuttings-power law flow.

Table 1. Simulation data for cuttings-fluid flow.

Rheological and Drilling Parameter	Case 1	Case 2
	Water	*Power Law*
Fluid density (kg/m^3)	998.5	1006.3
Cuttings density (kg/m^3)	2761.4	2761.4
Avg. cuttings size (m)	0.00201	0.00201
Flow behaviour index, n	1.00	0.51
Viscosity consistency, K, (Pa s^n)	0.001	0.289
Fluid velocity (m/s)	1.524-2.7432	1.524-2.7432
Inner pipe rotation speed (rpm)	0, 80, 120	0, 80, 120
Hole diameter (mm)	50.8	50.8
Wellbore inclination (degree)	90	90
Diameter ratio (κ=Di/Do)	0.64 - 0.90	0.64 - 0.90
Eccentricity (e)	0.623	0.623
ROP (m/s)	0.00508	0.00508

3. Results and discussion

3.1. Frictional pressure loss model

The frictional pressure loss can be expressed as a function of the developed dimensionless groups from Eq. 10 to 17. For cuttings-water flow, with drillpipe rotation ranging from zero (0) to 120 rpm, the frictional pressure loss model can be expressed as:

$$\left.\frac{\Delta P}{\Delta L}\right|_{cut-water} = \exp[a_1 \Pi_1 + a_2 \Pi_2 + a_3 \Pi_3 + a_4 \Pi_4 + a_5 \Pi_5 + a_6 \Pi_6 + a_7 \Pi_7 + a_8 \Pi_8 + a_9] \quad (18)$$

For cuttings-power law flow, with drillpipe rotation ranging from 80 to 120 rpm, the frictional pressure loss model can be expressed as:

$$\left.\frac{\Delta P}{\Delta L}\right|_{cut-PL} = a_1[\Pi_1{}^{a_2} \cdot \Pi_2{}^{a_3} \cdot \Pi_3{}^{a_4} \cdot \Pi_4{}^{a_5} \cdot \Pi_5{}^{a_6} \cdot \Pi_6{}^{a_7} \cdot \Pi_7{}^{a_8} \cdot \Pi_8{}^{a_9}] \quad (19)$$

In this study, the wellbore inclination and cuttings size are kept constant, hence, Π_7 and Π_8 are excluded from the analysis. Regression analysis is carried out to predict the constants in Eqs. 18 and 19 using the database generated from the simulation study. The final frictional pressure loss model for cuttings-water flow with drillpipe rotation ranging from zero (0) to 120 rpm is:

$$\left.\frac{\Delta P}{\Delta L}\right|_{cut-water} = \exp\left[23.06\Pi_1 - 23.47\Pi_2 + 2.643\frac{\Pi_3}{10^5} + 8.438\frac{\Pi_4}{10^5} + 4.417\frac{\Pi_5}{10^3} - 3.424\Pi_6 - 9.585\right] \quad (20)$$

Similarly, for cuttings-power law flow with drillpipe rotation ranging from 80 to 120 rpm, the final frictional pressure loss model gives:

$$\left.\frac{\Delta P}{\Delta L}\right|_{cut-PL} = 0.255[10^5\Pi_1{}^{-2.021} \cdot \Pi_2{}^{0.175} \cdot \Pi_3{}^{-1.51} \cdot \Pi_4{}^{1.105} \cdot \Pi_5{}^{0.967} \cdot \Pi_6{}^{-1.118}] \quad (21)$$

3.2. Relationship between measured and predicted data

The frictional pressure loss model for cuttings-water flow in Eq. 20 was developed from a total of 45 data points, while for cuttings-power law flow model in Eq. 21, a total of 40 data points were used. The comparison between the predicted and measured frictional pressure loss for both cuttings-water and cuttings-power law flow are respectively presented in Figures 3a and 3b.

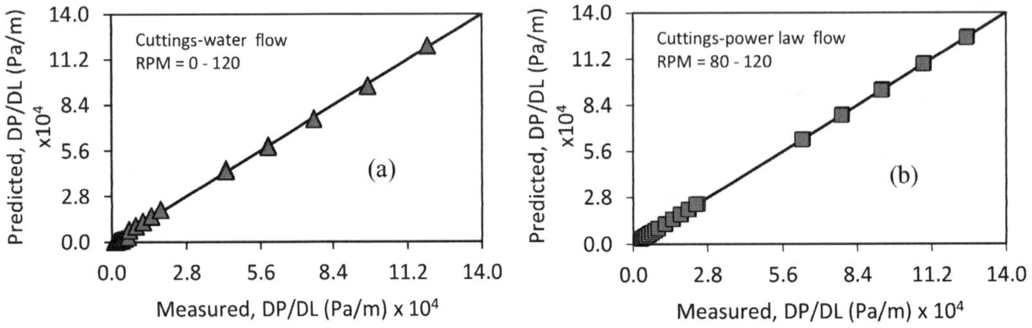

Figure 3. Measured and predicted frictional pressure loss for (a) cuttings-water flow & (b) cuttings-power law flow.

In Figures 3a and 3b, the performances of the frictional pressure loss models were measured by a statistical parameter: regression success index, R^2. Here, the R^2 value for Figure 3a is 0.9985. Similarly, the data in Figure 3b correlated to R^2 value of 0.9999, which are reasonably accurate.

3.3. Comparison of models with existing correlations

The proposed models in Equations 20 and 21 were compared with existing models in literature. It should be noted that the data range from these existing models [16-17], [26] are similar to that used to develop Equations 20 and 21, with the exception that the diameter ratio κ, of the annular setup for the existing models ranges from 0.50 to 0.64, whereas for the present study, the diameter ratio ranges from 0.64 to 0.90. Figures 5a and 5b present these comparisons for cuttings-water and cuttings power law flow respectively.

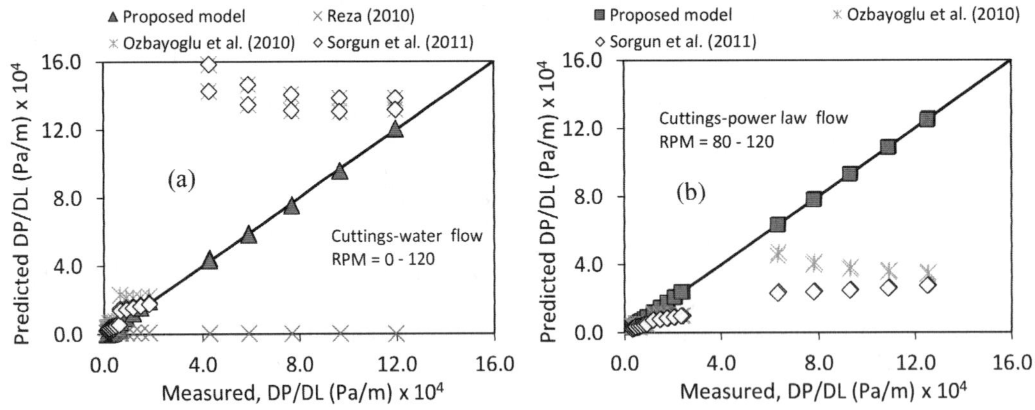

Figure 5. Model comparison with existing correlations for (a) cuttings-water flow & (b) cuttings-power law flow.

Figures 5a and 5b show that the performances of the existing frictional pressure loss models are quite satisfactory, especially, at low diameter ratios of $\kappa = 0.64$ and 0.70. However, as the diameter ratio increases above 0.70, the existing models become unreliable for predicting the frictional pressure loss. It is imperative to note that these existing models were developed for a specific range of diameter ratios, that is, $\kappa = 0.57 - 0.64$ hence, cannot be used for general purposes. Table 2 shows the performances of the proposed models and existing correlations using statistical measurements.

Table 2. Regression success index, R^2 for present and previous models

Models	R^2 for Cuttings-water flow	R^2 for Cuttings-power law flow
Equation 20	0.9985	-
Equation 21	-	0.9999
Sorgun et al. (2011)	0.757	0.337
Ozbayoglu et al. (2010)	0.750	0.280
Reza (2010)	0.208	-

4. Conclusions

The present study has presented frictional pressure loss models for both cuttings-water and cuttings-power law flow in horizontal narrow annuli with a wide range of diameter ratios from 0.64 to 0.90, and drillpipe rotation from 0 to 120 rpm. A full-scale experimental data were employed in a CFD simulation to predict annular frictional losses for validation purposes. The proposed models were developed based on dimensional analysis and their constants were fitted by regression analysis from the simulated data. The present models performed better than other existing models in literature, particularly, where the diameter ratio exceeds 0.70. The study has shown the deficiencies of existing frictional pressure loss models when they are applied, especially, to narrow horizontal wellbores, such as casing-while-drilling and slimhole applications.

Acknowledgement
The authors appreciate the financial support from the department of petroleum engineering, Universiti Teknologi PETRONAS (UTP), Malaysia.

References

[1] Delwiche R A, Stratabit D B, Lejeune M W D, Mawet P F B N and Vighetto R 1992 "Slimhole drilling hydraulics" *Soc. of Petrol. Eng.* DOI: 10.2118/24596-MS

[2] Marken C D, He X and Saasen A 1992 "The influence of drilling conditions on annular pressure losses" *Soc of Petrol Eng.* DOI:10.2118/24598-MS

[3] Ahmed R M and Miska S Z 2008 "Experimental study and modeling of Yield Power-Law fluid flow in annuli with drillpipe rotation" *Soc. of Petrol. Eng.* DOI:10.2118/112604-MS

[4] McCann R C, Quigley M S, Zamora M and Slater K S 1995 "Effects of high-speed pipe rotation on pressures in narrow annuli. *Soc. of Petrol. Eng.* DOI:10.2118/26343-PA

[5] Nouri J M and Whitelaw J H 1997 "Flow of Newtonian and non-Newtonian fluids in an eccentric annulus with drillpipe rotation of the inner cylinder" *Int. J. Heat Fluid Flow.* **18** (2): 236-46. DOI: 10.1016/S0142-727X(96)00086-0

[6] Wei X, Miska S Z, Takach N E, Bern P and Kenny P 1998 "The effect of drillpipe rotation on annular frictional pressure loss" *J. Energy Resour. Technol.* **120** (1): 61-66. DOI: 10.1115/1.2795011

[7] Ooms G, Burgerscentrum J M and Kampman-Reinhartz B E 1999 "Influence of drillpipe rotation and eccentricity on pressure drop over borehole during drilling" *Soc. of Petrol. Eng.* DOI:10.2118/56638-MS

[8] Wang H, Su Y, Bai Y, Gao Z and Zhang F 2000 "Experimental study of slimhole annular pressure loss and its field applications" *Soc. of Petrol. Eng.* DOI:10.2118/59265-MS

[9] Wan S, Morrison D and Bryden I G 2000 "The flow of Newtonian and inelastic non-Newtonian fluids in eccentric annuli with inner-cylinder rotation" *Theoretical and Comput. Fluid Dyna.* **13** (5): 349-59. DOI: 10.1007/S001620050004

[10] Escudier M P, Oliveira P J and Pinho F T 2002 "Fully developed laminar flow of purely viscous non-Newtonian liquids through annuli, including the effects of eccentricity and inner-cylinder rotation" *Int. J. Heat Fluid Flow.* **23** (1): 52-71. DOI: 10.1016/S0142-727X(01)00135-7

[11] Woo N, Kim Y and Hwang Y 2006 "Experimental study on the helical flow in a concentric annulus with rotating inner cylinder" *J. Fluid Eng.* **128** (1): 113-117. DOI: 10.1115/1.2136923

[12] Ozbayoglu E M and Sorgun M 2010 "Frictional pressure loss estimation of non-Newtonian fluids in realistic annulus with pipe rotation" *Soc. of Petrol. Eng.* DOI:10.2118/141518-PA

[13] Mokhtari M, Ermila M, Tutuncu A N and Karimi M 2012 "Computational modelling of drilling fluids dynamics in casing drilling" *Soc. of Petrol. Eng.* DOI:10.2118/161301-MS.

[14] Ozbayoglu M E, Saasen A, Sorgun M and Svanes K 2008 "Effect of pipe rotation on hole cleaning for water-based drilling fluids in horizontal and deviated wells. *Soc. of Petrol. Eng.* DOI:10.2118/114965-MS

[15] Ofei T N, Irawan S and Pao W 2015 "Modified yield power law fluid flow in narrow annuli with inner rotating pipe" *Can J of Chem. Eng.* DOI/10.1002/cjce.22100

[16] Ozbayoglu M E and Sorgun M 2010 "Frictional pressure loss estimation of water-based drilling fluids at horizontal and inclined drilling with pipe rotation and presence of cuttings" *Soc. of Petrol. Eng.* DOI:10.2118/127300-MS

[17] Sorgun M Aydin I and Ozbayoglu M E 2011 "Friction factor hydraulics calculations considering presence of cuttings and pipe rotation in horizontal/highly-inclined wellbore" *J. of Petrol. Sci. and Eng.* 78, 407-14.

[18] Ozbayoglu M E Ettehadi R O, Ozbayoglu A M and Yuksel E 2010 "Estimation of 'very-difficult-to-identify' data for hole cleaning, cuttings transport and pressure drop estimation in directional and horizontal drilling," *Soc. of Petrol. Eng.* DOI:10.2118/136304-MS.

[19] Ozbayoglu M E Saasen A Sorgun M and Svanes K 2008 "Effect of Pipe Rotation on Hole Cleaning for Water-Based Drilling Fluids in Horizontal and Deviated Wells" *Soc. of Petrol. Eng.* DOI:10.2118/114965-MS

[20] Ozbelge T A and Unal G C 2009 "A new correlation for two-phase pressure drops in fully developed dilute slurry up-flows through an annulus" *Chem. Eng. Comm.*, **196**:491-98. DOI: 10.1080/009864-40802483822

[21] Ofei T N Irawan S and Pao W 2014 "CFD method for predicting annular pressure losses and cuttings concentration in eccentric horizontal wells" *J. of Petrol. Eng.* DOI: 10.1155/2014/486423

[22] Han S-K, Hwang Y-K, Woo N-S and Kim Y-J 2010 "Solid-liquid hydrodynamics in a slim hole drilling annulus" *J. of Petrol. Sci. and Eng.* **70** 308-19

[23] Thomas D G 1965 "Transport Characteristics of Suspensions. Part VIII. A Note on Viscosity of Newtonian Suspensions of Uniform Spherical Particles" *J. Colloid Sci.* **20,** 267-77

IOP Conf. Series: Materials Science and Engineering **78** (2015) 012009 doi:10.1088/1757-899X/78/1/012009

[24] Slattery J C 1999 "Advanced Transport Phenomena" *Cambridge University Press*.
[25] ANSYS CFX version 14.0–Solver Modelling Guide, "Multiphase flow modelling".
[26] Osgouei R.E 2010 "Determination of cuttings transport properties of gasified drilling fluids" Ph.D Dissertation, Middle East Technical University, Ankara-Turkey.

IOP Conf. Series: Materials Science and Engineering **78** (2015) 012010 doi:10.1088/1757-899X/78/1/012010

Voltage harmonic elimination with RLC based interface smoothing filter

K Chandrasekaran[1] and V K Ramachandaramurthy[2]

[1] Department of Electronic Engineering, University Tunku Abdul Rahman, Perak, Malaysia
[2] Power Quality Research Group, Universiti Tenaga Nasional, Malaysia

E-mail: kchandra@utar.edu.my

Abstract-A method is proposed for designing a Dynamic Voltage Restorer (DVR) with RLC interface smoothing filter. The RLC filter connected between the IGBT based Voltage Source Inverter (VSI) is attempted to eliminate voltage harmonics in the busbar voltage and switching harmonics from VSI by producing a PWM controlled harmonic voltage. In this method, the DVR or series active filter produces PWM voltage that cancels the existing harmonic voltage due to any harmonic voltage source. The proposed method is valid for any distorted busbar voltage. The operating VSI handles no active power but only harmonic power. The DVR is able to suppress the lower order switching harmonics generated by the IGBT based VSI. Good dynamic and transient results obtained. The Total Harmonic Distortion (THD) is minimized to zero at the sensitive load end. Digital simulations are carried out using PSCAD/EMTDC to validate the performance of RLC filter. Simulated results are presented.

I. INTRODUCTION

Modern usage of static power converters has increased the problems of power system harmonics. Static power converters use power semiconductor devices as electronic switches to transfer and convert the dc to ac supply. The alternate switching actions of these power devices and its non-linearity characteristics result in a distorted input current. The power converter will behave like a current source, injecting harmonic.

This research paper discusses a dc source supplying the power electronic equipment connected to a series APF or DVR. Rectifier or dc supply with a large filter capacitor on its dc side is a voltage type harmonic source [1][2][3]. One of the problems of power system harmonics is the supply voltage distortion at the Point of Common Coupling (PCC).

The main purpose of the LC filters is to attenuate the voltage ripple from inverter switching [4][5]. To suppress harmonics, specifically the 5th, 7th and 9th orders, shunt LC filters had been used. [6[7][[8].This is not cost effective. The output voltage on the LC filter capacitor is controlled by the switching operation of the PWM inverter, where a time delay is introduced by the LC filters causing resonance in the output AC voltage. Number of research has been done on regulating the output voltage across the LC filter capacitor [9][10].Various control techniques has been introduced to control harmonics in inverter such as digital control strategies based on repetitive control, dead-beat control, and discrete-time sliding-mode control [11]. Although there are some advantages of the said techniques, these techniques do suffer from some drawbacks, such as complexity, sensitivity to parameter variations and loading conditions, and steady-state errors in distorting conditions.

This paper proposes a smoothing RLC filter design method that will minimize the transient current overshoot problem without decreasing the active damping performance of inverter systems. Large inductor minimizes current ripple of PWM inverter but increases the voltage drop [12]. Large inductor will contribute to the overall cost, weight and increase the voltage stress on the inverter switches.

This paper analyses the RLC filters characteristics used in PWM inverters. The supply harmonic source is a 6-pulse Bridge rectifier and a three-phase diode rectifier with a dc source. An improved filter design method offering good control dynamics and minimum transient current overshoot is considered. The proposed design method is verified in PSCAD/EMTDC.

A 12-pulse three-phase VSI model as shown in Figure 1 is used. The IGBT based VSI can operate with conventional sinusoidal pulse width modulation (SPWM) method. Simulated voltage harmonic elimination results are presented.

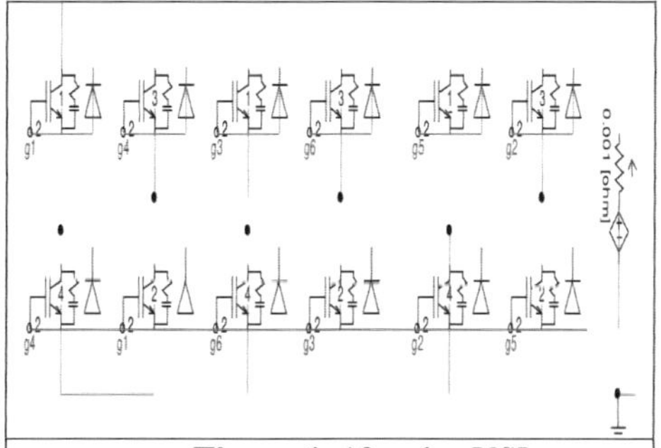

Figure 1. 12 pulse VSI

II. RLC FILTER SYSTEM

The single-phase one leg equivalent circuit of an inverter system is shown in Figure 2.

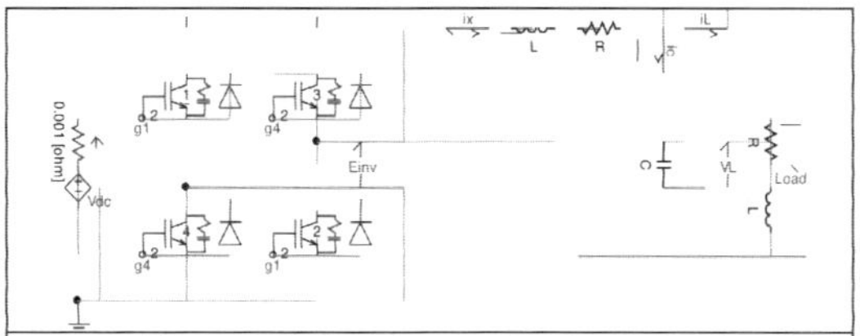

Figure 2. One leg equivalent circuit of PWM inverter

*E*inv is the VSI output voltage supply current is *i*x, current through capacitor is *ic*, load current is *i*L and *VL* is voltage across the load.

The dc supply voltage *V*dc is constant and the gate switch of VSI is ideal. The output voltage of the VSI is written as;

$$Einv = VL + R_f ix + L_f \frac{dix}{dt} \quad (1)$$

IOP Conf. Series: Materials Science and Engineering **78** (2015) 012010 doi:10.1088/1757-899X/78/1/012010

Separating both the capacitor voltage and the current through the inductor into average over a switching cycle and harmonics ripple component,

$$E\text{inv} = \overline{VL} + \tilde{V}L \tag{2}$$

$$ic = \bar{\imath}\text{x} + \tilde{\imath}\text{x} \tag{3}$$

Substituting eq. 2 and 3 in eq. 1 gives

$$E\text{inv} = \overline{VL} + \widetilde{VL} + R_f(\bar{\imath}\text{x} + \tilde{\imath}\text{x}) + L_f\frac{d}{dt}(\bar{\imath}\text{x} + \tilde{\imath}\text{x}) \tag{4}$$

Considering $\widetilde{VL}$ and $R_f\tilde{\imath}\text{x}$ is small as compared to
$L\frac{d}{dt}\tilde{\imath}\text{x}$, then the ripple component of the filter inductor current is

$$\tilde{\imath}\text{x} = \frac{1}{L_f}\int(E\text{inv} - \overline{E\text{inv}})dt \tag{5}$$

and the average value of output voltage is

$$E\text{inv} = \overline{VL} + R_f i\text{x} + L_f\frac{dix}{dt} \tag{6}$$

The capacitor voltage harmonic, with reference to Fig. 2;

$$ic = i\text{x} - i\text{L} \tag{7}$$

Splitting the capacitor and inverter output currents into the average and harmonic components,

$$ic = \bar{\imath}\text{c} + \tilde{\imath}\text{c} \tag{8}$$

$$i\text{x} = \bar{\imath}\text{x} + \tilde{\imath}\text{x} \tag{9}$$

Substituting eq. (3) and eq. (8)-(9) into eqn.(7), the following equation is obtained

$$\bar{\imath}\text{c} + \tilde{\imath}\text{c} = \bar{\imath}\text{x} + \tilde{\imath}\text{x} - \bar{\imath}\text{L} - \tilde{\imath}\text{L} \tag{10}$$

Thus,
The average and harmonic components of the filter capacitor current can be obtained as

$$\bar{\imath}\text{c} = \bar{\imath}\text{x} - \bar{\imath}\text{L} \tag{11}$$

$$\tilde{\imath}\text{c} = \tilde{\imath}\text{x} - \tilde{\imath}\text{L} \tag{12}$$

Considering, the filter capacitor is large, and all the inductor harmonic current flows through the capacitor. Thus,
$\tilde{\imath}\text{c} = \tilde{\imath}\text{x}$
The ripple component of the filter capacitor voltage is

$$\tilde{V} = \frac{1}{C_f}\int \tilde{\imath}\text{c}\, dt = \frac{1}{C_f}\tilde{\imath}\text{x}\, dt \tag{13}$$

III. LC FILTER VALUES

The reactive power of the LC filter is

$$P_{rec} = \omega_s L_f\left(\bar{I}_x^2 + \tilde{I}_{pav}^2\right) + \omega_s C_f\left(\bar{V}L^2 + \widetilde{VL}_{av}\right) \tag{14}$$

where,
$\omega_s = 2\pi f_s$ and f_s is fundamental frequency, $\bar{I}_X$ and $\overline{VL}$ are rms value of the fundamental component of the inductor current and load voltage. The harmonic components is much smaller than the fundamental components and eq. 14 is simplified as

$$P_{rec} = \omega_s L_f\left(\bar{I}_X^2\right) + \omega_s C_f\left(\overline{VL^2}\right) \tag{15}$$

The rms value of the current through the inductor is
Splitting the current Ix in terms of real and imaginary, eq. 15 is rewritten as

$$P_{rec} = \omega_s L_f\left[\bar{I}_{re}^2 + (\bar{I}_{im} - \omega_s C_f \bar{V}L)^2\right] + \omega_s C_f \overline{VL}^2 \tag{16}$$

Capacitance is given as

$$C_f = k\frac{Vdc}{L_f f_{sw}^2 \bar{V}_{av}} \tag{17}$$

Inductance is given as

$$L_f = \frac{\bar{V}}{\tilde{\imath}L f_s}\{k\frac{\bar{V}_{dc}}{\bar{v}_{AV}}[1 + 4\pi^2(\frac{f_s}{f_{sw}})^2 k\frac{V_{dc}}{\bar{V}_{ac}}]\}^{1/2} \tag{18}$$

where,

k is the modulation index.

To suppress the lower order harmonics, the cut-off frequency has to be considered in the RLC filter. To operate as an ideal voltage source, with no additional voltage distortion the output impedance of the inverter must be kept zero. Then, the capacitance value of capacitor is large and the inductance value is small at the selected cut-off frequency of the low-pass filter. In this design model, from extensive simulation done, the value of R_f is selected at 2.0 ohm. The selected value gives good dynamic performance and, actively damps the resonance oscillations, during disturbances.

The value of L and C component is determined such as to minimize the reactive power consumption in these components. From eq. 17 and 18, the capacitance and inductance values are 55μF and 0.99 mH respectively, for this model. The selection of control scheme parameters, play a crucial role in the design of this DVR model. In this model, an improvised sinusoidal pulse width modulation (SPWM) technique is used to generate switching pulses to the VSI switches.

IV. SIMULATION RESULT

Modeling and simulation is done using PSCAD/EMTDC software. The DVR power system consists of a three-phase at 415 V, 50 Hz, three single-phase injection transformer, and three single pole VSI as shown in Figure 3. The simulation conditions are shown in table 1: The sensitive load on the feeder is a 3-phase R-L circuit rated at (R = 10 ohms, L = 0.05 mH) and a parallel connected 3-phase diode bridge rectifier non-linear load. Similarly, to introduce distortion in supply voltages of feeder, 2 harmonic voltage sources, consisting of a 6-pulse Bridge rectifier and three-phase diode Bridge rectifier are connected to the input supply voltages.

Table 1. Simulation condition

Supply voltage	415 V
*F*s	50 Hz
*F*sw	15 kHz
*f*cut-off	650 Hz

To demonstrate the performance of DVR, exhaustive studies is done on the voltage waveforms of the distribution system by taking into consideration factors that have significant influence on the feeder voltage harmonics and the load voltage THD.

Case study 1: Voltage harmonic distortion is introduced in the supply voltage and both the supply voltage and load voltage waveform are distorted with DVR in off state. The THD of load voltage is greater than 55 %. The simulated waveform without DVR is shown in Figure 4.

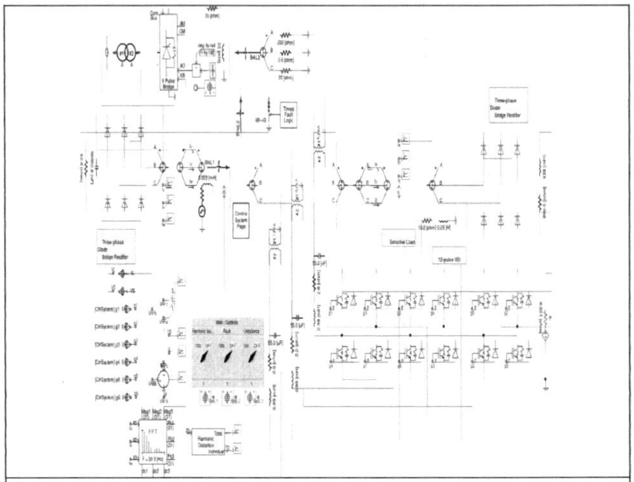

Figure 3. Power circuit of DVR

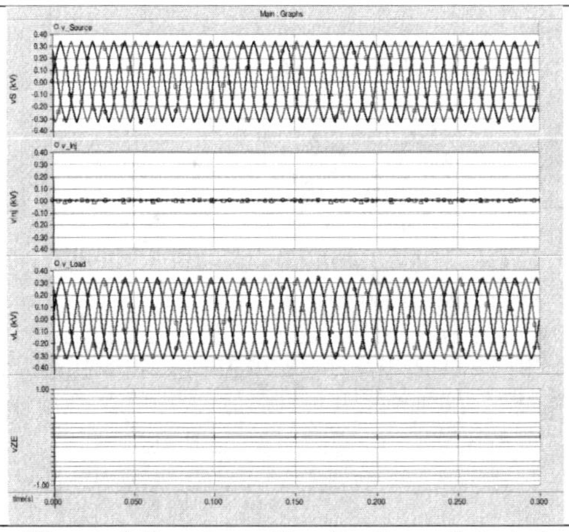

Figure 4. Supply and load voltage distorted

Case study 2: Compensation of voltage harmonics.

To consider the DVR performance, harmonic voltage sources are introduced in the supply voltage. The effectiveness of DVR to compensate the voltage distortion is shown in Figure5. With DVR in operation, the load voltage waveform is sinusoidal as shown in Figure 5.

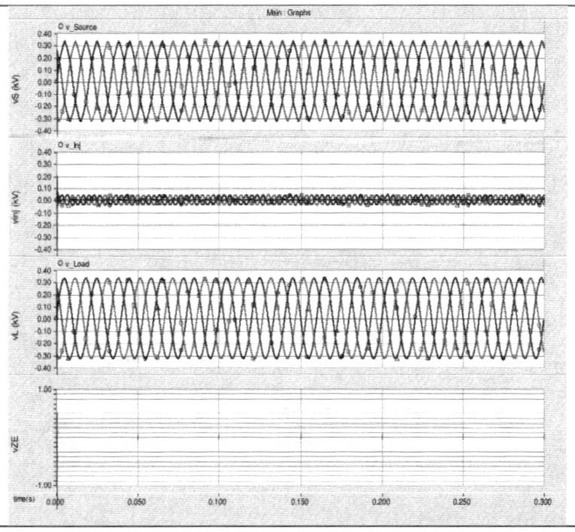

Figure 5. Load voltage compensated.

Figure 6 shows .the distorted source voltage, injection voltage and compensated load voltage for phase a. With the DVR adopting the proposed control approach, the load voltage as in Figure 6 is sinusoidal voltage. When the DVR is in operation, the load voltage at sensitive load is sinusoidal and is at rated value, at 1.0 pu, thus achieving the objective. The source and load voltage harmonic spectrum and THD after compensation are shown as in Figure 7(a) (b) and (c), respectively. Simulation is based on 50 Hz base frequency and the THD is considered up to the 31st harmonics order.

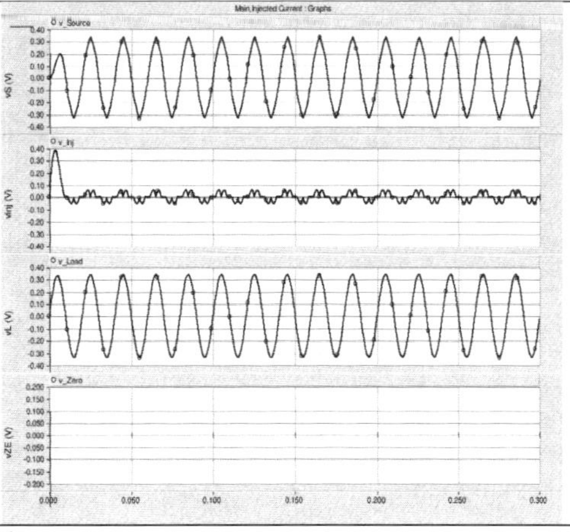

Figure 6. Load voltage compensated on phase a.

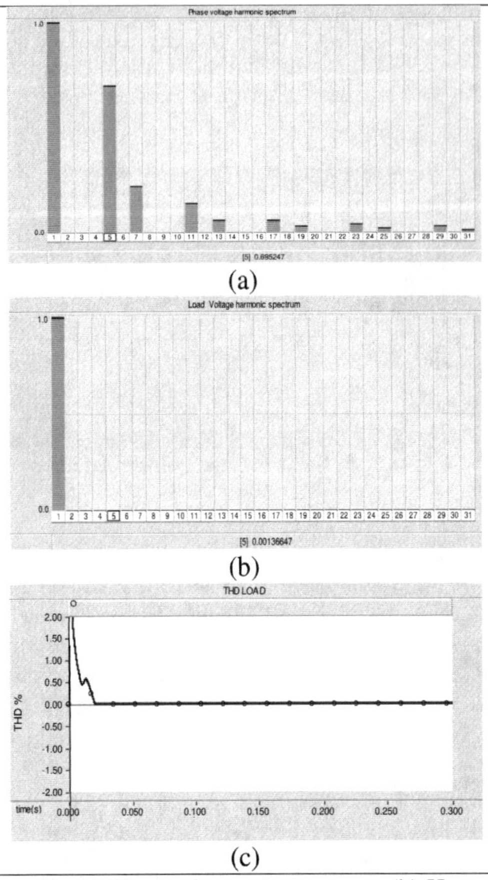

Figure 7. (a) Harmonic spectrum source (b) Harmonic spectrum load (c) THD load.

It is evident from Figure 5 and Figure 6 the effectiveness of DVR to restore sinusoidal load voltage at the required level at the sensitive load.

V. CONCLUSION

In this research paper, analysis and design methods for the RLC filter for a three-phase PWM inverter for DVR is presented. The interface filter is able to provide the compensating voltage for the DVR performance suppressing the voltage harmonic. The DVR is considered as the controlled voltage source to generate a compensation voltage for the harmonic source. The DVR is very suitable for compensating the voltage type harmonic source. The compensation performance of DVR is established by the simulation results test run on a three-phase 415 V distribution system. The simulation results have demonstrated excellent suppression of voltage harmonic capabilities of DVR using the proposed new RLC filter carrier based PWM technique. The proposed DVR is suitable to be used for high voltage systems to compensate various power quality problems. The DVR is an attractive custom power device for power quality improvement specifically to suppress voltage harmonics and also to mitigate voltage sag and combination both as a hybrid unit to protect sensitive load.

VI. REFERENCES

[1] Dayi Li, Jun Tian, "A Novel Active Power Filter for the Voltage-Source Type Harmonic Source", Electrical Machines and Systems, International Conference, pp. 2077-2080, 17-20 Oct. 2008.

[2] R. B. Gonzatti, "Hybrid active power filter applied to harmonic compensation of current-source type and voltage-source type nonlinear loads", Power Electronics Conference, pp. 1257-1262, 27-31 Oct, 2013.

[3] R. Chan BahaudinI. Daut, "Investigation of Influence of Voltage Source Type on Harmonic Characteristics", The 4th International Power Engineering and Optimization Conf. (PEOCO2010), Shah Alam, Selangor, MALAYSIA: pp. 391-395, 23-24 June 2010.

[4] B. Swathi, "Minimization of Resonant Frequency Oscillations in Terminal Voltages of AC Motors Using Active Damping Technique", International Journal of Engineering Research and Development, Volume 5, Issue 6 (December 2012), PP. 20-31.

[5] Kamalesh Hatua, "Active Damping of Output LC Filter Resonance for Vector-Controlled VSI-Fed AC Motor Drives", IEEE Transactions on Industrial Electronics, Vol. 59, N0. 1, pp. 334-342, January 2012.

[6] Nisha G. K, "Harmonic Elimination of Space Vector Modulated Three Phase Inverter", IMECS, March 14-16, 2012.

[7] Ab. Hamadi, "A New Hybrid Series Active Filter Configuration to Compensate Voltage Sag, Swell, Voltage and Current Harmonics and Reactive Power", IEEE International Symposium on Industrial Electronics, pp.286-291, 2009.

[8] M. A. Mulla, "A Simplified Control Strategy for Series Hybrid Active Power Filter that Compensat Voltage Sag, Swell, Unbalance and Harmonics" IEEE, pp. 1-6, 6-8 Dec. 2012.

[9] Patricio Salmer, "A Control Strategy for Hybrid Power Filter to Compensate Four-Wires Three-phase Systems", IEEE Transactions on Power Electronics, Vol. 25, No. 7, pp.1923-1931, July 2010.

[10] A.Z.M. Shahriar Muttalib, "Design and Simulation of an Inverter with High Frequency Sinusoidal PWM Switching Technique for Harmonic Reduction in a Standalone/ Utility Grid Synchronized Photovoltaic System", International Conference on Informatics, Electronics & Vision, pp. 1168-1173, 2012.

[11] Mohammad Monfared, "Analysis, Design, and Experimental Verification of a Synchronous Reference Frame Voltage Control for Single-Phase Inverters", IEEE Transactions on Industrial Electronics, Vol. 61, No. 1, pp. 258-269, January 2014.

[12] Khaled H. Ahmed, "Passive Filter Design for Three-Phase Inverter Interfacing in Distributed Generation", Electrical Power Quality and Utilisation, Journal Vol. XIII, No. 2, pp. 49-58, 2007.

IOP Conf. Series: Materials Science and Engineering **78** (2015) 012011 doi:10.1088/1757-899X/78/1/012011

Cmos spdt switch for wlan applications

M A S Bhuiyan, M B I Reaz, L F Rahman and K N Minhad
Electrical, Electronic and Systems Engineering, Universiti Kebangsaan Malaysia, 43600 Bangi, Selangor, Malaysia

E-mail: asobhan@eng.ukm.my

Abstract. WLAN has become an essential part of our today's life. The advancement of CMOS technology let the researchers contribute low power, size and cost effective WLAN devices. This paper proposes a single pole double through transmit/receive (T/R) switch for WLAN applications in 0.13 µm CMOS technology. The proposed switch exhibit 1.36 dB insertion loss, 25.3 dB isolation and 24.3 dBm power handling capacity. Moreover, it only dissipates 786.7 nW power per cycle. The switch utilizes only transistor aspect ratio optimization and resistive body floating technique to achieve such desired performance. In this design the use of bulky inductor and capacitor is avoided to evade imposition of unwanted nonlinearities to the communication signal.

1. Introduction
The increased popularity of WLAN applications, in the recent years, has raised the demand for small power and less off-chip components which tends to achieve integration of multiple transmitter/receiver chains on the same die [1], which further increases the demand for highly integrated and area efficient designs.

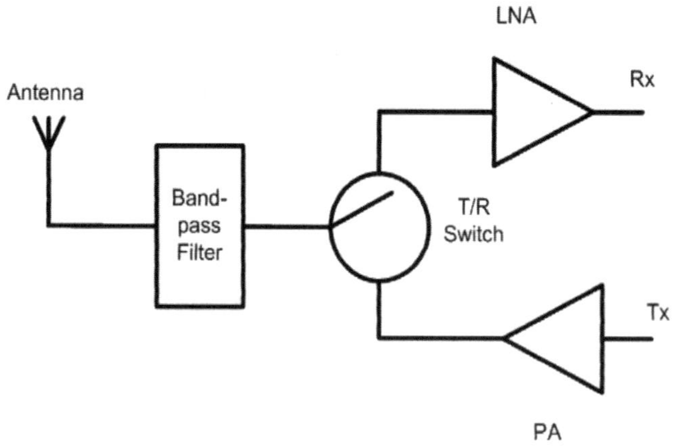

Figure 1. T/R switch at the WLAN transceiver front-end.

All WLAN transceivers essentially need an efficient T/R switch for sharing a single antenna between its transmitter and receiver to compensate the size, cost and portability. It is one of the very crucial parts of a transceiver which deals with both high power transmitter and low power receiver circuit as shown in Figure 1. During transmission stage, the switch connects the antenna to the power amplifier of the transmitter to radiate large signal to the surroundings and at the same time protects the low power receiver circuit from being get damaged [2]. On the other hand, during reception stage, the switch ensures that the low noise amplifier receives the intended signal with minimum loss. However, the major requirements for a T/R switch are low insertion loss, high power handling capability, high linearity, good isolation and acceptable reliability [3-4].

Advances in CMOS technology allows the researchers to fabricate fully on-chip transceivers without compromising the performance issues [5]. This results in low power, small size and low cost wireless terminals for different applications such as RFID, Bluetooth, Zigbee, Wi-Fi and WLAN devices [6-7]. Consequently, to a great extent, this attracts more subscribers to enjoy affordable wireless communication services through WLAN.

The CMOS T/R switch presented, in this paper, can withstand more than 24 dBm PA output power with the normal 0.13 μm CMOS devices, by employing a novel floating device scheme. Over the 802.11g operation frequency band, the insertion loss due to the switch is less than 1.4 dB and it provides isolation of more than 25 dB to the receiver. This switch will be very useful for 2.4 GHz WLAN devices.

2. Methodology

The objective of this study is to design and simulate a CMOS switch to obtain better performance. The software named Mentor Graphics by Emerald Systems is used for the design and simulation of the switch. This software has two parts; first part is Design Architect IC (DA-IC) and second part is IC-Station. Design Architect has been used to design the schematics of the T/R switch circuit and for the simulation to obtain the result.

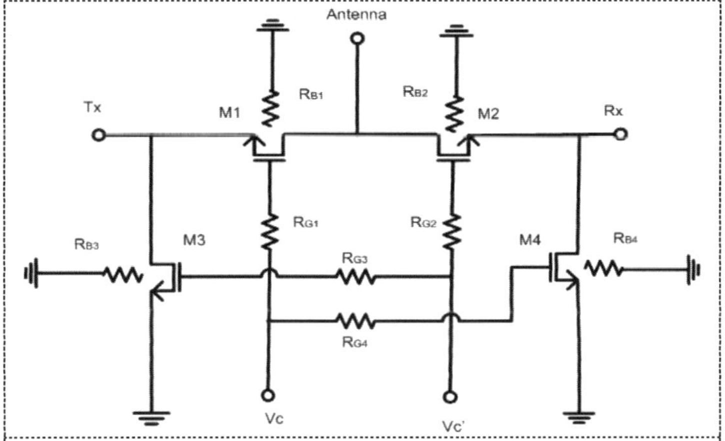

Figure 2. Schematic of the proposed CMOS SPDT switch.

The schematic of the proposed series-shunt T/R switch is illustrated in Figure 2. It consists of one voltage source, six transistors and eight resistors only. The transistors M5 and M6 form an inverter circuit. Control signals Vc and Vc′ are used to switch the transistors M1, M2, M3 and M4 from their ON to OFF states and vice versa. The series transistors, M1 and M2, perform main switching task while the shunt transistors, M3 and M4, provide low-impedance paths for the unwanted signals to the RF ground. Therefore, the two shunt arms make the T/R switch achieve relatively better isolation between transmitter and receiver port but at the cost of high insertion loss.

Transistor ON state resistance is the dominant factor responsible for insertion loss. Because of this, only n-channel MOS transistors are utilized in the design as their conductivity is higher than the p-channel MOSs. The drain-to-body and source-to-body junction capacitances of the transistors are also critical factors determining insertion loss. Therefore, optimization of the transistor width-length is very important [8-9].

The gate resistances help to improve the isolation by decreasing the effect of capacitive coupling of the OFF transistors [10] whereas the body floating resistors helps to meliorate power handling capacity by retaining the high input impedance of the transistor by offering a high resistance to the source/Drain to body diode [11-12]. The components used in this study are given in Table 1.

Table 1. T/R switch elements and their values.

Elements	Values
M1 – M4	60/0.13 μm
RG1 – RG4	5 Ω
RB1 – RB4	50Ω
Vc and Vc′	1.2 V and 0 V

3. Results and discussions

The switch proposed in this paper has been designed in 130 nm CMOS technology. Mentor graphics Design Architect has been used for the simulation study. The isolation, insertion loss and power handling capacity are the three main characteristics of a transmit/receive switch. These parameters are also interrelated and tradeoffs among them cannot be ignored [9].

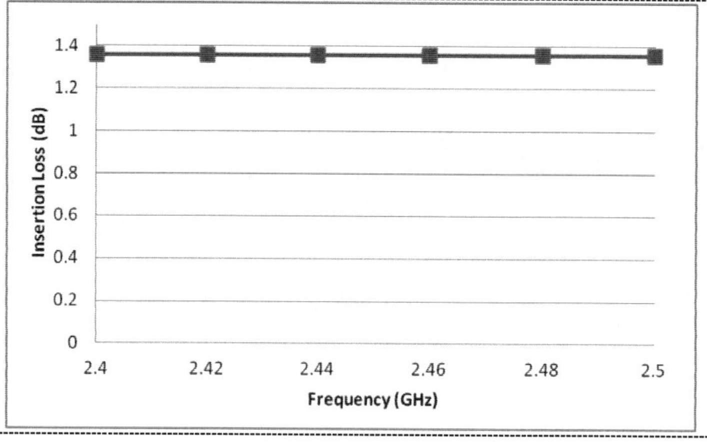

Figure 3. Insertion loss of the switch at 2.4 GHz band.

In this paper, the insertion loss, isolation and power handling capacity of the proposed switch is measured in transient analysis mode. Figure 3 and Figure 4 show the insertion loss and isolation of the switch. From the simulation it is obvious that the switch exhibit 25.1 dB isolation and 1.36 dB insertion loss. It is also apparent that the values of insertion loss and isolation for the 2.4 GHz WLAN remains almost constant for the proposed switch.

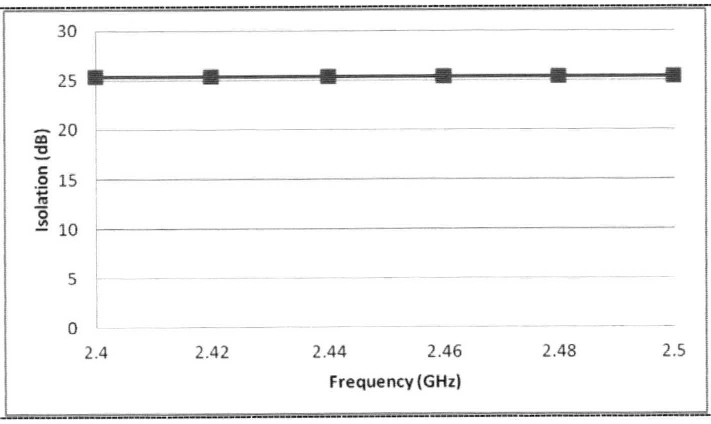

Figure 4. Isolation of the switch at 2.4 GHz band.

Figure 5 shows the insertion loss and isolation of the switch as a function of input power (Pin). The insertion loss of the switch is found to increase exponentially with input power but the isolation is found to decrease steadily. At P1dB point the values of the insertion loss and isolation are 1.36 dB and 25.3 dB respectively.

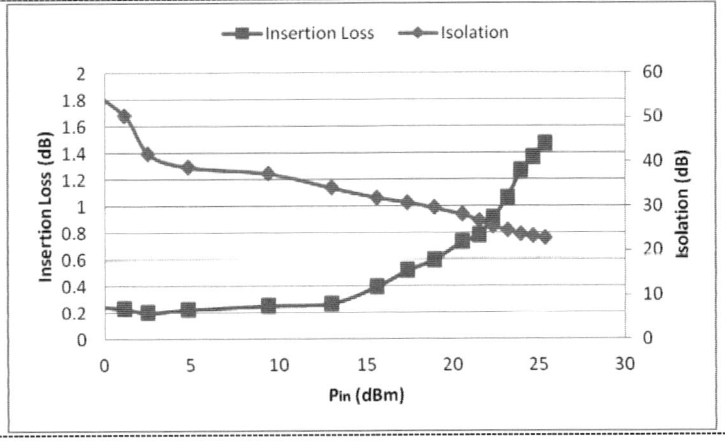

Figure 5. Insertion loss and isolation of the switch with input power.

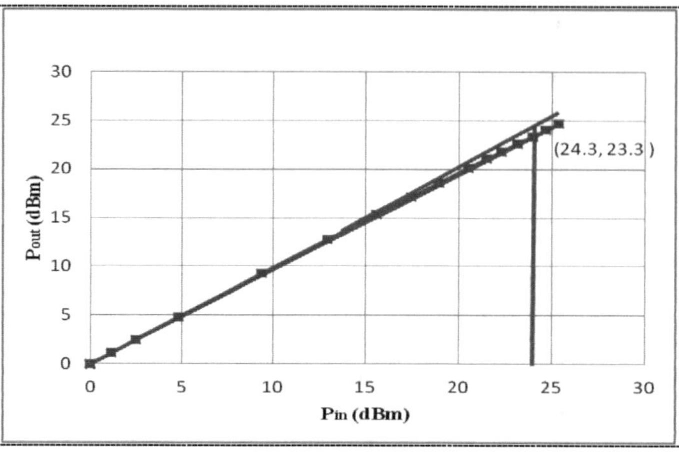

Figure 6. Power handling capacity of the switch.

The linearity as well as the power handling capacity of a switch is usually determined by the 1 dB compression (P1dB) point. The higher the P1dB point, the more the switch handles RF power within acceptable level of nonlinearity. Therefore, this characteristic can be considered as stability of the switch at high input power during the transmission period. Figure 6 shows the large signal linearity of the switch and the value of the P1db found is 24.3 dBm.

Therefore, the proposed switch can handle large amount of power during transmission mode with a very small insertion loss and better isolation of low power receiver circuit. It also dissipates very small amount of power which is only 786.7 nW per cycle. Table 2 gives the summery of the performance of the switch.

Table 2. T/R switch performance summery

Parameters	**values**
Insertion loss	1.36 dB
Isolation	25.3 dB
P1dB	24. 3 dBm
Power dissipation	786.7 nW

4. Conclusion

This paper proposes a single pole double through transmit/receive (T/R) switch for compact WLAN devices. For the simulation, Mentor graphics 0.13 μm CMOS technology is used. The proposed switch exhibit 1.36 dB insertion loss, 25.3 dB isolation and 24.3 dBm power handling capacity. Transistor aspect ratio optimization and resistive body floating technique are utilized in this circuit to achieve such desired performance. It only dissipates 786.9 nW power per cycle. In this design the use of bulky inductor and capacitor is avoided and therefore, the size and cost of the devices will be reduced. Such a switch will be very useful for cost effective WLAN devices.

5. References

[1] Behzad A, Carter K A, Ming C H, Wu S, An P M, Lee C P, Qiang L, Leete J C, Au S, Kappes M S, Zhimin Z, Ojo D, Lijun Z, Zolfaghari A, Castanada J, Darabi H, Benson Y, Rofougaran A, trachewsky J, Moorti T, Gaikwad R, Bagchi A, Hammerschmidt J S, Pattin J,

Rael J J and Marholev B 2007 a fully integrated MIMO multiband direct conversion CMOS transceiver for WLAN applications (802.11n) *IEEE J. Solid-St. Circ.* **42** 2795

[2] Kidwai A A, Fu C T, Sadhwani R, Chi D C, Jensen J C, and Taylor S 2008 An ultra-low insertion loss T/R switch fully integrated with 802.11b/g/n transceiver in 90nm CMOS *Proc. 2008 IEEE Radio Frequency Integrated Circuits Symposium* (Atlanta, GA, June 2008) pp. 313-316

[3] Bhuiyan M A S, Reaz M B I, Jalil J, Rahman L F and Chang T G 2014 Design trends in fully integrated 2.4 GHz CMOS SPDT switches *Curr. Nanosci.* **10** pp. 334-343

[4] Kidwai A A, Tsung F C, Jensen J C and Taylor S S 2009 A fully integrated ultra-low insertion loss T/R switch for 802.11b/g/n application in 90 nm CMOS process *IEEE J. Solid-St. Circ.* **44** pp. 1352-1360

[5] Uddin M J, Nordin A N, Reaz M B I and Bhuiyan M A S 2013 A CMOS power splitter for 2, 45 GHz ISM band RFID reader in 0, 18 μm CMOS technology *Teh. vjesnik* **20** pp. 125-129

[6] Idris M I B, Reaz M B I and Bhuiyan M A S 2013 A low voltage VGA for RFID receivers *Proc. 2013 IEEE International Conference on RFID Technologies and Applications* (Johor, Malaysia, September 2013) pp. 1-4

[7] Uddin M J, Ibrahimy M I, Reaz M B I and Nordin A N 2009 Design and application of radio frequency identification systems *Euro. J. Sci. Res.* **33** pp. 438-453

[8] Jung H F and O K K 2004 Single-pole double-throw CMOS switches for 900-MHz and 2.4-GHz applications on p- silicon substrates *IEEE J. Solid-St. Circ.* **39** pp. 35-41

[9] Li X J and Zhang Y P 2010 Flipping the CMOS switch *IEEE Microw. Mag.* **11** pp. 86-96

[10] Yamamoto K, Heima T, Furukawa A, Ono M, Hashizume Y, Komurasaki H, Shigenobu M, Sato H and Kato N 2001 A 2.4-GHz-band 1.8-V operation single-chip Si-CMOS T/R-MMIC front-end with a low insertion loss switch *IEEE J. Solid-St. Circ.* **36** pp. 1186-1197

[11] Zhang Y P, Qiang L, Wei F, Hoe A C and He L 2006 A Differential CMOS T/R Switch for Multistandard Applications *IEEE Trans. Circ. Systems II: Express Briefs* **53** pp. 782-786

[12] Yeh M C, Tsai Z M, Liu R C, Lin K Y, Chang Y T and Wang H 2006 Design and analysis for a miniature CMOS SPDT switch using body-floating technique to improve power performance *IEEE Trans. Microw. Theory Tech.* **54** pp. 31-39

An analytical study of various telecomminication networks using markov models

M Ramakrishnan[1], **E Jayamani**[2] and **P Ezhumalai**[3]
Faculty of Engineering, Computing and Science
Swinburne University of Technology Sarawak Campus
Kuching, Sarawak, Malaysia

E-mail:[1]mramakrishnan@swinburne.edu.my, [2]ejayamani@swinburne.edu.my, [3]pezhumalai@swinburne.edu.my

Abstract. The main aim of this paper is to examine issues relating to the performance of various Telecommunication networks, and applied queuing theory for better design and improved efficiency. Firstly, giving an analytical study of queues deals with quantifying the phenomenon of waiting lines using representative measures of performances, such as average queue length (on average number of customers in the queue), average waiting time in queue (on average time to wait) and average facility utilization (proportion of time the service facility is in use). In the second, using Matlab simulator, summarizes the finding of the investigations, from which and where we obtain results and describing methodology for a) compare the waiting time and average number of messages in the queue in M/M/1 and M/M/2 queues b) Compare the performance of M/M/1 and M/D/1 queues and study the effect of increasing the number of servers on the blocking probability M/M/k/k queue model.

1. Introduction

This paper represents the analytical study of different queuing Markov models' a type of single server performances with suitable applications on telecommunication networks. In general, queuing systems may be characterized by complex input process, service time and distribution and queue disciplines.[1] paper also analysying the development of multivariant models based on the Markov chains applications in the field of reliability,economics, survival analysis, engineering, social sciences, environmental studies, biological sciences, etc. In practice, such queuing processes and disciplines are often amendable to analysis.

2. Literature Review

In telecommunication networks queuing theories are modelled in a variety of network engineering principles and techniques. [2] consider the discrete-time process that the length of off-periods property taken into account which are idependent interval distribution with an arbitrary distribution. [3] emerging Value at Risk(VaR) as one of the measure in markt place to examine the high –order effect of Mark chain through backtesting. There are different testing made in Markov process, [4] is a statistic test with the stationary that to divide whole sequence of events into subintervals and then

[1 2 3] Faculty of Engineering, Computing and Science, Swinburne University of Technology, Sarawak Campus.

compute and comparing the transition probability matrix amongst each intervals. Testing and analysis plays a vital role to improve the performances of the network systems. [5] defines through specification test constructing a non-parametric simultaneous confidence band for conditional regression functions.[6] proposing the new model that Diffential Output-ports Choosing Probability scheme , it is applied to optical burst switching networks that the blocking probability and the delay time will change along with the changeable ports choosing probability and the ratio between different length burst.

3. Basic Notation Representing Queuing Models

A commonly used shorthand notation, called Kendall notation, for such single queue models describes the arrival process, service distribution, the number of servers and the buffer size(waiting line) as follows:

Arrival process/service distribution/ number of servers / waiting room

The complete notation expressed as (a/b/c): (d/e/f) where,

a =arrival distribution,
b=departure distribution,
c=number of parallel service channel in the system,
d=service discipline,
e=maximum number of customers allowed in the system,
f=calling source or population

Commonly used characters for the first two positions in the shorthand notations are: M(Markovian – Poision for the arrival or Exponential for the service time) , G(General), D(Deterministic), GI (General and Independent) and Geom(Geomentric). The fourth positions used for the number of buffer places in addition to the number of servers and its usually not used if the waiting room is unlimited.

Symbols for d,

FCFS=First Come, First Served,
LCFS=Last Come, First Served,
SIRO=Service In Random Order,
GD=General service Discipline.

The symbols e and f represent a finite (N) or infinite () number of customers in the system and calling source respectively. For instance, (M/E/1):(FCFS/N/∞) represents a Poisson arrival (exponential interarrival), Erlangen departure, single server, First Come First Served discipline, maximum allowable customers N in the system and infinite population model.

4. Telecommunication System Analysis

In this telecommunication queuing system, exploring the effect of having more than one server in a queuing system.Queuing networks have been analysed and solved under different assumptions and constraints [7]. To get a good feel for how does such a system behave, comparing the waiting time and average number of messages in the single server system, such as M/M/1 with that of a Multiple server's system such as M/M/2, M/M/D and M/M/k/k systems. As like [2], the number of iterations taken into account in the MatLab simulator to measure the performance and analyse it.

4.1 M/M/1 vs M/M/2

To compare the performances of an M/M/1 and M/M/2, plot the mean total time in the system against the utilization factor in both systems. The total time a customer spends in a system(Ws) is given by,

$$Ws = Wq + \frac{1}{\mu} \quad (1)$$

While considering,λ is arrival rate,μ is service rate and ρ is traffic intensity or utilization factor.

For M/M/1, The total time a customer spends in the queue(Wq) is given by,

$$Wq = \frac{\rho}{\mu(1-\rho)} \tag{2}$$

For μ =1, Equation (2) becomes,

$$Wq = \frac{\rho}{1-\rho} \tag{3}$$

For M/M/2, and μ=1, Wq is given by,

$$Wq = \frac{\rho^2}{1-\rho^2} \tag{4}$$

Use MatLab simulator, plotting Equation (1), the values of that are of interest are ρ<0.2 and ρ> 0.9

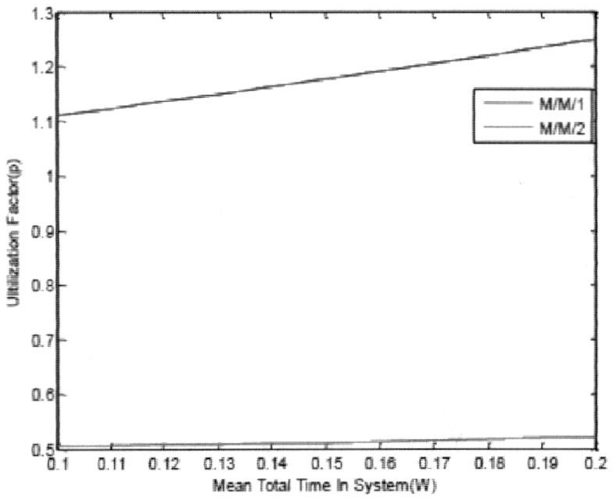

Figure 1. Mean total against Utilization Factor ρ<0.2

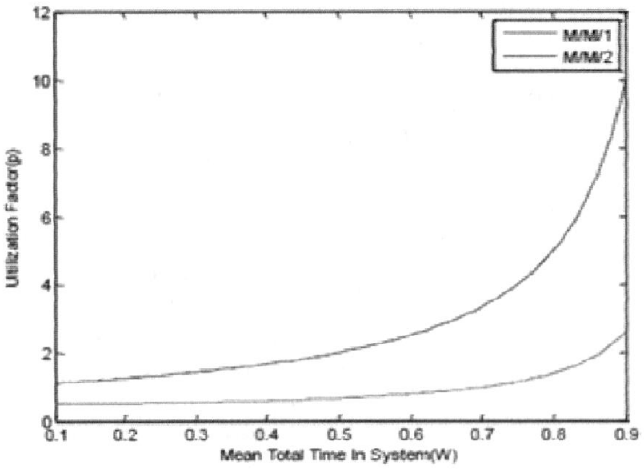

Figure 2. Mean total against Utilization Factor ρ >0.9

The above two figures, Figure 1 represents the utilization factor against mean total time for the value taken as greater than 0.2 and Figure 2 represents the utilization factor against mean total time for value taken as less than 0.9. Utilization factor represents the performance of any system being one by the system. For a lower utilization factor, the mean time in the system is very low. As the utilization factor increases, in the M/M/1 queue, the mean total time in the system increases greatly while in the M/M/2 queue, the mean total time in the system only has a slight increment. Therefore, M/M/2 queue has a superior performance than M/M/1 queue. Matlab simulator working controls are given below,

```
p = 0:0.01:0.99; % the utilization factor
w1 = []; % an array to store the values
w2 = []; % an array to store the values
lamda1 = 0:0.01:0.99;
lamda2 = lamda1 * 2;
% a for loop to do the calculations for more than a single value
for k = 1:100
w1 = [w1 (lamda1(k)/(1-lamda1(k)))+1];
w2 = [w2 (lamda2(k)^2)/(4 - (lamda2(k)^2))+0.5];
end
%plotting the graph
plot(p,w1,'r',p,w2,'b');
Title('Average time in the system for M/M/1 And M/M/2 queues when \mu = 1');
Xlabel('Utilization \rho');
Ylabel('Average Time In System');
tleg = legend('M/M/1','M/M/2');
```

4.2 M/M/1 vs M/M/1 and λ=1

Repeating the same process, however, this time needs different equations to work with, since the service rate in an M/M/2 system is twice that of an M/M/1 system, use the following equations;

For M/M/1, Ws is given by,

$$Ws = \frac{\rho^2}{2(1-\rho)} + \frac{1}{2} \tag{5}$$

For M/M/2, Ws is given by,

$$Ws = \frac{\rho^3}{(1-\rho^2)} \tag{6}$$

Plotting graph on the simulator will get $\rho < 0.2$ and $\rho > 0.9$ as below, Figure 3 represents the utilization factor against mean total time for the value taken less than 0.2 and Figure 4 represents the utilization factor against mean total time for the value taken greater than 0.9.

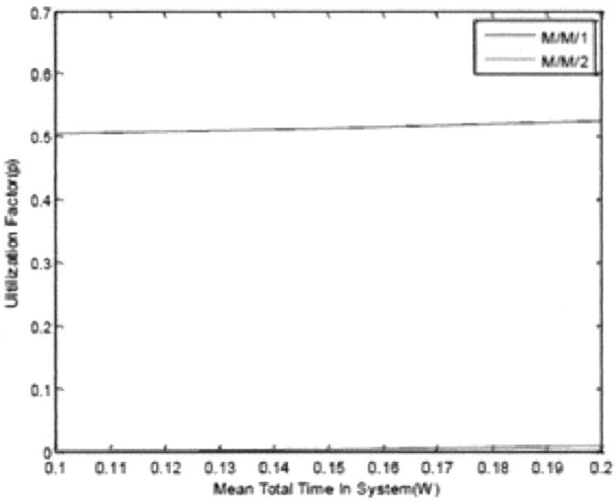

Figure 3. Utilization factor against Mean total time in the system for ρ<0.2

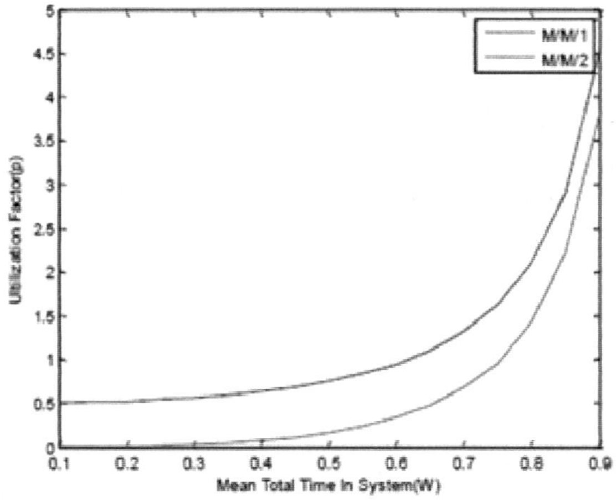

Figure 4. Utilization factor against Mean total time in the system for ρ > 0.9

The expected waiting time for the systems is almost same with increasing values of utilization factor. Although the M/M/2 system performed better, both systems relatively equal, because of the arrival rate is 1 for both systems. Nevertheless, the biggest difference can be observed for smaller values ofρ . In Matlab simulator working controls are given below,

```
p = 0:0.01:0.99;%utilisation factor
W1 = []; % an array to store the values
W2 = [];% an array to store the values
mu = 50:-0.5:0.5;
%a for loop to repeat the calculations for more than a single value
for k = 1:100
W1 = [W1 (1/(2*mu(k)*((2*mu(k)) -1))) + (1/(2*mu(k)))];
```

```
W2 = [W2 (1/(mu(k)*((4*mu(k)^2) -1))) + (1/(2*mu(k)))];
end
%plotting the graph
plot(p ,W1,'r', p, W2, 'g');
Title('Average time in system for M/M/1 and M/M/2 Queues when \lambda = 1, M/M/1 \mu = 2
(M/M/2 \mu)');
Xlabel('Utilization \rho');
Ylabel('Average time tn system');
tleg = legend('M/M/1','M/M/2');
set(tleg, 'Location', 'NorthWest');
```

4.3 M/M/1 vs M/M/2 and μ =1, Number of customers in the system(Ls)

By applying the Little's formula:

$$Ls = \lambda Ws$$

Using Equations(1), (3) and (4) and using the fact that μ=1, get the following expressions:
For M/M/1, Ls is given by,

$$Ls = \frac{\rho}{1-\rho} \tag{7}$$

For M/M/2, Ls is given by,

$$Ls = \frac{\rho}{1-\rho^2} \tag{8}$$

Again, plotting the number of customers in the system against utilization factor reveals that at low values of utilization factor, the number of customers in the system is almost zero for both M/M/1 and M/M/2 queue. With increasing utilization factor, the number of customers also increases. It's an seen that the M/M/1 model has more customers in the system that the M/M/2 model, hence the number of servers, the best in performance. In Matlab simulator working controls are given below,

```
p = 0:0.01:0.99; %utilisation factor
W1 = []; % an array to store the values
W2 = []; % an array to store the values
lamda1 = 0:0.01:0.99;
lamda2 = lamda1 * 2;
N1 = [];
N2 = [];

%a for loop to repeat the calculations for more than a single value
for k = 1:100
W1 = [W1 (lamda1(k)/(1-lamda1(k)))+1];
W2 = [W2 (lamda2(k)^2)/(4 - (lamda2(k)^2))+0.5];
N1(k) = lamda1(k) * W1(k);
N2(k) = lamda2(k)* W2(k);
end

%plotting the graph
plot(p,N1,'r',p,N2,'b');
Title('Average Number of Customers In System for M/M/1 And M/M/2 Queues when \mu = 1');
Xlabel('Utilisation \rho');
Ylabel('Average Number of Customers In System');
tleg = legend('M/M/1','M/M/2');
set(tleg, 'Location', 'NorthWest');
```

4.4 M/M/1 vs M/D/1 and μ=1

Comparing the performances of M/M/1 with M/D/1 in terms of average waiting time in the queue. An expression for M/M/1 was derived earlier, namely, equation(3), and for M/D/1 and μ=1, will have,

$$W_q = \frac{\rho}{2\mu(1-\rho)} \tag{9}$$

A graph containing the two plots is shown below,

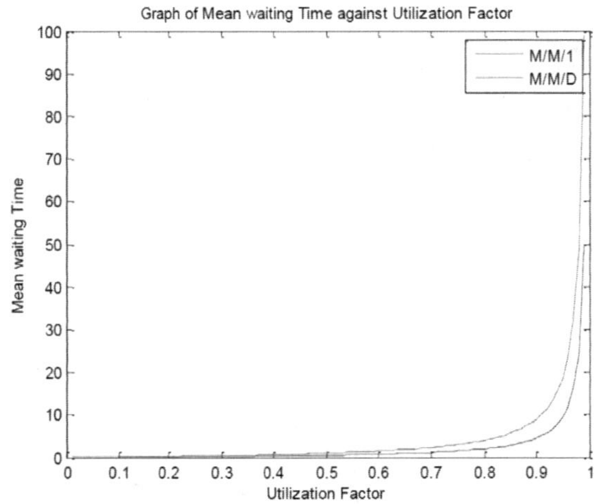

Figure 5. Utilization factor against Average waiting time

From the graph, understanding that for the both M/M/1 and M/M/D queuing models, the average waiting time in the queue is approximately until about ρ=0.3 for M/M/1 and ρ= 0.5 for M/M/2 after which it starts increasing exponentially. Also adding that as utilization factor increases, the average waiting time for M/M/1 is greater than M/M/D. Therefore M/M/D has a better performance than M/M/1 model. In Matlab simulator working controls are given below,

```
p = 0:0.01:0.99;
W1 = [];
W2 = [];
lamda1 = 0:0.01:0.99;
for k = 1:100
W1 = [W1 (lamda1(k)/(1-lamda1(k)))+1];
W2 = [W2 (lamda1(k)/(2*(1-lamda1(k))))+1];
end
plot(p,W1,'r',p,W2,'b');
Title('Average Time In System for M/M/1 And M/D/1 Queues when \mu = 1');
Xlabel('Utilization \rho');
Ylabel('Average Time In System');
tleg = legend('M/M/1','M/D/1');
set(tleg, 'Location', 'NorthWest');
buffer = plot(pD_Array,WD_Array,'b');
legend('M/M/1','M/M/D');
title('Graph of Mean waiting Time against Utilization Factor')
xlabel('Utilization Factor');
ylabel('Mean waiting Time');
```

4.5. M/M/k/k, Probability of Loss

In this session, investigates the effect of increasing the load of an M/M/k/k model while the number of servers remains constant. A graph is generated with different plots for 5,10,15 servers describing the probability loss(Erlang B) for traffic intensity per server ranging between 0 and 1. The Erlang B formula is given by,

$$\text{Erlang } B_k = \text{load} * \text{Erlang } B_k - \frac{1}{k + (Load * ErlangB_k - 1)}$$

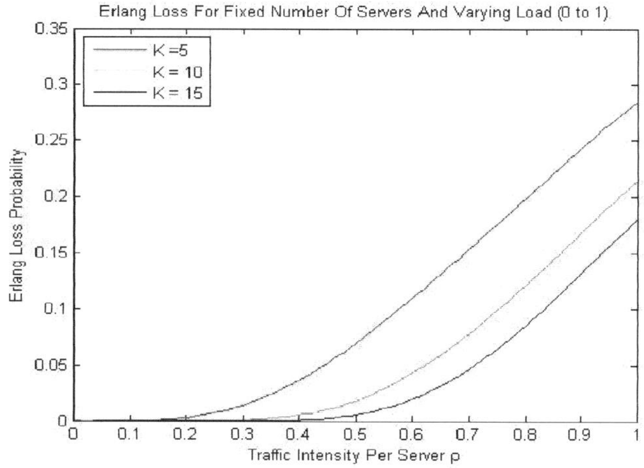

Figure 6. Erlang Loss Probability against Traffic Intensity

Based on the graph, the lower the blocking probability is seen by the system with the highest if servers ρ=A/k, A being the load and k being the number of servers. As the number of servers increases, then the load also increases by the same factor. Given the explanation refers to the Erlang B formula given above in he procedure, it can be deduced that as the number of servers, k increases, then the denominator will keep increasing therefore the probability of the system being full at any time $E_k(A)$ keeps reducing. [8] using blocking probability entend analysis with single layer unicast to multilayer cast call. Using Erlang 's fixed point approximation [9] finds new way to estimate blocking probability in overflow loss networks.Blocking probability is a basic quantity that is used in many network design and application tasks[10]. In Matlab simulator working controls are given below,

```
FILE 1:
function y = erlangB (A, k)
e = 1;
for i = 1:k
e = (A * e)/(i + (A * e));
end
y = e;
FILE 2:
load1 = 0:0.05:5;
load2 = 0:0.1:10;
load3 = 0:0.15:15;
fixedK1 = 5;
fixedK2 = 10;
fixedK3 = 15;
p = 0:0.01:1;
for k = 1:101
Bm1(k) = erlangB(load1(k), fixedK1);
Bm2(k) = erlangB(load2(k), fixedK2);
Bm3(k) = erlangB(load3(k), fixedK3);
end
clf;
plot(p, Bm1, 'r', p, Bm2, 'g', p, Bm3, 'b');
Title('Erlang Loss For Fixed Number Of Servers And Varying Load (0 to 1)');
Xlabel('Traffic Intensity Per Server \rho');Ylabel('Erlang Loss Probability');
tleg = legend('K =5','K = 10', 'K = 15');
set(tleg, 'Location', 'NorthWest');
```

4.6. M/M/k/k and Probability of Loss with ρ ϵ [0,10]

In this session, repeating the investigation effects of increasing the load A of an M/M/k/k model while the number of servers remains constant. The graph from Matlab simulator shown below,

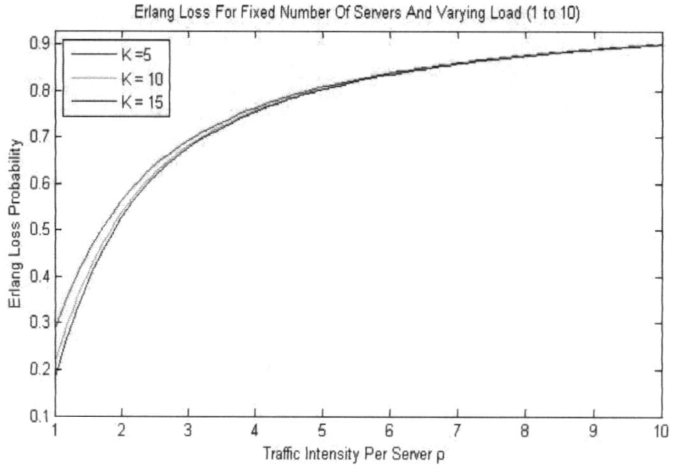

Figure 7. Erlang Loss for fixed servers against fixed traffic intensity

From the graph, it shows that the traffic intensity per server increases, the Erlang loss plots tend to converge. It is also seen that as the load increases, the blocking probability also increases, therefore reducing the performance of the system. Thus, the system with five servers has the last performance as load increases. Never prefer such a design of the system because of its poor performance that comes as a result of the high blocking probability. In Matlab simulator working controls are given below,

```
load1 = 5:0.45:50;
load2 = 10:0.9:100;
load3 = 15:1.35:150;
fixedK1 = 5;
fixedK2 = 10;
fixedK3 = 15;
p = 1:0.09:10;
for k = 1:101
Bm1(k) = erlangB(load1(k), fixedK1);
Bm2(k) = erlangB(load2(k), fixedK2);
Bm3(k) = erlangB(load3(k), fixedK3);
end
clf;
plot(p, Bm1, 'r', p, Bm2, 'g', p, Bm3, 'b');
Title('Erlang Loss For Fixed Number Of Servers And Varying Load (1 to 10)');
Xlabel('Traffic Intensity Per Server \rho');
Ylabel('Erlang Loss Probability');
tleg = legend('K =5','K = 10', 'K = 15');
set(tleg, 'Location', 'NorthWest');
```

5. Conclusions

From results analysis, understanding that M/M/2 queuing model has a better performance than M/M/1 model based on average waiting time and utilization factors of both systems. A M/M/D model given more performance measure than an M/M/1 model using the same variable as above. The M/M/D performance can be approximated to that of M/M/2 model. Therefore, the more the number of servers in a system, get more and better performance. It can also be deduced that while keeping the utilization factor constant and increasing the number of servers, the blocking probability reduces thus better performances.

REFERENCES

[1] WKChing, MKNg, and ESFung, "Higher-order multivariate Markov chains and their applications," *Linear Algebra and its Applications,* vol. 428, pp. 492-507, 1/15/ 2008.

[2] RGClegg, "A discrete-time Markov-modulated queuing system with batched arrivals," *Performance Evaluation,* vol. 67, pp. 376-385, 5// 2010.

[3] TKSiu, WKChing, EFung, MNg, and XLi, "A high-order Markov-switching model for risk measurement," *Computers & Mathematics with Applications,* vol. 58, pp. 1-10, 7// 2009.

[4] HYang, YLi, LLu, and RQi, "First order multivariate Markov chain model for generating annual weather data for Hong Kong," *Energy and Buildings,* vol. 43, pp. 2371-2377, 9// 2011.

[5] SKim and ZZhao, "Specification test for Markov models with measurement errors," *Journal of Multivariate Analysis,* vol. 130, pp. 118-133, 9// 2014.

[6] RHou, "Blocking and delay performance for differential output-ports choosing probability scheme applied optical burst switching network," *Optik - International Journal for Light and Electron Optics,* vol. 121, pp. 1295-1299, 8// 2010.

[7] PRParthasarathy, NSelvaraju, and GManimaran, "Exact transient analysis of a circulant queuing network," *Computers & Mathematics with Applications,* vol. 44, pp. 1297-1306, 11// 2002.

[8] AAlyatama, "Wavelength decomposition approach for computing blocking probabilities in multicast WDM optical networks," *Optical Switching and Networking,* vol. 12, pp. 24-33, 4// 2014.

[9] EWMWong, AZalesky, ZRosberg, and MZukerman, "A new method for approximating blocking probability in overflow loss networks," *Computer Networks,* vol. 51, pp. 2958-2975, 8/8/ 2007.

[10] AFaragó, "Efficient blocking probability computation of complex traffic flows for network dimensioning," *Computers & Operations Research,* vol. 35, pp. 3834-3847, 12// 2008.

Transmission Line Series Compensation for Wind Energy Transmission

Prof Dr C Palanichamy and Y C Wong

Dept. of Electrical and Computer Engineering, School of Engineering and Science, Curtin University, Miri, Sarawak, Malaysia.

E-mail: drcpc@curtin.edu.my, 7e0b7889@student.curtin.edu.my

Abstract. Wind energy has demonstrated to be a clean, copious and absolutely renewable source of energy, and the large penetration of it into the power grid indicates that wind energy is considered an effective means of power generation, Transmission of wind energy from remote locations to load centers necessitates long transmission lines. Series compensation is a proven and economical transmission solution to address system power transfer strength, grid stability, and voltage profile issues of long transmission lines. In this paper, a programmable approach to determine the capacitive reactance of series capacitor and optimum location for its placement to achieve maximum power transfer gas been presented. The respective program with sample solutions has been provided for real-time applications.

1. Introduction

Transmission grids across the glove have experienced noteworthy increase in renewable energy penetration primarily driven by utility scale wind turbines [1-3]. In more cases than not, regions concomitant with plentiful resource availability for wind power generation are far from the lad centers [4-6]. Therefore long transmission networks are essential to deliver the renewable energy generation to the load centers. Transmission systems must be flexible to react to more diverse generation and load patterns. In addition, the economical utilization of transmission system assets is of vital importance to enable electric utilities in industrialized countries to remain competitive and to survive [7,8]. In developing countries, the optimized use of transmission system investments is also important to support industry, create employment, and utilize efficiently the scarce economic resources.

In order to confirm passable system performance in terms of large power transfers over long distances, system operators are considering series compensated transmission facilities to integrate remotely located wind generation and deliver power to load centers [9,10]. Series compensation on transmission lines is a proven and established technology in terms of adding series capacitors along the transmission line to reduce the "effective line impedance" thereby improving power transfer capability [11-13]. Series capacitors may be installed at one or both line ends of the transmission line. Line ends are archetypal capacitor locations, because it is usually possible to use space available in the substation. Consequently, this reduces installation/ capital cost. Alternatively series capacitors shall be installed at some central location on the line. Series capacitors placed at the line ends generate added complex protection problems than those installed at the center of the line.

2. Wind Energy Transmission

2.1. Remote Locations
One of the prevalent dares to wind power grid integration in many parts of the world is the necessity of developing new transmission lines to transfer power from wind farms, usually in remote lowly populated areas due to availability of wind, to high load locations, where population density is higher. The existing transmission lines in remote locations were not designed for the transport of large amounts of wind energy. As transmission lines become longer the losses associated with power transmission increase, as modes of losses at lower line lengths are exacerbated and new modes of losses are no longer negligible as the length is increased, making it tougher transferring huge loads over large distances.

2.2. Transmission Congestion
Transmission congestion occurs when power flow over a power transmission line exceeds its capacity, resulting a higher local marginal electricity prices in the electricity demand center, since more electricity must be supplied by local generators rather than less expensive, distant generators such as wind turbines. Escalating line capacity does not remove the transmission line congestion externality and only temporarily eases congestion. As far as wind power is concerned, they are more susceptible to transmission congestion than conventional power plants for two reasons. First, wind power must be transported over greater distances to meet sufficient demand, since it is typically windy at night when electricity demand is low. Second, conventional power plants are better connected with demand centers.

3. Compensated Transmission Lines
Electrical power transmission lines have three main parameters, such as resistance, inductance, and capacitance. The inductive reactance of a transmission line, which is directly proportional to the line inductance and the system frequency, is by far the most important line parameter because for normal line design, the power transfer capability is highly affected by its magnitude [14]. Hence, the inductive reactance of transmission lines becomes a problem of increasing importance when transmission line length continues to increase. A decrease in the line inductive reactance increases the power transfer capability.

3.1. Series Compensation
Series compensation schemes permit the connection of capacitors in series with the transmission lines, thus opposing directly the effect of the series inductive reactance of the lines. Series capacitors virtually reduce the length of the line, making it easier to keep all parts of the power system running in synchronism, and maintain a constant voltage level throughout the system. This method of improving the power transmission system performance is effective and inexpensive [9-15].

3.2. Thyristor Controlled Series Capacitor
The Thyristor Controlled Series Capacitor (TCSC) presented in Figure 1 is one such popular series compensation scheme. By controlling the firing of thyristor [14,15], one can change its apparent reactance smoothly and rapidly to manage the active power flow in the transmission line.

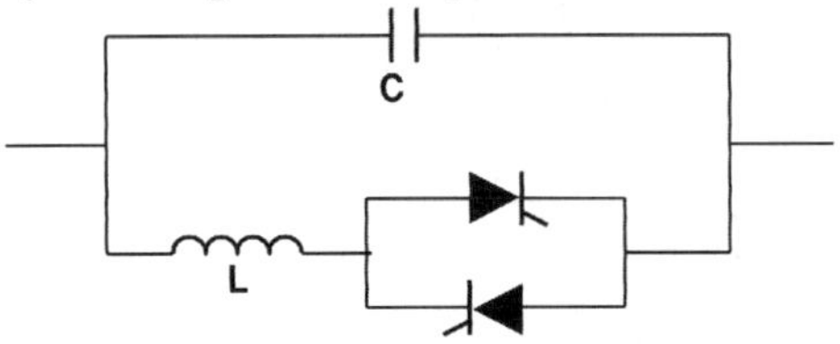

Figure 1. A Simple Circuit Model of TCSC

Assuming series capacitor compensation at a distance of x-km from the sending end as shown in Figure 2, the resultant generalized circuit constants of the compensated transmission line are generally given by equation (1) as

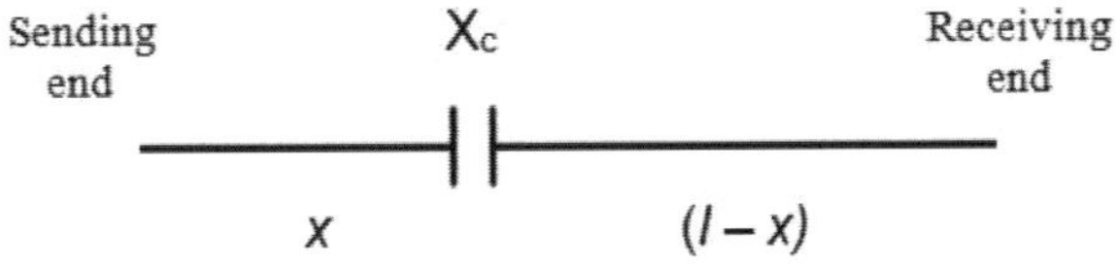

Figure 2. Compensated Transmission Line

$$\begin{vmatrix} A_c & B_c \\ C_c & D_c \end{vmatrix} = \begin{vmatrix} A_1 & B_1 \\ C_1 & D_1 \end{vmatrix} \begin{vmatrix} 1 & -iX_c \\ 0 & 1 \end{vmatrix} \begin{vmatrix} A_2 & B_2 \\ C_2 & D_2 \end{vmatrix} \quad (1)$$

where X_c = capacitive reactance of series capacitor

$A_1 = \cosh \gamma x$; $A_2 = \cosh \gamma(l-x)$;
$B_1 = Z_c \sinh \gamma x$; $B_2 = Z_c \sinh \gamma(l-x)$;
$C_1 = Y_c \sinh \gamma x$; $C_2 = Y_c \sinh \gamma(l-x)$;
$D_1 = \cosh \gamma x$; and $D_2 = \cosh \gamma(l-x)$;

With E_s and E_r as sending end and receiving end voltages, the received power, P_r can be written in the same form as that of an uncompensated as

$$Pr = \left(\frac{E_s E_r}{B_c}\right) \cos(\beta_o - \delta_o) - \left(\frac{A_c E_r^2}{B_c}\right) \cos(\beta_o - \alpha_o) \quad (2)$$

where E_s = magnitude of sending end voltage,
E_r = magnitude of receiving end voltage,
B_c = magnitude of the constant B_c
A_c = magnitude of the constant A_c
β_o = Argument of B_c
α_o = Argument of A_c
δ_o = Phase angle between E_s and E_r,

and the maximum received power of the compensated line, P_{max} is also given as

$$P_{max} = \left(\frac{E_s E_r}{B_c}\right) - \left(\frac{A_c E_r^2}{B_c}\right) \cos(\beta_o - \alpha_o) \quad (3)$$

3.3. Compensation Effect on Power Transfer

Among the four constants of the compensated line, only constants A_c and B_c are involved in the received power equation as seen in equation (2), hence any change in these will affect the received power [14,15].

From equation (1), the constant A_c can be written in a simplified form as

$$A_c = \cosh \gamma l - iX_c \left(\frac{1}{Z_c}\right) \cosh \gamma x \sinh \gamma(l - x) \quad (4)$$

One can see from equation (4) that the length of the transmission line (l), the location of the series capacitor from the sending end (x), the capacitive reactance of the series capacitor (X_c), the characteristic impedance of the transmission line (Z_c), and the propagation constant of the transmission line (γ) are the quantities that affect the value of the complex constant, A_c.

Likewise, the complex constant B_c (which is an impedance function) can be written in a simplified form after proper manipulation as

$$B_c = Z_c \sinh \gamma l - iX_c \cosh \gamma x \cosh \gamma(l - x) \quad (5)$$

Here also, once can see from equation (5) that the length of the transmission line, the location of the compensating device (i.e., the location of the series capacitor in this case), the capacitive reactance of the series capacitor, the characteristic impedance of the transmission line, and the propagation constant of the transmission line affect the value of the complex constant, B_c. Since for a transmission line of specified length the characteristic impedance and the propagation constant are known quantities, the unknown quantities in equation (5) are the capacitive reactance of the series capacitor and the capacitor location from the sending end. Or in other words, particularly with respect to series compensation, the impedance of the compensated transmission line is affected by the location and the value of the compensating device; therefore, the changes in the line impedance because of series compensation, affect the power transferring capability of the compensated line.

4. Programmable Algorithm

4.1. Computational Flow

The functional steps of the proposed algorithm are listed out along with a flow chart as in Figure 3:

Step 1: Determine the degree of percentage compensation

Step 2: Compute the value of X_c as degree of compensation (%) = X_c/X_L.

Step 3: Establish an arbitrary location for the series capacitor, say x=0 (from sending end).

Step 4: Compute the values of A_c and B_c by using equations (4) and (5) respectively.

Step 5: With the computed values of A_c and B_c, determine the power transfer for the specified location, x by using equation (2).

Step 6: Likewise, assume different values of x until x = l, and compute the power transfer.

Step 7: Compare the power transfers; the one with the highest magnitude is the optimum value of received power. The corresponding value of x is the optimum location for the placement of the series capacitor.

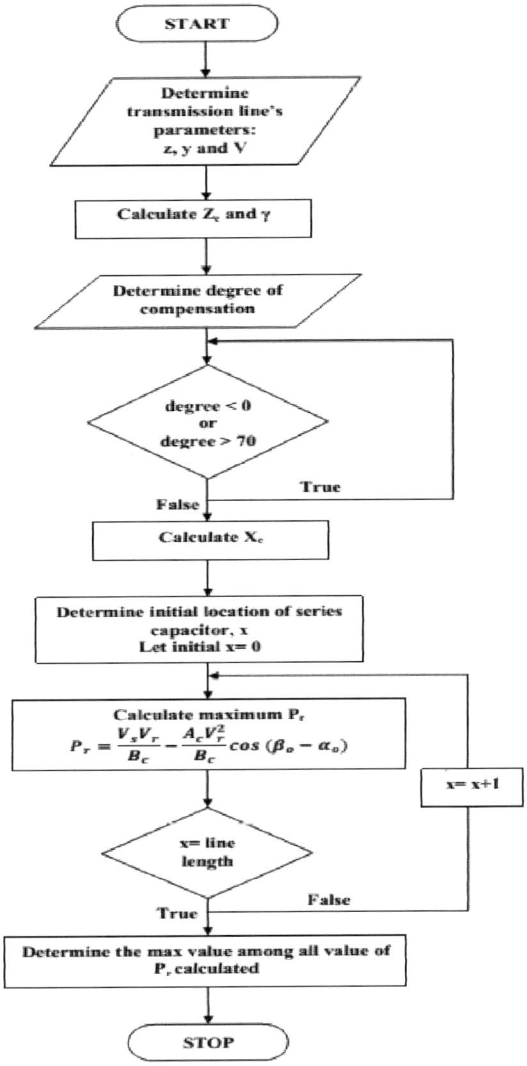

$$P_r = \frac{V_s V_r}{B_c} - \frac{A_c V_r^2}{B_c} \cos(\beta_o - \alpha_o)$$

Figure 3. Flow Chart

4.2. C++ Programming

```
#include <cstdlib>
#include <iostream>
#include <stdio.h>
#include <math.h>
#define SIZE 99999
#define pi 3.141592654

using namespace std;

int main(int argc, char *argv[])
{
    int i,k;

    double length, Vs, Vr;
    double resistance, inductive, susceptance, cImpR, cImpI;
    double propagationR, propagationI, icImpR, icImpI;
    double a, b, c, d, m, n;
    double AcR, AcI, Ac, alpha, BcR, BcI, Bc, beta, min=1, max=0;
    double degree, Pmax[SIZE];
    double x,Xc;

    cout<<endl<<"PERCENTAGE COMPENSATION "<<endl<<endl<<endl;
    cout<<"Please enter the values for the parameters below: "<<endl<<endl;

    cout<<"Line length (km): ";
    cin>>length;
    cout<<"Resistance per unit length (Ohm/km): ";
    cin>>resistance;
    cout<<"Reactance per unit length (Ohm/km): ";
    cin>>inductive;
    cout<<"Susceptance per unit length (Mho/km): ";
    cin>>susceptance;
    cout<<"Sending end voltage (kV): ";
    cin>>Vs;
    cout<<"Receiving end voltage (kV): ";
    cin>>Vr;

    cImpR=sqrt(sqrt(pow(resistance,2)+pow(inductive,2))/(susceptance)) *cos
((atan(inductive/resistance)-(pi/2))/2);

    cImpI=sqrt(sqrt(pow(resistance,2)+pow(inductive,2))/(susceptance)) *sin
((atan(inductive/resistance)-(pi/2))/2);


    propagationR=sqrt(sqrt(pow(resistance,2)+pow(inductive,2))*(susceptance)) *cos
((atan(inductive/resistance)+(pi/2))/2);

    propagationI=sqrt(sqrt(pow(resistance,2)+pow(inductive,2))*(susceptance)) *sin
((atan(inductive/resistance)+(pi/2))/2);


    icImpR=1/(sqrt(pow(cImpR,2)+pow(cImpI,2)))*cos(-atan(cImpI/cImpR));
    icImpI=1/(sqrt(pow(cImpR,2)+pow(cImpI,2)))*sin(-atan(cImpI/cImpR));

    do
    {
    cout<<"Degree of Compensation (%)= ";

    cin>>degree;
              if (degree<0||degree>70){
                                    cout<<"Invalid Entry."<<endl<<"Range: 0
< Degree of Compensation < 70"<<endl;
                                    }
    }while(degree<0||degree>70);

    Xc=degree/100*(inductive*length);

    a=propagationR*length;
    b=propagationI*length;

    for (x=1;x<=length;x++)
    {
      c=propagationR*x;
      d=propagationI*x;
      m=a-c;
      n=b-d;


      AcR=(1.0/2.0)*(exp(a)+exp(-a))*cos(b)+(Xc/4.0)*icImpR*((exp(c+m))*sin(d+n)-
(exp(c-m))*sin(d-n)+(exp(-c+m))*sin(-d+n)-(exp(-c-m))*sin(-d-
n))+(Xc/4.0)*icImpI*((exp(c+m))*cos(d+n)-(exp(c-m))*cos(d-n)+(exp(-c+m))*cos(-d+n)-
(exp(-c-m))*cos(-d-n));

      AcI=(1.0/2.0)*(exp(a)-exp(-a))*sin(b)-(Xc/4.0)*icImpR*((exp(c+m))*cos(d+n)-
(exp(c-m))*cos(d-n)+(exp(-c+m))*cos(-d+n)-(exp(-c-m))*cos(-d-
n))+(Xc/4.0)*icImpI*((exp(c+m))*sin(d+n)-(exp(c-m))*sin(d-n)+(exp(-c+m))*sin(-d+n)-
(exp(-c-m))*sin(-d-n));

      Ac=sqrt(pow(AcR,2.0)+pow(AcI,2.0));

      alpha=atan(AcI/AcR);


      BcR=(1.0/2.0)*cImpR*(exp(a)-exp(-a))*cos(b)-(1.0/2.0)*cImpI*(exp(a)+exp(-
a))*sin(b)+(Xc/4.0)*((exp(c+m))*sin(d+n)+(exp(c-m))*sin(d-n)+(exp(-c+m))*sin(-
d+n)+(exp(-c-m))*sin(-d-n));

      BcI=(1.0/2.0)*cImpR*(exp(a)+exp(-a))*sin(b)+(1.0/2.0)*cImpI*(exp(a)-exp(-
a))*cos(b)-(Xc/4.0)*((exp(c+m))*cos(d+n)+(exp(c-m))*cos(d-n)+(exp(-c+m))*cos(-
d+n)+(exp(-c-m))*cos(-d-n));

      Bc=sqrt((BcR*BcR)+(BcI*BcI));

      beta=atan(BcI/BcR);


      for(i=0;i<=(length+1);i++){
                              Pmax[i]=(Vs*Vr)/Bc-((Ac*pow(Vr,2.0))/Bc)*cos(beta-
alpha);
                              if (Pmax[i]>max){
                                            max=Pmax[i];
                                            k=x;
                                           }
```

IOP Conf. Series: Materials Science and Engineering **78** (2015) 012013 doi:10.1088/1757-899X/78/1/012013

```
            }

    }
    cout<<endl<<"Series Capacitor Reactance Value= "<<Xc<<" Ohm"<<endl;
    cout<<endl<<"Maximum Power Transfer= "<<max<<" MW"<<endl;
    cout<<endl<<"Optimum Series Capacitor Location= "<<k<<" km away from sending
end"<<endl<<endl;

    system("PAUSE");
    return EXIT_SUCCESS;
}
```

Figure 4. C++ Program

5. Application To A Prototype System

The suitability of the proposed programmable algorithm has been ascertained by applications to various transmission study systems and the proposed algorithm is found to be technically sound. The application to a prototype wind energy transmission system is presented here.

5.1. Prototype System

A remote wind farm is connected to a load center by means of a 320 km ling 345 kV high voltage transmission line. TCSC compensation has been incorporated with a maximum permissible compensation of 50%. Figure 5 depicts the interconnections.

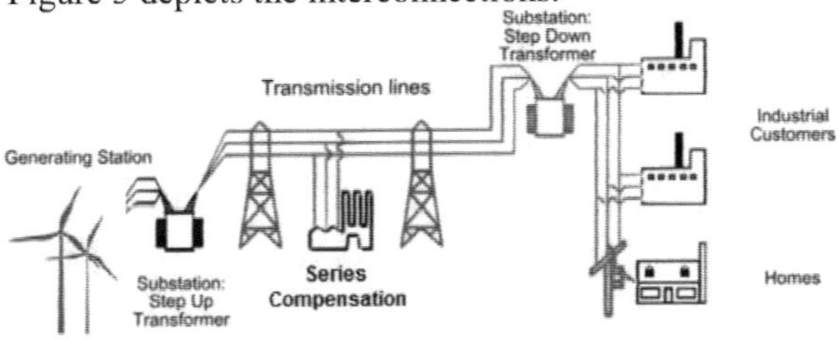

Figure 5. Wind Energy Transmission System

The parameters of the transmission line are:

Resistance	0.032 ohm/km
Inductive reactance	0.320 ohm/km
Susceptance	0.0000042 mho/km
Line length	320 km
Sending end voltage	345 kV
Receiving end voltage	345 kV (desired level)

5.2. Program Output

The line parameters are entered in the program through keyboard and mouse control. The degree of compensation is varied from 10% to 50% for equal sending end and receiving end voltages of 345 kV and the results are shown in Table 1.

For the given transmission line configuration, for a degree compensation of 25%, the maximum received power is found to be 1266.19 MW with the series capacitor reactance value of 28.0 ohms placed at a distance of 176 km from the sending end. The optimum location is not at the center of the transmission line but nearer to the middle towards the receiving end. Likewise, for every degree of compensation, the maximum power transfer is different with different capacitor reactance value. However, there is not much difference in the optimum location of the capacitor placement.

Table 1. Equal sending and receiving end voltages

Sending End Voltage: 345 kV Receiving End Voltage: 345 kV			
Degree of Compensation (%)	Series Capacitor Reactance (Ohms)	Maximum Power Transfer (MW)	Optimum Series Capacitor Location from Sending End (km)
10	11.2	1084.22	176
15	16.8	1138.93	176
20	22.4	1199.30	176
25	28.0	1266.19	176
30	33.6	1340.69	177
35	39.2	1424.08	177
40	44.8	1517.92	177
45	50.4	1624.09	178
50	56.0	1744.85	179

The screen shot of 25% compensation with equal sending and receiving end voltages is presented as in Figure 6.

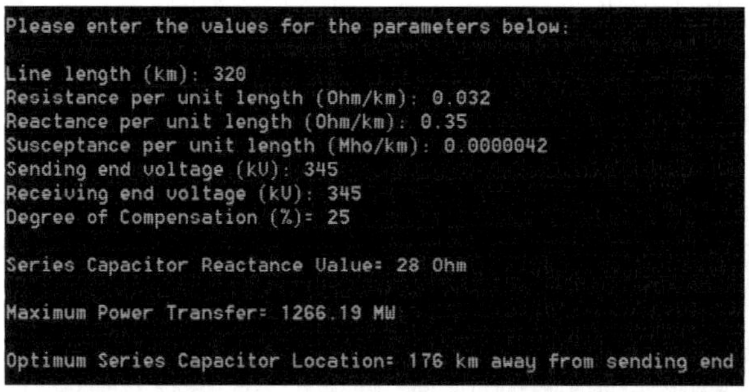

Figure 6. Screen shot with equal voltages

In order to know the effect of receiving end voltage on maximum power transfer and the capacitor location, the degree of compensation is varied from 10% to 50% for a sending end voltage of 345 kV and a receiving end voltage of 330 kV and the results are shown in Table 2.

Table 2. Unequal sending and receiving end voltages

Sending End Voltage: 345 kV Receiving End Voltage: 330 kV			
Degree of Compensation (%)	Series Capacitor Reactance (Ohms)	Maximum Power Transfer (MW)	Optimum Series Capacitor Location from Sending End (km)
10	11.2	1042.02	175
15	16.8	1094.95	175
20	22.4	1153.38	175
25	28.0	1218.22	176
30	33.6	1290.50	176
35	39.2	1371.52	176
40	44.8	1462.84	177
45	50.4	1566.38	177
50	56.0	1684.46	178

The screen shot of 25% compensation with unequal sending and receiving end voltages is presented in Figure 7.

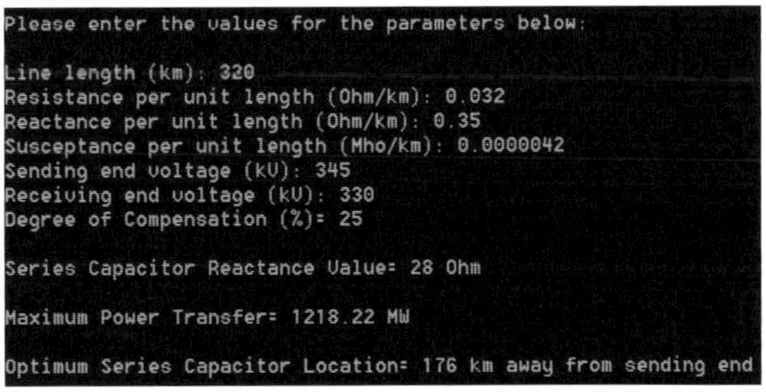

Figure 7. Screen shot with unequal voltages

While comparing the results of Tables 1 and 2, for the same amount of 25% compensation, due to the reduction in receiving end voltage to 330 kV, the maximum power transfer becomes 1218.22 MW at the same location. That means, a reduction of 47.83 MW is noticed due to the dip in the receiving end voltage. Similar situations happen for the different degrees of compensation.

Considering the magnitude of wind power generated, the degree of series compensation shall be adjusted to accommodate the excess generated power for transferring without upgrading the existing transmission facilities or constructing additional transmission lines. The developed program is highly flexible and it suitable for real-time applications.

6. Conclusions

This paper has deliberated the application of series compensation of long transmission lines intended for transferring wind energy from remote locations to the load centers. A programmable approach to determine the capacitive reactance of the series capacitor and the optimum location for its placement to achieve maximum power transfer has been presented. The respective program is suitable for real-time applications. From the prototype system test result, it is evident that series compensation techniques increase the power transferring capacity of existing transmission systems, and avoid the formation of new transmission lines to cope up with additional power transfer. Since formation of new transmission lines is very expensive compared to installing series capacitors for performing the same task, the series capacitor compensation is said to be cost effective. However, series compensation results in transmission line relaying and protection issues, an adverse effect on the power system stability, etc. and all these issues should be considered whole implementing series compensation.

References

[1] Palanichamy C and Naveen P 2014 A sustainable renewable energy mix option for secluded society *Journal of Renewable and Sustainable Energy* **6**

[2] Palanichamy C 2010 Economic performance of wind turbines *International Journal of Distributed Energy Resources* **6** 41-60

[3] Palanichamy C 2004 Renewable energy investment opportunities in Mauritius – an investor's perspective *Renewable Energy – An international journal* **29** 703-16

[4] Shrestha G B and Wang F 2005 Effects of series compensation on spot price power markets *Electrical Power and Energy Systems* **27** 428-36

[5] Duong T L, Yao J G and Truong V A 2014 Application of min cut algorithm for optimal location of FACTS devices considering system loadability and cost of installation *Electrical Power and Energy Systems* **63** 979-87

[6] Alabduljabbar A A and Milanovic J V 2010 Assesment of techno-ecomonic contribuion of

FATCS devices to power system operation *Electric Power Systems Research* **80** 1247-55
[7] Palanichamy C and Sundar B N 2006 Transmission system upgrading – analyzing the alternatives for final decision making *International Journal of Emerging Electric Power Systems* **6**
[8] Palanichamy C and Krishna K S 1991 Simple algorithm for economic power dispatch *Electrical Power Systems Research* **121** 147-53
[9] Elanders Sverige AB 2010 Series compensation *ABB developments and achievements A02-0135E*
[10] Basu M 2011 Multi-objective optimal power flow with FACTS devices *Energy Conversion and Management* **52** 903-10
[11] Leonidaki E A, Manos G A and Hatziargyriou N D 2005 An effective method to locate series compensation for voltage stability enhancement *Electric Power Systems Research* **74** 73-81
[12] Safari A, Shayanfar H A and Kazemi A 2013 Optimal location of PWM based series compensator in a power system *Electrical Power and Energy Systems* **51** 127-33
[13] Gerbex S, Cherkaoui R and Germond A J 2001 Optimal location of multi-type FACTS devices in a power system by means of genetic algorithms *IEEE Trans. Power Systems* **16** 537-44
[14] Saadat H 1999 *Power System Analysis* (New York: McGraw-Hill)
[15] Palanichamy C and Sundar B N 2005 A visual package for educating preparatory transmission line series compensation *IEEE Transactions on Education* **48** 16-22

IOP Conf. Series: Materials Science and Engineering **78** (2015) 012014 doi:10.1088/1757-899X/78/1/012014

Enhanced Electric Power Transmission by Hybrid Compensation Technique

C Palanichamy and GQ Kiu

Dept. of Electrical and Computer Engineering, School of Engineering and Science, Curtin University, Miri, Sarawak, Malaysia.

E-mail: drcpc@curtin.edu.my, 7e1b8907@student.curtin.edu.my

Abstract. In today's competitive environment, new power system engineers are likely to contribute immediately to the task, without years of seasoning via on-the-job training, mentoring, and rotation assignments. At the same time it is becoming obligatory to train power system engineering graduates for an increasingly quality-minded corporate environment. In order to achieve this, there is a need to make available better-quality tools for educating and training power system engineering students and in-service system engineers too. As a result of the swift advances in computer hardware and software, many windows-based computer software packages were developed for the purpose of educating and training. In line with those packages, a simulation package called Hybrid Series-Shunt Compensators (HSSC) has been developed and presented in this paper for educational purposes.

1. Introduction

The increasing load demand has necessitated the development of new generating facilities, transmission and distribution networks. However the construction of new transmission lines [1, 2] has been limited due to economic and environmental constraints. Therefore, in a competitive market milieu due to the challenges associated with the construction of new transmission lines, it is desirable to apply the existing transmission systems to the fullest possible extent without compromising quality and reliability [3-6]. Moreover, growth of different commercial schemes for the electric power industry, particularly deregulation of market, has made this requirement even stronger [7-9]. In developing countries, the optimized use of transmission system investments is also important to support industry, create employment, and employ efficiently the scarce economic resources [10-12].

The use of series and shunt compensation schemes [5, 6, 13-16] to increase the power transfer capability of long transmission line is well-known. The advent and development of power electronics resulted in the introduction of Flexible AC Transmission System (FACTS) controllers such as Thyristor Controlled Series Compensators (TCSC), Unified Power Flow Controllers (UPFC) and Static Var Compensators (SVC) for transmission line compensation [17-19]. The TCSC is a series connected device used to regulate the reactance of a transmission line thereby controls the real and reactive power flow and redistributes power flow even under highly loaded conditions. Hence FACTS controllers can be used to increase system load ability and the available transfer capacity as situation demands. Steady state and transient stability is also improved with the help of FACTS controller. SVC is a shunt connected device which regulates the transmission system voltage by reactive power injections. UPFC is a combination of static synchronous compensators (STATCOM) and static synchronous series compensators (SSSC) coupled through a common DC voltage link used for controlling active and reactive power flow through the transmission lines.

In recent years, computer-aided instruction has been extensively used in electrical and power engineering education [20]. Computer simulation models have been employed to support and improve power engineering courses [21-23]. In this paper, an educational software package called Hybrid Series-Shunt Compensators (HSSC) for enhanced electric power transmission has been developed. The design and simulations of HSSC are proposed by using the Matlab R2013b and Matlab GUI. Windows based Graphical User Interface (GUI) concept is used, which gives the advantages of interactive visual communication between users and computer processes and quick interpretation of test results.

2. Compensated Transmission Line

Transmitted real and reactive power of the transmission line, which is shown in Fig. 1, can be derived in terms of the ABCD parameters of the line using the following notations:

l = *length of transmission line*
Z_0 = *characteristic impedance of transmission line*
V_S = *sending end voltage*
I_S = *sending end current*
V_R = *receiving end voltage*
I_R = *receiving end current*
Z_{se} = *series capacitor compensation*
Y_{sh} = *shunt reactor/capacitor compensation*
P_S = *sending end real power/operating power*
P_{S-max} = *maximum sending end real power*
Q_S= *sending end reactive power*
S_S = *sending end apparent power*
α = *argument of TL parameter A*
β = *argument of TL parameter B*
δ = *power angle*

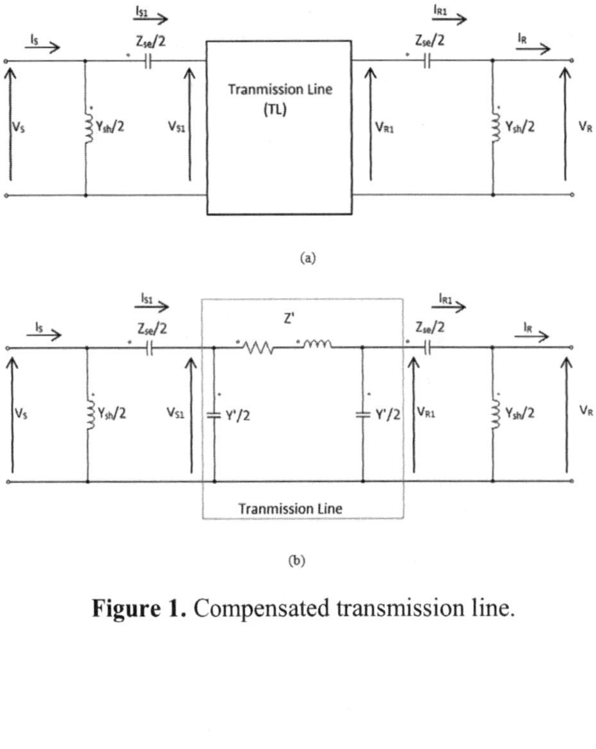

Figure 1. Compensated transmission line.

For the compensated lines, the degree of series (K_{se}) and shunt (K_{sh}) compensation, respectively, is defined by

$$K_{se} = \frac{Total\ capacitive\ reactance\ of\ series\ compensators}{Total\ inductive\ reactance\ of\ the\ line}$$

And

$$K_{sh} = \frac{Total\ inductive\ susceptance\ of\ shunt\ compensators}{Total\ charging\ susceptance\ of\ the\ line}$$

Following the notations defined above, the ABCD constants of the two port network representation of the compensated transmission line as depicted in Fig.2 are derived as

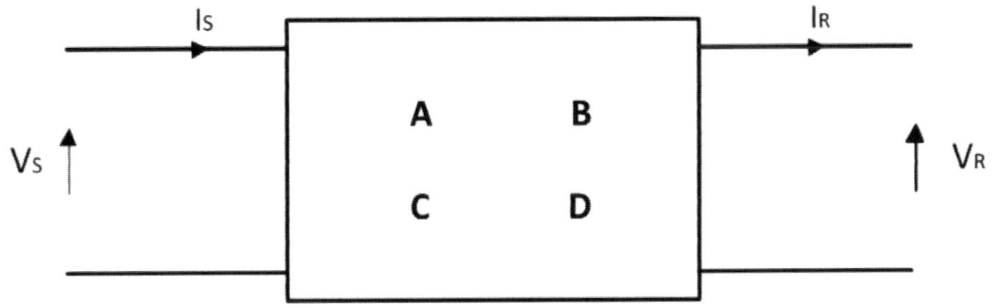

Figure 2. Two port network.

$$V_{S1} = (\cosh \gamma l)V_{R1} + (Z_0 \sinh \gamma l)I_{R1} \tag{1}$$

$$I_{S1} = \left(\frac{1}{Z_0}\sinh \gamma l\right)V_{R1} + (\cosh \gamma l)I_{R1} \tag{2}$$

$$\begin{aligned} V_S &= (\cosh \gamma l)V_{R1} + (Z_0 \sinh \gamma l)I_{R1} + \left(\frac{Z_{se}}{2}\right)I_{S1} \\ &= (\cosh \gamma l)V_{R1} + (Z_0 \sinh \gamma l)I_{R1} + \left(\frac{Z_{se}}{2}\right)\left(\left(\frac{1}{Z_0}\sinh \gamma l\right)V_{R1} + (\cosh \gamma l)I_{R1}\right) \\ &= (\cosh \gamma l)V_{R1} + (Z_0 \sinh \gamma l)I_{R1} + \left(\frac{Z_{se}}{2Z_0}\sinh \gamma l\right)V_{R1} + \left(\frac{Z_{se}}{2}\cosh \gamma l\right)I_{R1} \\ &= \left(\frac{Z_{se}}{2Z_0}\sinh \gamma l + \cosh \gamma l\right)V_{R1} + \left(\frac{Z_{se}}{2}\cosh \gamma l + Z_0 \sinh \gamma l\right)I_{R1} \end{aligned} \tag{3}$$

$$\begin{aligned} I_S &= \frac{Y_{sh}}{2}V_S + I_{S1} \\ &= \frac{Y_{sh}}{2}V_S + \left(\frac{1}{Z_0}\sinh \gamma l\right)V_{R1} + (\cosh \gamma l)I_{R1} \\ &= \frac{Y_{sh}}{2}\left(\left(\frac{Z_{se}}{2Z_0}\sinh \gamma l + \cosh \gamma l\right)V_{R1} + \left(\frac{Z_{se}}{2}\cosh \gamma l + Z_0 \sinh \gamma l\right)I_{R1}\right) + \left(\frac{1}{Z_0}\sinh \gamma l\right)V_{R1} + (\cosh \gamma l)I_{R1} \\ &= \left(\frac{Y_{sh}Z_{se}}{4Z_0}\sinh \gamma l + \frac{Y_{sh}}{2}\cosh \gamma l\right)V_{R1} + \left(\frac{Y_{sh}Z_{se}}{4}\cosh \gamma l + \frac{Y_{sh}Z_0}{2}\sinh \gamma l\right)I_{R1} + \left(\frac{1}{Z_0}\sinh \gamma l\right)V_{R1} + (\cosh \gamma l)I_{R1} \\ &= \left(\frac{Y_{sh}Z_{se}}{4Z_0}\sinh \gamma l + \frac{Y_{sh}}{2}\cosh \gamma l + \frac{1}{Z_0}\sinh \gamma l\right)V_{R1} + \left(\frac{Y_{sh}Z_{se}}{4}\cosh \gamma l + \frac{Y_{sh}Z_0}{2}\sinh \gamma l + \cosh \gamma l\right)I_{R1} \end{aligned} \tag{4}$$

$$\begin{aligned} I_R &= I_{R1} - \frac{Y_{sh}}{2}V_R \\ I_{R1} &= I_R + \frac{Y_{sh}}{2}V_R \end{aligned} \tag{5}$$

$$\begin{aligned} V_R &= V_{R1} - \frac{Z_{se}}{2} I_{R1} \\ &= V_{R1} - \frac{Z_{se}}{2}\left(I_R + \frac{Y_{sh}}{2} V_R\right) \\ V_{R1} &= V_R + \frac{Z_{se}Y_{sh}}{4} V_R + \frac{Z_{se}}{2} I_R \end{aligned} \tag{6}$$

$$\begin{aligned} V_S &= \left(\frac{Z_{se}}{2Z_0}\sinh\gamma l + \cosh\gamma l\right)\left(V_R + \frac{Z_{se}Y_{sh}}{4}V_R + \frac{Z_{se}}{2}I_R\right) + \left(\frac{Z_{se}}{2}\cosh\gamma l + Z_0\sinh\gamma l\right)\left(I_R + \frac{Y_{sh}}{2}V_R\right) \\ &= \left(\frac{Z_{se}}{2Z_0}\sinh\gamma l + \frac{Z_{se}{}^2Y_{sh}}{8Z_0}\sinh\gamma l + \frac{Z_0Y_{sh}}{2}\sinh\gamma l + \cosh\gamma l + \frac{Z_{se}Y_{sh}}{2}\cosh\gamma l\right)V_R \\ &\quad + \left(\frac{Z_{se}{}^2}{4Z_0}\sinh\gamma l + Z_0\sinh\gamma l + Z_{se}\cosh\gamma l\right)I_R \\ &= \left(\left(1 + \frac{Z_{se}Y_{sh}}{2}\right)\cosh\gamma l + \left(\frac{Z_{se}}{2Z_0} + \frac{Z_{se}{}^2Y_{sh}}{8Z_0} + \frac{Z_0Y_{sh}}{2}\right)\sinh\gamma l\right)V_R + \left(Z_{se}\cosh\gamma l + \left(\frac{Z_{se}{}^2}{4Z_0} + Z_0\right)\sinh\gamma l\right)I_R \end{aligned} \tag{7}$$

$$\begin{aligned} I_S &= \left(\frac{Y_{sh}Z_{se}}{4Z_0}\sinh\gamma l + \frac{Y_{sh}}{2}\cosh\gamma l + \frac{1}{Z_0}\sinh\gamma l\right)\left(V_R + \frac{Z_{se}Y_{sh}}{4}V_R + \frac{Z_{se}}{2}I_R\right) \\ &\quad + \left(\frac{Y_{sh}Z_{se}}{4}\cosh\gamma l + \frac{Y_{sh}Z_0}{2}\sinh\gamma l + \cosh\gamma l\right)\left(I_R + \frac{Y_{sh}}{2}V_R\right) \\ &= \frac{Y_{sh}Z_{se}}{4Z_0}\sinh\gamma l\,V_R + \frac{Y_{sh}}{2}\cosh\gamma l\,V_R + \frac{1}{Z_0}\sinh\gamma l\,V_R + \frac{Y_{sh}{}^2Z_{se}{}^2}{16Z_0}\sinh\gamma l\,V_R + \frac{Z_{se}Y_{sh}{}^2}{8}\cosh\gamma l\,V_R + \frac{Y_{sh}Z_{se}}{4Z_0}\sinh\gamma l\,V_R \\ &\quad + \frac{Y_{sh}Z_{se}{}^2}{8Z_0}\sinh\gamma l\,I_R + \frac{Z_{se}Y_{sh}}{4}\cosh\gamma l\,I_R + \frac{Z_{se}}{2Z_0}\sinh\gamma l\,I_R + \frac{Z_{se}Y_{sh}}{4}\cosh\gamma l\,I_R + \frac{Y_{sh}Z_0}{2}\sinh\gamma l\,I_R + \cosh\gamma l\,I_R \\ &\quad + \frac{Z_{se}Y_{sh}{}^2}{8}\cosh\gamma l\,V_R + \frac{Z_0Y_{sh}{}^2}{4}\sinh\gamma l\,V_R + \frac{Y_{sh}}{2}\cosh\gamma l\,V_R \\ &= \left(\left(Y_{sh} + \frac{Z_{se}Y_{sh}{}^2}{4}\right)\cosh\gamma l + \left(\frac{Y_{sh}Z_{se}}{2Z_0} + \frac{1}{Z_0} + \frac{Y_{sh}{}^2Z_{se}{}^2}{16Z_0} + \frac{Z_0Y_{sh}{}^2}{4}\right)\sinh\gamma l\right)V_R \\ &\quad + \left(\left(\frac{Z_{se}Y_{sh}}{2} + 1\right)\cosh\gamma l + \left(\frac{Z_{se}}{2Z_0} + \frac{Y_{sh}Z_0}{2} + \frac{Y_{sh}Z_{se}{}^2}{8Z_0}\right)\sinh\gamma l\right)I_R \end{aligned} \tag{8}$$

Comparing with the standard equations of the two port network, we get the A B C D constants as

$$A = \left(1 + \frac{Z_{se}Y_{sh}}{2}\right)\cosh\gamma l + \left(\frac{Z_{se}}{2Z_0} + \frac{Z_{se}{}^2Y_{sh}}{8Z_0} + \frac{Z_0Y_{sh}}{2}\right)\sinh\gamma l \tag{9}$$

$$B = Z_{se}\cosh\gamma l + \left(\frac{Z_{se}{}^2}{4Z_0} + Z_0\right)\sinh\gamma l \tag{10}$$

$$C = \left(Y_{sh} + \frac{Z_{se}Y_{sh}{}^2}{4}\right)\cosh\gamma l + \left(\frac{Y_{sh}Z_{se}}{2Z_0} + \frac{1}{Z_0} + \frac{Y_{sh}{}^2Z_{se}{}^2}{16Z_0} + \frac{Z_0Y_{sh}{}^2}{4}\right)\sinh\gamma l \tag{11}$$

$$D = A \tag{12}$$

Also the equations of the operating power P_S and the maximum power $P_{S\text{-max}}$ of the compensated transmission line are obtained as

$$P_S = \frac{|A||V_S|^2}{|B|}\cos(\beta - \alpha) - \frac{|V_S|^2}{|B|}\cos(\delta + \beta) \tag{13}$$

When $\delta + \beta = \pi$, $P_S = P_{S-max}$.

$$P_{S-max} = \frac{|A||V_R|^2}{|B|}\cos(\beta - \alpha) + \frac{|V_S||V_R|}{|B|} \tag{14}$$

3. Hybrid Series-Shunt Compensators Package

3.1. HSSC Working Main Screen & User Manual

The Working Main Screen of the HSSC package and the HSSC User Manual are shown in Fig. 3 & 4 respectively.

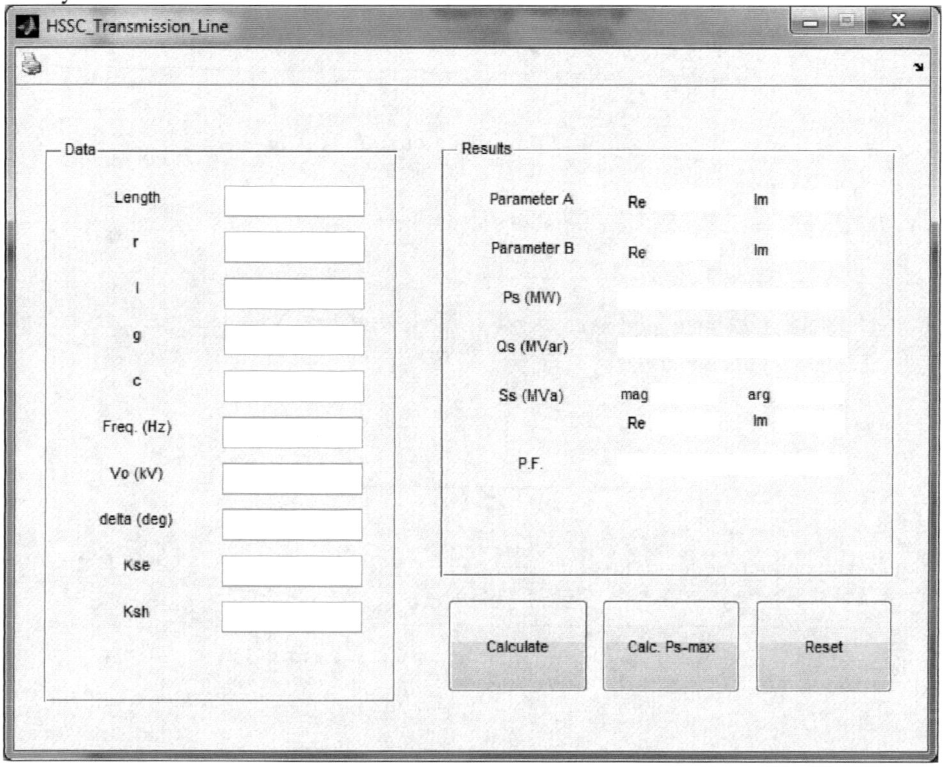

Figure 3. HSSC working main screen.

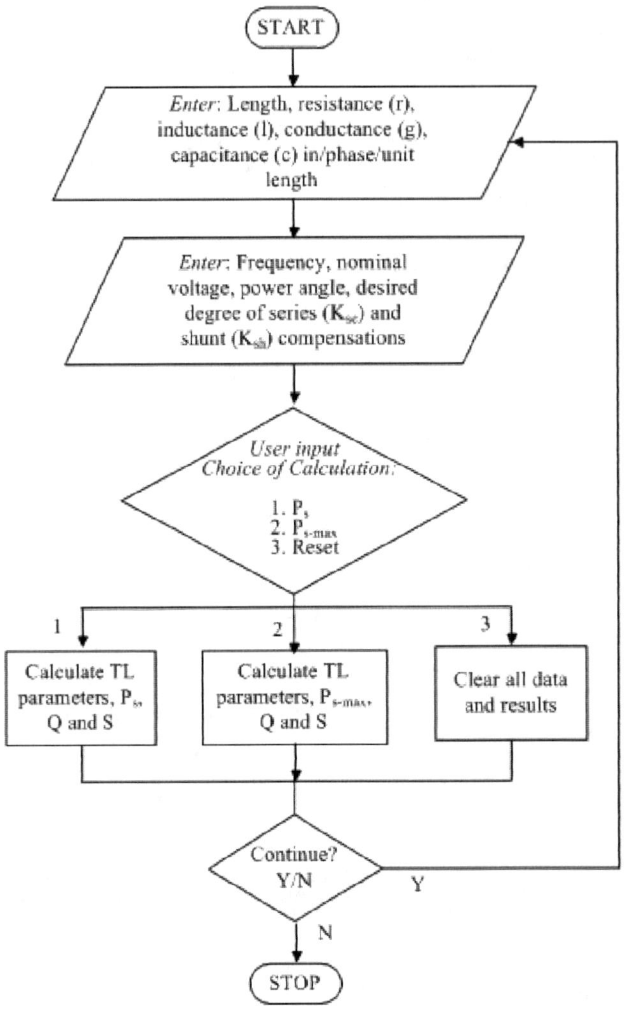

Figure 4. HSSC user manual.

3.2. Program Coding

The program coding of HSSC by using the Matlab R2013b and Matlab GUI is given in two segments as shown below.

```
function calc_Callback(hObject, eventdata, handles)
% hObject    handle to calc (see GCBO)
% eventdata  reserved - to be defined in a future version of MATLAB
% handles    structure with handles and user data (see GUIDATA)
length=str2double(get(handles.length,'String'));
r=str2double(get(handles.resistance,'String'));
l=str2double(get(handles.inductance,'String'));
g=str2double(get(handles.conductance,'String'));
c=str2double(get(handles.capacitance,'String'));
f=str2double(get(handles.freq,'String'));
Vo=str2double(get(handles.volt_nom,'String'));
d=str2double(get(handles.delta,'String'));
Kse=str2double(get(handles.Kse,'String'));
Ksh=str2double(get(handles.Ksh,'String'));
R=length*r;
L=length*l;
G=length*g;
C=length*c;
d=d*pi/180;
i=sqrt(-1);
XL=2*pi*f*L;
XC=1/(2*pi*f*C);
BC=1/XC;
Z=R+XL*i;
Y=G+BC*i;
Z0=sqrt(Z/Y);
gamma1=sqrt(Z*Y);
Xc=Kse*XL;
Zc=-Xc*i;
Bsh=Ksh*BC;
Ysh=-Bsh*i;
paraA=((Zc/(2*Z0))+((Zc^2)*Ysh/(8*Z0))+(Z0*Ysh/2))*sinh(gamma1+1+(Zc*Ysh/2)
)*cosh(gamma1);
Ar=real(paraA);
Ai=imag(paraA);
A1=abs(paraA);
alpha=angle(paraA);
paraB=(((Zc^2)/(4*Z0))+Z0)*sinh(gamma1)+Zc*cosh(gamma1);
Br=real(paraB);
Bi=imag(paraB);
B1=abs(paraB);
beta=angle(paraB);
Ps=(A1*(Vo^2)/B1)*cos(beta-alpha)-((Vo^2)/B1)*cos(d+beta);
Qs=(A1*(Vo^2)/B1)*sin(beta-alpha)-((Vo^2)/B1)*sin(d+beta);
Ss=(A1*(Vo^2)/B1)*exp(i*(beta-alpha))-((Vo^2)/B1)*exp(i*(d+beta));
Ssr=real(Ss);
Ssi=imag(Ss);
Ss1=abs(Ss);
theta=angle(Ss);
pf=cos(theta);
% 'if' condition below define that if error% of d is < 0.5% then 'Ps-max
% (MW)' will show, else 'Ps (MW)' will replace the textbox
if (abs((d+beta)/pi-1)<0.005)
    set(handles.text_ps,'String','Ps-max (MW)');
else
    set(handles.text_ps,'String','Ps (MW)');
end
set(handles.Ar,'String',Ar);
set(handles.Ai,'String',Ai);
set(handles.Br,'String',Br);
set(handles.Bi,'String',Bi);
set(handles.Ps,'String',Ps);
set(handles.Qs,'String',Qs);
set(handles.Ss1,'String',Ss1);
theta=theta*180/pi;
theta=sprintf('%.2f%c',theta,char(176));
set(handles.theta,'String',theta);
set(handles.Ssr,'String',Ssr);
set(handles.Ssi,'String',Ssi);
set(handles.pf,'String',pf);
```

IOP Conf. Series: Materials Science and Engineering **78** (2015) 012014 doi:10.1088/1757-899X/78/1/012014

The above section is intended for calculating the sending end real power/operating power, the reactive power, the apparent power and the power factor. The apparent power is represented in rectangular as well as polar forms for through understanding of the concept. The user input choice of calculation 1 displays the said results.

```
function calc_psmax_Callback(hObject, eventdata, handles)
% hObject    handle to calc_psmax (see GCBO)
% eventdata  reserved - to be defined in a future version of MATLAB
% handles    structure with handles and user data (see GUIDATA)
length=str2double(get(handles.length,'String'));
r=str2double(get(handles.resistance,'String'));
l=str2double(get(handles.inductance,'String'));
g=str2double(get(handles.conductance,'String'));
c=str2double(get(handles.capacitance,'String'));
f=str2double(get(handles.freq,'String'));
Vo=str2double(get(handles.volt_nom,'String'));
% d=str2double(get(handles.delta,'String'));
Kse=str2double(get(handles.Kse,'String'));
Ksh=str2double(get(handles.Ksh,'String'));
R=length*r;
L=length*l;
G=length*g;
C=length*c;
% d=d*pi/180;
i=sqrt(-1);
XL=2*pi*f*L;
XC=1/(2*pi*f*C);
BC=1/XC;
Z=R+XL*i;
Y=G+BC*i;
Z0=sqrt(Z/Y);
gammal=sqrt(Z*Y);
Xc=Kse*XL;
Zc=-Xc*i;
Bsh=Ksh*BC;
Ysh=-Bsh*i;
paraA=((Zc/(2*Z0))+((Zc^2)*Ysh/(8*Z0))+(Z0*Ysh/2))*sinh(gammal)+
(1+(Zc*Ysh/2))*cosh(gammal);
Ar=real(paraA);
Ai=imag(paraA);
A1=abs(paraA);
alpha=angle(paraA);
paraB=(((Zc^2)/(4*Z0))+Z0)*sinh(gammal)+Zc*cosh(gammal);
Br=real(paraB);
Bi=imag(paraB);
B1=abs(paraB);
beta=angle(paraB);
d=pi-beta;
Psmax=(A1*(Vo^2)/B1)*cos(beta-alpha)-((Vo^2)/B1)*cos(d+beta);
Qs=(A1*(Vo^2)/B1)*sin(beta-alpha)-((Vo^2)/B1)*sin(d+beta);
Ss=(A1*(Vo^2)/B1)*exp(i*(beta-alpha))-((Vo^2)/B1)*exp(i*(d+beta));
Ssr=real(Ss);
Ssi=imag(Ss);
Ss1=abs(Ss);
theta=angle(Ss);
pf=cos(theta);
d=d*180/pi;
set(handles.delta,'String',d);
set(handles.Ar,'String',Ar);
set(handles.Ai,'String',Ai);
set(handles.Br,'String',Br);
set(handles.Bi,'String',Bi);
set(handles.text_ps,'String','Ps-max (MW)');
set(handles.Ps,'String',Psmax);
set(handles.Qs,'String',Qs);
set(handles.Ss1,'String',Ss1);
theta=theta*180/pi;
theta=sprintf('%.2f%c',theta,char(176));
set(handles.theta,'String',theta);
set(handles.Ssr,'String',Ssr);
set(handles.Ssi,'String',Ssi);
set(handles.pf,'String',pf);
```

IOP Conf. Series: Materials Science and Engineering **78** (2015) 012014 doi:10.1088/1757-899X/78/1/012014

The above program section is intended for calculating the maximum sending end real power/operating power in addition to displaying the already calculated reactive power, the apparent power and the power factor as in the previous section. Here also the apparent power is represented in rectangular as well as polar forms for through understanding of the concept. The user input choice of calculation 2 displays the said results. The user input choice of calculation 3 resets all the data to facilitate new computations.

4. An Illustrative Example

The suitability of the developed package has been ascertained by applications to various transmission study systems and it is found to be technically sound. The application to a prototype transmission system is presented here.

4.1. System Data

The sample transmission line considered [24] in this study has the following data:

$l = 550$ *miles*
$V_0 = 500$ *kV (line to line), double circuit*
$f = 60$ *Hz*
$R = 0.02495$ *ohm/mile/phase/circuit*
$G = 2x10^{-12}$ *mho/mile/circuit*
$L = 1.3925$ *mH/mile/phase/circuit*
$C = 0.020885$ *μF/mile/phase to neutral/circuit*

4.2. Program Output

The line parameters are entered in the HSSC Working Main Screen Data column. The package is designed such that both series and shunt compensation is possible at a time. The degree of series as well as shunt compensation can be varied from 0% to 100% (i.e. 0 to 1 p.u.); however due to stability reasons, it is limited to a maximum of 70%. As an illustration, the degree of shunt compensation is kept fixed at 50% and the series compensation is varied from 0% to 70% for a power angle of 30 degrees and the results are shown in Table 1.

Table 1. Fixed Shunt & Variable Series Compensation.

Degree of Compensation (p.u.)		Ps (MW)	Qs (MVar)	Ps_{max} (MW)
k_{se}	k_{sh}			
0.0	0.5	546.10	-210.83	1118.26
0.1	0.5	579.53	-185.93	1191.91
0.2	0.5	621.79	-161.54	1285.46
0.3	0.5	675.98	-137.52	1406.28
0.4	0.5	746.89	-113.84	1566.06
0.5	0.5	842.41	-90.84	1784.54
0.6	0.5	976.45	-69.91	2097.63
0.7	0.5	1175.73	-55.64	2577.88

From the results of Table 1, for the given transmission line configuration, for every degree of compensation, the real power, reactive power and the maximum power transfer are increasing as the degree of series compensation increases. For series and shunt compensation of each 0.5 p.u. (50%),

Fig. 5 shows the screen shot of "*choosing user input choice of calculation 1*" and Fig. 6 shows the "*screen shot of choosing user input choice of calculation 2*".

As a second illustration, the degree of series compensation is kept fixed at 50% and the shunt compensation is varied from 0% to 70% for a power angle of 30 degrees and the results are shown in Table 2. From the results of Table 2, for the same transmission line configuration, for every degree of shunt compensation, the reactive power gets varied whereas the real power and the maximum power transfer remain constant irrespective of the variations in shunt compensation. That means, reactive shunt compensation affects the reactive power and the series compensation affects the real power transfer.

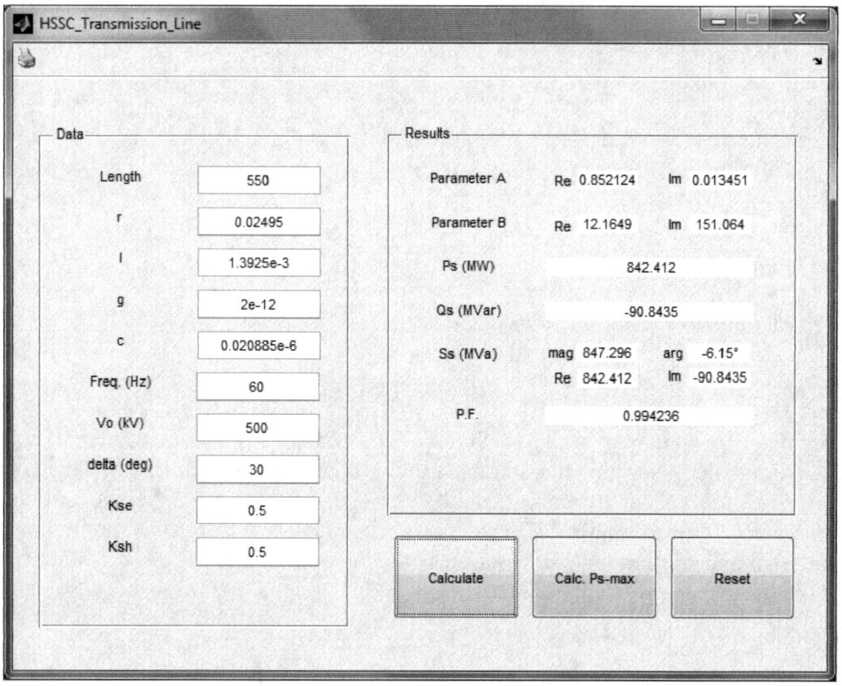

Figure 5. User input choice of calculation 1.

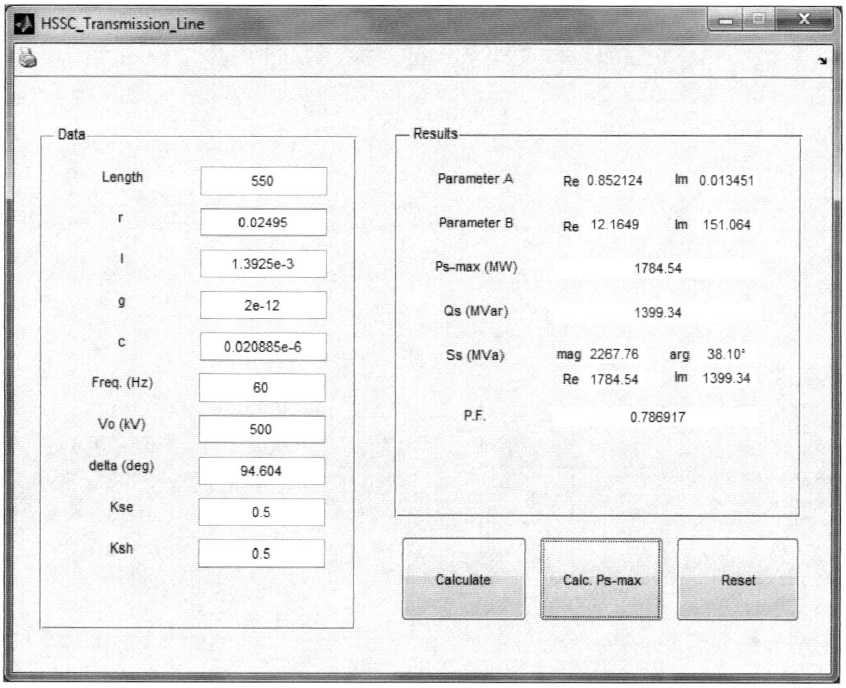

Figure 6. User input choice of calculation 2.

Table 2. Fixed Series & Variable Shunt Compensation.

Degree of Compensation (p.u.) k_{se}	k_{sh}	Ps (MW)	Qs (MVar)	Ps_{max} (MW)
0.5	0.0	842.41	-361.49	1784.54
0.5	0.1	842.41	-307.36	1784.54
0.5	0.2	842.41	-253.23	1784.54
0.5	0.3	842.41	-199.10	1784.54
0.5	0.4	842.41	-144.97	1784.54
0.5	0.5	842.41	-90.84	1784.54
0.5	0.6	842.41	-36.71	1784.54
0.5	0.7	842.41	17.42	1784.54

As a third illustration, the degrees of series and shunt compensations are varied from 0% to 70% for a power angle of 30 degrees and the results are shown in Table 3. From the results of Table 3, it is seen that, for every degree of compensation, the real, reactive and maximum power transfer gets varied following the variations in compensation.

Table 3. Variable Series & Shunt Compensation.

Degree of Compensation (p.u.)		Ps (MW)	Qs (MVar)	Ps_{max} (MW)
k_{se}	k_{sh}			
0.0	0.0	546.10	-481.48	1118.26
0.1	0.1	579.53	-402.44	1191.91
0.2	0.2	621.79	-323.93	1285.46
0.3	0.3	675.98	-245.78	1406.28
0.4	0.4	746.89	-167.97	1566.06
0.5	0.5	842.41	-90.84	1784.54
0.6	0.6	976.45	-15.78	2097.63
0.7	0.7	1175.73	52.62	2577.88

From the above three illustrations, it could be seen that compensation techniques especially series and shunt compensation (hybrid) is quite effective for enhancing electric power transfer of transmission lines and simple packages such as the developed one will be very useful for educating students as well as power engineers.

5. Conclusions

This paper has deliberated the application of series and shunt compensation of transmission lines for enhanced real and reactive power transfer. A programmable approach is used to quantify the amount of power transfers for various compensation levels and based on that a user friendly educative software package is developed and presented. Due to the graphical user interface approach, the data manipulation task and output task have been made simple and the program part is also very flexible for modifications when demanded. The series and shunt compensation techniques result in transmission line relaying and protection issues, and adverse effect on the power system stability, etc. and all these issues are not under the concern of this paper.

References

[1] Roberto Benato and Antonio Paolucci 2008 Operating capability of ac EHV mixed lines with overhead and cables links *Electric Power Systems Research* **78** 584–594

[2] Planichamy C et al 2006 Transmission System Upgrading - Analyzing the Alternatives for Final Decision Making *International Journal of Emerging Electric Power Systems* **6**

[3] Series Compensation | ABB developments and achievements A02-0135 E, 2010-12, Elanders Sverige AB, www.abb.com/FACTS

[4] Palanichamy C and Sri Krishna K 1991 Simple Algorithm for Economic Power Dispatch *Electrical Power Systems Research* **121** 147-153

[5] Shafiul Alam M, Md Abdur Razzak, Md Nazmul Hasan and Hasib Chowdhury A 2012 Transmission Capacity Enhancement of East-West Interconnectors Using Series-Shunt Compensation *7th International Conference on Electrical and Computer Engineering*, 20-22 December, 2012, Dhaka, Bangladesh, 579-582

[6] Shubhanga KN and Anil Kulkarni 2002 Application of Structure Preserving Energy Margin Sensitivity to Determine the Effectiveness of Shunt and Series FACTS Devices *IEEE Transactions on Power Systems* **17** 730-738

[7] Naveen P and Palanichamy C 2014 A Sustainable Renewable Energy Mix Option for the Secluded Society *Journal of Renewable and Sustainable Energy* **6** 023124

[8] Palanichamy C 2004 Renewable energy investment opportunities in Mauritius-an investor's perspective *Renewable Energy - An international journal* **29** 703-716

[9] Palanichamy C et al 2001 Budget Constrained Energy Conservation-An Experience with a Textile Industry *IEEE Transactions on Energy Conversion* **16** 340-345

[10] ThanhLong Duong, Yao JianGang and VietAnh Truong 2014 Application of min cut algorithm for optimal location of FACTS devices considering system loadability and cost of installation *Electrical Power and Energy Systems* **63** 979–987

[11] Alabduljabbar AA and Milanovic JV 2010 Assessment of techno-economic contribution of FACTS devices to power system operation *Electric Power Systems Research* **80** 1247-1255

[12] Palanichamy C et al 1999 Restructuring the Indian power sector with energy conservation as the motive for economic and environmental benefits *IEEE Transactions Energy Conversion* **14** 1589-1596

[13] Taghavi R and Seifi A 2012 Optimal reactive power control in hybrid power systems *Electr Power Compon Syst* **40** 741–58

[14] Shrestha GB and Wang Feng 2005 Effects of series compensation on spot price power markets *Electrical Power and Energy Systems* **27** 428-436

[15] Safari A, Shayanfar HA and Kazemi A 2013 Optimal location of PWM based series compensator in a power system *Electrical Power and Energy Systems* **51** 127-133

[16] Leonidaki EA, Manos GA and Hatziargyriou ND 2005 An effective method to locate series compensation for voltage stability enhancement *Electric Power Systems Research* **74** 73-81

[17] Basu M 2011 Multi-objective optimal power flow with FACTS devices *Energy Conversion and Management* **52** 903-910

[18] Orfanogianni T and Bacher R 2003 Steady-state optimization in power systems with series FACTS devices *IEEE Trans Power Syst* **18** 19-26

[19] Stéphane Gerbex, Rachid Cherkaoui and Alain Germond 2001 Optimal Location of Multi-Type FACTS Devices in a Power System by Means of Genetic Algorithms *IEEE Trans. Power Systems* **16** 537-544

[20] Palanichamy C et al 2003 An educational package for environmentally friendly, economic operation of power systems *International Journal of Electrical Engineering Education* 40 130-143

[21] Shaheen HI, Rashed GI and Cheng SJ 2010 Application and comparison of computational intelligence techniques for optimal location and parameter setting of UPFC *Eng Appl Artif Intell. Elsevier* **23** 203-216

[22] Palanichamy C et al 2005 A Visual Package for Educating Preparatory Transmission Line Series Compensation *IEEE Transactions on Education* 48 16-22

[23] Ayetül Gelen and Tankut Yalcinoz 2010 An educational software package for Thyristor Switched Reactive Power Compensators using Matlab/Simulink *Simulation Modelling Practice and Theory* **18** 366-377

[24] Magdy El Marsafaway 1991 Economical Design of Series Capacitor Compensation for Long Transmission Power Systems *Electrical Power & Energy Systems* **13**

Adaptive Precoded MIMO for LTE Wireless Communication

A F Nabilla and T C Tiong

Department of Electrical and Computer Engineering, Faculty of Engineering and Science, Curtin University, Miri, Sarawak, Malaysia

E-mail : tiong.teck.chai@curtin.edu.my

Abstract. Long-Term Evolution (LTE) and Long Term Evolution-Advanced (ATE-A) have provided a major step forward in mobile communication capability. The objectives to be achieved are high peak data rates in high spectrum bandwidth and high spectral efficiencies. Technically, pre-coding means that multiple data streams are emitted from the transmit antenna with independent and appropriate weightings such that the link throughput is maximized at the receiver output thus increasing or equalizing the received signal to interference and noise (SINR) across the multiple receiver terminals. However, it is not reliable enough to fully utilize the information transfer rate to fit the condition of channel according to the bandwidth size. Thus, adaptive pre-coding is proposed. It applies pre-coding matrix indicator (PMI) channel state making it possible to change the pre-coding codebook accordingly thus improving the data rate higher than fixed pre-coding.

1. Introduction

Long Term Evolution (LTE) and LTE-Advanced (LTE-A) are the development of 3GPP that should fulfill the current necessities and to accomplish the global broadband mobile communication [1]. These are

1. Higher data rates
2. Better system capacity and coverage area
3. Superb bandwidth operation
4. Improving spectral efficiency
5. Low latency
6. Low operating cost
7. Supporting multiple antennas
8. seamless integration with the Internet and existing mobile communication systems

The development of the mobile communication system had come a long way. From 2G mobile standard that support mobile telephony and voice application to 3G system that is the starting point of using packet-based data and Internet application and now the 4G mobile communication system that is based on IP-packet networks and supports the high bandwidth consumption applications.

It has been demonstrated by Shannon's fundamental work on channel capacity that the data rates are not enough for the existing received signal power. Thus, the usage of orthogonal frequency division multiplexing (OFDM) system and multiple-input multiple-output (MIMO) system that allows the multi-carrier transmission and the usage of multi antennas at transmitter and receiver end are important to increase the data rates and improve the system performance [2]-[4].

With the existence of broadband multi-media services such as smartphones, laptop computers, we currently live in an era where mobile data is very important hence leading to the development of wireless network standard. Telecommunication operators are constrained with limited frequency spectrum purchased from government [5]-[7].

Fixed MIMO modes and fixed precoding cannot adapt to the channel condition to achieve both high data rate and low bit error rate.

In this paper, we are investigating adaptive pre-coder in increasing spectral efficiency.

2. System description

2.1. Adaptive Precoding

Adaptive precoding implements precoding matrix indicator (PMI) channel state and by using PMI the system could modify the precoding matrix index accordingly. A parameter is used to enable the precoding matrix index so that it could be modified according to the channel condition instead of providing a fixed codebook index for the transmission. The PMI index will then be selected at the receiver and is passed to the transmitter to be used at the next step. PMI channel state designated the most suitable precoding matrix codebook for the corresponding frequency unit. This channel state is only available in closed-loop single-user and multi-user spatial multiplexing transmission modes [2].

2.2. Adaptive MIMO

Adaptive MIMO implements rank indicator (RI) channel state and by using RI the system could switch the transmission mode between transmit diversity and spatial multiplexing mode. The system will undergo spatial multiplexing transmission if the RI is equal to the transmitting antennas number and transmit diversity mode if the RI is less than it. The antennas number and data rate remained constant while the manipulated parameter will be the threshold that trigger the switching. Thus, a wideband rank value is used for the simulation. RI channel state designated the number of layers that the spatial multiplexing channel can adapt to [3]. RI channel state does not affect the performance of the system much for the frequency selective thus only one wideband RI is reported for the whole bandwidth [3].

3. Simulation results and discussion

3.1. Adaptive Precoding

This section describes the implementation of PMI and RI in Adaptive Pre-coded MIMO. The PMI channel state designated the most suitable pre-coding matrix codebook for the corresponding frequency unit. RI channel state designates the number of layers that the spatial multiplexing channel can adapt to. RI channel state does not affect the performance of the system much for the frequency selective thus only one wideband RI is reported for the whole bandwidth. For Part A of this section, a comparison between baseline simulation and adaptive pre-coder based on PMI feedback is done to investigate the effect of PMI.

Figure 1 shows the spectrum graph of adaptive precoding with SNR value of 30dBm, coding rate 1/3 for 4x4 antenna configuration with PMI being enabled. The transmitted signal and received signal is shown in it. From the graph, it could be seen that the received signals and transmitted signal are quite similar. The received signal before equalization is showing the effects of frequency selective fading. After equalization, the frequency selective fading effect is equalized by the closed loop spatial multiplexing and is showing a more frequency flat nature that resembles the transmitted signal spectrum. Figure 2 shows the fading channel effect (left) is reduced after the equalization is done. The equalizer compensates the multipath fading effect and rotate and scales back the effected constellation (left) to a constellation symbols can be more easily demodulated (right) as the received signal cluster is closer to the ideal constellation after equalizing.

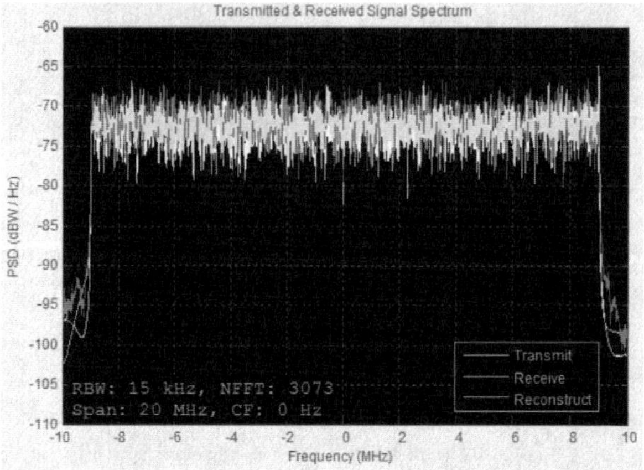

Figure 1. Spectrum of adaptive precoding simulation for 4x4 antenna configuration.

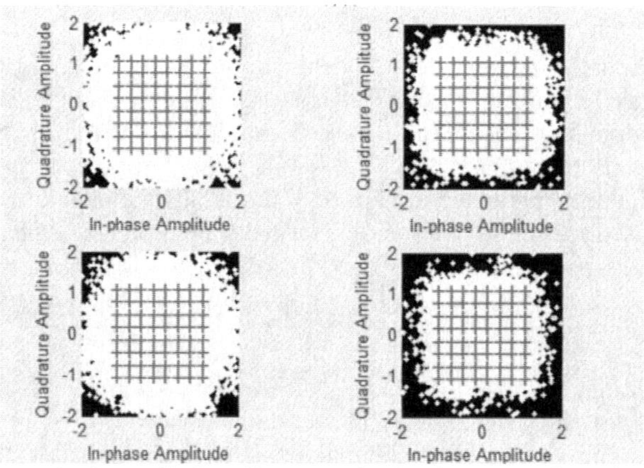

Figure 2. Constellation diagram of adaptive precoding simulation for 4x4 antenna configuration.

The average data rate for the baseline simulation with no adaptation is the same as adaptive precoding result albeit with higher BER can be seen in Table 1. Adaptive Precoder and Baseline are simulated with the same modulation rate and coding rate. The table also illustrates that the adaptive precoder index based on PMI feedback simulation has lower bit error rate (BER) compared to the Baseline. This proves that by varying the precoding codebook according to the channel condition, the system will produce less BER, thus improving the transmission of data rate hence improving the system. Hence, it can be concluded that the performance for adaptive precoder index based on PMI feedback simulation is better than the baseline simulation with no adaptation.

Table 1. Adaptive precoding for 4X4 antenna configuration.

Type of Modulation	Average Data Rate (Mbps)	Modulation Rate	Coding Rate	Bit error rate
Baseline simulation with no adaptation	124.46	6	1/3	0.151606
Adaptive precoder based on PMI feedback	124.46	6	1/3	0.148546

3.2. Adaptive MIMO

For Part B of this section, a comparison between MIMO techniques; transmits diversity and closed-loop spatial multiplexing is done. Next, a closed-loop spatial multiplexing is compared with adaptive MIMO based on RI feedback.

Figure 3 shows the spectrum with SNR value of 30dBm, coding rate 1/3 for 4x4 antenna configuration with PMI being enabled. The received signal and transmitted signal spectrum are quite similar.

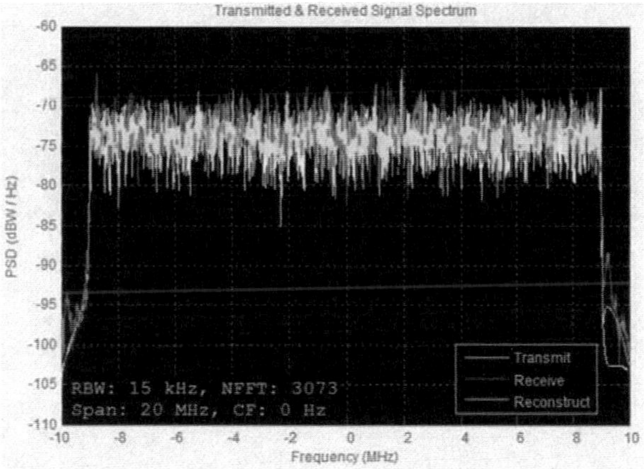

Figure 3. Spectrum graph of adaptive MIMO simulation for 4x4 antenna configuration.

The received signal before equalization is showing frequency selective fading. After equalization, the frequency selective fading effect is equalized by spatial multiplexing and is showing a more frequency flat nature that resembles the transmitted signal spectrum. Figure 4 is a constellation diagram. The equalizer mitigates the multipath fading effect and symbols can now be more easily demodulated as the received signal cluster is closer to the constellation after equalizing.

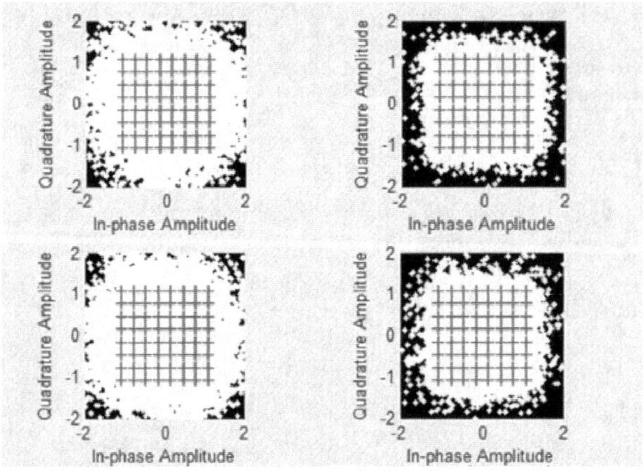

Figure 4. Constellation diagram of adaptive MIMO simulation for 4x4 antenna configuration.

Table 2 shows that the average rate for closed-loop spatial diversity mode is higher than transmit diversity mode. This is because in closed-loop spatial multiplexing, the bit stream is divided into multiple substreams and then modulated separately thus increasing the data rate. Other than that, closed-loop spatial multiplexing mode also has lower BER, which is 0.000631 compared to transmit diversity BER (0.000772) Thus, it could be concluded that the performance for closed-loop spatial multiplexing mode is better than the transmit diversity mode.

Table 2. Adaptive MIMO – transmit diversity mode and closed-loop spatial multiplexing mode for 4x4 antenna configuration.

Type of Modulation	Average Data Rate (Mbps)	Modulation Rate	Coding Rate	Bit error rate
Fixed mode: transmit diversity	48.06143	6	1/3	0.000772
Fixed mode: closed loop spatial multiplexing	50.635338	6	1/3	0.000631

In Table 3, it could be seen that average data rate for closed-loop spatial multiplexing mode is higher than adaptive MIMO based on RI feedback by 0.39857 Mbps. This is because in spatial multiplexing, the bit stream is divided into multiple substreams and then modulated separately thus increasing the data rate. However, when comparing the BER for closed-loop spatial multiplexing mode and adaptive MIMO based on RI feedback mode, it could be observed that the adaptive MIMO based on RI feedback mode has lower BER which is 0.000612. It is because spatial multiplexing is sensitive to the rank deficiencies of the matrix characterizing the equation of MIMO.

Proportionately, the bit error rate for the adaptive MIMO based on RI feedback mode is

$$\frac{47.66286}{48.06143} \times 0.000631 = 6.23 \times 10^{-4} \approx 0.000623$$

The Adaptive mode based on RI feedback has lower BER (0.000612) than pro rata value (0.000623).

Table 3. Adaptive MIMO –closed-loop spatial multiplexing mode and RI feedback for 4x4 antenna.

Type of Modulation	Average Data Rate (Mbps)	Modulation Rate	Coding Rate	Bit error rate
Fixed mode: closed-loop spatial multiplexing	48.06143	6	1/3	0.000631
Adaptive mode based on RI feedback	47.66286	6	1/3	0.000612

4. Conclusion

By using PMI, adaptive precoding is able to increase spectral efficiencies while keeping the bit error rate at acceptable low value.

Adaptive MIMO based on RI can switch between spatial multiplexing and transmit diversity to achieve high data rate at reasonable BER.

References

[1] Becker R T 2009 Precoding and spatially multiplexed mimo in 3gpp long-term evolution *High Frequency Electronics* pp 18-26

[2] David A E , Furuskär A, Jading Y, Lindström M and Parkvall S 2009 LTE: the evolution of mobile broadband *IEEE Communications Magazine* p 45

[3] Akyildiz I F, Gutierrez-Estevez D M, Balakrishnan R and Chavarria-Reyes E 2014 LTE-Advanced and the evolution to beyond 4G (B4G) systems *Physical Communication,* vol 10 pp 31-60

[4] Hanzo L, Akhtman Y, Jiang M and Wang L 2010 *Mimo-Ofdm for LTE, Wifi And Wimax: Coherent Versus Non-Coherent and Cooperative Turbo-Transceivers* (Wiley-IEEE Press)

[5] Erik Dahlman S P and Johan Sköld 2011 *4G LTE/LTE-Advanced for Mobile Broadband* USA: Elsevier Ltd

[6] Erik Dahlman SP, Sköld J and Beming P 2008 *3G Evolution: HSPA and LTE for Mobile Broadband* Second Edition ed.: Elsevier Science & Technology

[7] Házy L 1997 *Introduction to OFDM* Carleton University Canada

[8] Li Q, Li G, Lee W, Lee M i, Mazzarcsem D, Clerckx B and Li Z 2010 MIMO techniques in wimax and LTE: a feature overview *IEEE Communications Magazine* 48 pp 86-92

[9] Juho L, Han J-K and Zhang J 2009 MIMO technologies in 3GPP LTE and LTE-Advanced *EURASIP J. on Wireless Communications and Networking* pp 1-10

IOP Conf. Series: Materials Science and Engineering **78** (2015) 012016 doi:10.1088/1757-899X/78/1/012016

Adaptive Modulation and Coding for LTE Wireless Communication

S S Hadi and T C Tiong

Department of Electrical and Computer Engineering, Faculty of Engineering and Science, Curtin University, Miri, Sarawak, Malaysia

E-mail : tiong.teck.chai@curtin.edu.my

Abstract. Long Term Evolution (LTE) is the new upgrade path for carrier with both GSM/UMTS networks and CDMA2000 networks. The LTE is targeting to become the first global mobile phone standard regardless of the different LTE frequencies and bands use in other countries barrier. Adaptive Modulation and Coding (AMC) is used to increase the network capacity or downlink data rates. Various modulation types are discussed such as Quadrature Phase Shift Keying (QPSK), Quadrature Amplitude Modulation (QAM). Spatial multiplexing techniques for 4×4 MIMO antenna configuration is studied. With channel station information feedback from the mobile receiver to the base station transmitter, adaptive modulation and coding can be applied to adapt to the mobile wireless channels condition to increase spectral efficiencies without increasing bit error rate in noisy channels. In High-Speed Downlink Packet Access (HSDPA) in Universal Mobile Telecommunications System (UMTS), AMC can be used to choose modulation types and forward error correction (FEC) coding rate.

1. Introduction

Long Term Evolution (LTE) is a standard for wireless data communications technology and an evolution of the GSM/UMTS standards. The goals of LTE was to increase the capacity and the speed of wireless data networks by utilizing a new Digital Signal Processing (DSP) techniques and modulations. Adaptive modulation and coding is used to increase the network capacity or data rates [1]. Network architectures used to redesign significantly reduce transfer latency compared to the 3G architectures. The LTE wireless interface is incompatible with 2G and 3G networks, so that it must be operated on a separate wireless spectrum. The physical layer is simulated using Matlab. Channel Quality Information estimation is used to improve the data rates and bit error rate (BER). Various modulation types are discussed.

Increasing broadband multi-media services including VoIP, mobile TV, audio and video streaming, video conferencing, FTP and internet access require larger wireless cellular network and reliability.

Multiple-Input-Multiple-Output (MIMO) antenna systems can provide enormous capacity by spatial multiplexing [2], [3]. The performance of a MIMO system is highly dependent on the characteristics of the matrix channel. Link adaptation can be employed to any time varying systems to improve the transmission throughput by dynamically changing transmission parameters. Therefore the combination of link adaptation and MIMO is promising to realize higher spectral efficiency and higher throughput for mobile communication systems [4].

Fixed modulation cannot make maximum use of information transfer rate to suit changing channel conditions within given bandwidth allocation. Spectral efficiency is the average data rate per bandwidth unit Hertz per cell. Tradeoffs are required in designing mobile communication systems. For a given bandwidth allocation, spectral efficiency can be increased by increasing data rate using higher order modulation or larger multiple-input-multiple-output MIMO techniques. However, this can increase the probability of error in noisy channel

The problem is to search feasible solution to increase spectral efficiency by matching modulation and coding to conditions on the radio link. Spatial multiplexing techniques for 4×4 MIMO antenna configuration is studied. With channel station information feedback from the mobile receiver to the base station transmitter, adaptive modulation and coding can be applied to adapt to the mobile wireless channels condition to increase spectral efficiencies without increasing bit error rate in noisy channels.

2. System description

2.1. Adaptive Modulation

Modulation is the approach by which a carrier wave has the ability to convey the message or digital signal. There are three essential methods to this: amplitude, frequency and phase shift keying. More bits per symbol or period can be encoded using a high order of modulation. Distinctive order modulation permits to send more bits per symbol and in this way attain higher throughputs or better spectral efficiencies [5]. The utilization of adaptive modulation (AM) permits a wireless system to pick the highest order modulation upon the channel conditions.

2.2. Adaptive modulation and coding

Nowadays, Adaptive Modulation and Coding (AMC) which is a key innovation to progressively respond to the channel change of fading channels [6]. The quintessence of the plan is that, with accessible channel state information (CSI) estimation at the transmitter, link adaptation is empowered by the AMC for compelling usage of the available channel capacity for resulting transmissions while provisioning focused on at the receiver in fighting remote channel fluctuations.

3. Simulation results

3.1. Adaptive Modulation

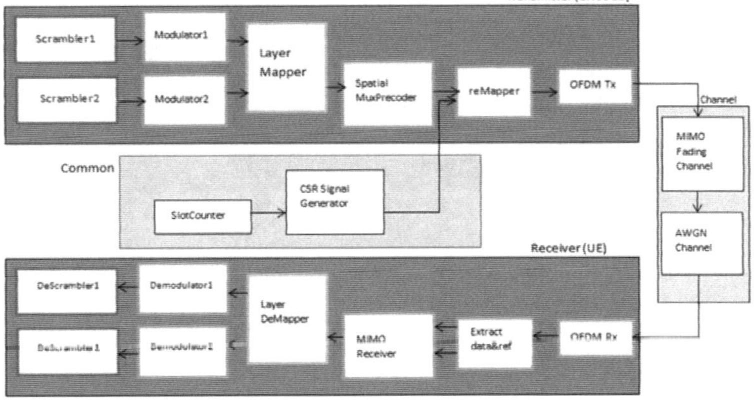

Figure 1. Physical Downlink Shared Channel Processing.

Figure 1 shows the physical downlink shared channel processing (PDSCH). The Downlink Physical Channel processing using spatial multiplexing codebook-based transmission by firstly scrambling the transport channel encoded bits by a bit-level scrambling sequence. This depends on the physical layer cell identity and ensures the randomization of cells. Secondly, the data modulation is converting the scrambled bits into complex modulated symbols. Layer Mapping is then used to map the complex modulated symbols received. The numbers of layers are same as the number of transmitted antennas. After that, Codebook-based precoding is used to precode the modulated symbols per layer. Then, the resource antenna used to map the precoded symbols are transmitted on each antenna and mapped to the resource elements of the resource blocks available for the transmission. Then, the Cell specific reference signal (CSR) is used to specify number of antenna in a cell and utilized for channel estimation at the receiver. Then, for the resource element carrying the reference signal for an antenna, the corresponding resource elements in other antennas have null transmissions. It permits the CSR signals to transmit without interference from the other antenna transmissions. Then, OFDM transmission is used to create the complex-valued time-domain OFDM signal per antenna from the fully populous resource grid, using the OFDM Modulator block. Then, in the MIMO channel, the higher mobility profiles are avoided as the closed-loop spatial multiplexing mode would be appropriate to high data rate and low mobility situations. Lastly is the Receiver UE processing that comprise of OFDM receiver and MIMO receiver.

The following will cover the channel quality indicator (CQI) channel-state report effect to change the modulation scheme in the subframe. Hence, to understand the design trade-offs, there will be comparison on: firstly, baseline –no adaptation, secondly, random selection of adaptation mode, and lastly adaptive modulation –CQI based.

Figure 2 shows the spectrum of user data acquired from the receive antenna. The receive signal shows the effects of frequency-selective fading effectively balanced by the close-loop spatial multiplexing used in transmission mode. Figure 3 demonstrate the pattern diagrams before (first column) and after (second column) equalization of information achieved from the receive antenna. The receive cluster signal shows that the equalizer can reduce the consequence of fading channel and produce pattern that more closely represents that of the 16QAM constellation.

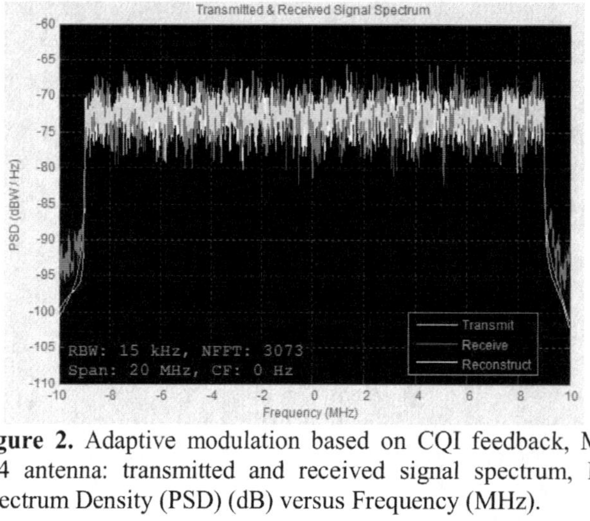

Figure 2. Adaptive modulation based on CQI feedback, MIMO 4x4 antenna: transmitted and received signal spectrum, Power Spectrum Density (PSD) (dB) versus Frequency (MHz).

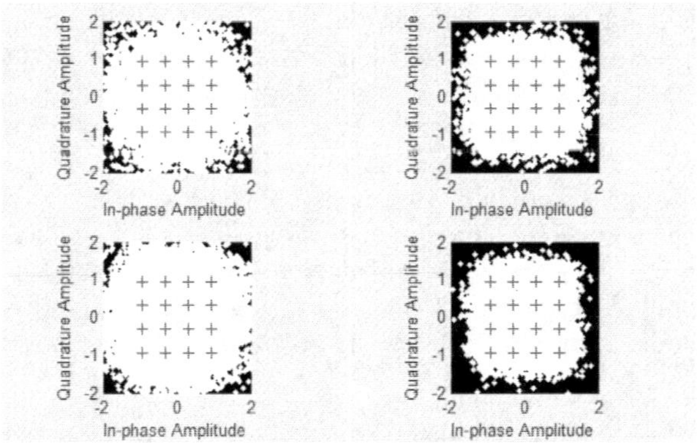

Figure 3. Adaptive modulation based on CQI feedback, MIMO 4x4 antennas: pattern diagram before and after equalization, left (before equalization), right- (after equalization).

Table 1 summarize the results for Adaptive Modulation MIMO 4. Table 1 shows that for fixed modulation, a higher modulation order produces higher data rate at the cost of higher BERs. As the modulation change to lower modulation order QPSK, it shows a lower data rate and lower BERs. Moreover, when randomly selecting the modulation scheme with no consideration of channel quality, both the average data rate and the BER are average value.

Table 1. Adaptive modulation MIMO 4×4.

Type of Modulation	Average Data Rate (Mbps)	Modulation Rate	Coding Rate	Bit Error Rate (BER)
QPSK	57.34	2	0.5	0.00001
16QAM	112.4	4	0.5	0.12183
64QAM	157.43	6	0.5	0.21584
Random selection	102.214	2 or 4 or 6	0.5	0.11461
Adaptive Modulation	99.24	2 or 4 or 6	0.5	0.11239

For AM scheme with channel quality estimation, the result comes out close to random selection. Random Selection scenario has data rate of 102.214 Mbps and BER of 0.11461 while for AM these are 99.24 Mbps and 0.11239 for data rates and BER respectively. Thus Adaptive Modulation can ensure a high data rate at reasonable Bit Error Rate compared to fixed modulation.

3.2. Adaptive modulation and coding

CQI channel-state report effect is used to modify the modulation scheme and coding rate in the sub-frame. Hence, to understand the design trade-offs, comparison is based on the baseline –no adaptation simulation and comparing the result with the random changing of both adaptive modulation and coding rate simulation.

Figure 4 demonstrates the spectrum acquired from the receive antenna in a subframe for adaptive modulation and coding rate. The receive signal shows the impact of frequency-selective fading effectively balanced by the close-loop MIMO spatial multiplexing.

Figure 5 demonstrate the pattern diagrams before (first column) and after (second column) equalization of data achieved from both of the receive antenna. The cluster region before equalization scatters farther than the constellation point. However, after equalization, the cluster region starts to surround the nearest constellation point. The received cluster signals form a smaller square region.

Table 2 shows the results for MIMO 4×4 simulation for AMC. For fixed modulation, a higher modulation order produces higher data rate at the cost of higher BERs However, when switching to random modulation selection, it gives the average results of the three fixed modulation cases (QPSK, 16QAM and 64QAM).

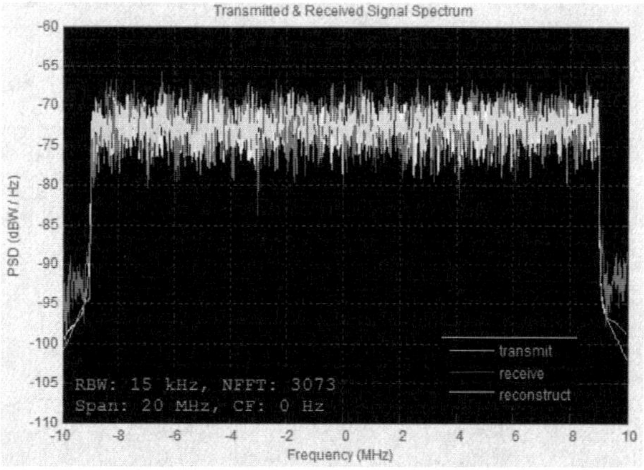

Figure 4. Adaptive modulation and coding based on CQI feedback, MIMO 4x4 antennas: transmitted and received signal spectrum, PSD (dB) versus Frequency (MHz).

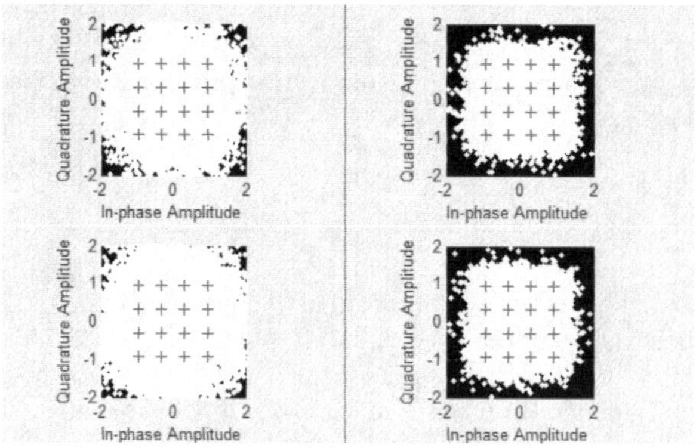

Figure 5. Adaptive modulation and coding based on CQI feedback, MIMO 4x4 antennas: pattern diagram before and after equalization, left (before equalization), right (after equalization).

Table 2. Adaptive modulation and coding MIMO 4×4.

Type of Modulation	Average Data Rate (Mbps)	Modulation Rate	Coding Rate	Bit Error Rate (BER)
QPSK	57.34	2	0.5	0.00001
16QAM	112.4	4	0.5	0.12183
64QAM	157.43	6	0.5	0.21584
Random selection	130.91	2 or 4 or 6	0.4958	0.18133
Adaptive Modulation and Coding	85.83	2 or 4 or 6	0.39-0.5	0.08375

For AMC, with channel quality estimation, the result comes out close to Random Selection. Random selection scheme has data rate of 130.91 Mbps and BER of 0.18133 while for AMC these are 85.83 Mbps and 0.08375 for data rates and BER respectively. Proportionally, by using a calculation:

$$\frac{85.93}{99.24}\times 0.11 = 0.0972$$

By comparing with the value in Table 1 and Table 2, the pro rata value obtained is 0.0972. The BER in adaptive modulation and coding is 0.08375 which is lower than the proportional value (0.0972). Thus AMC can further reduce BER compared to AM.

Overall, the CQI estimation techniques have been implemented and It also proves that AMC can increase data rate or spectra efficiency and reduce bit error rate better compared to fixed modulation. The user data rates and efficiency spectrum can be altered by using the selection measures for link adaptation. The main objective of LTE to experience high speed and high capacity can be easily achieved by using these techniques thus providing a faster and better performance.

4. Conclusion

Adaptive Modulation and Coding can increase data rate or spectra efficiency and reduce BER better compared to fixed modulation.

References

[1] Yamindi J-B, Hong J and Wu M-Q 2012 The optimization capacity of the MU-MIMO with channel quality information *IEEE 16th International Symposium on Consumer Electronics (ISCE)* pp 1-5 doi: 10.1109/ISCE.2012.6241685

[2] Foschini G J and Gans M J 1998 On limits of wireless communications in a fading environment when using muptiple antennas *Wireless Personal Comm.* vol 6

[3] Telatar E *Capacity of Multi-Antenna Gaussian Channels* June 1995 Technical Report, AT&T Bell Labs

[4] Catreux S, Gesbert D and Heath R W June 2002 Adaptive modulation and MIMO coding for broadband wireless data networks *IEEE Commun. Mag.* vol 40 pp 108-115

[5] Goldsmith A J and Chua S-G 1999 Adaptive modulation and coding for fading channels *Proc. IEEE Information Theory and Communications Workshop* pp 24-26 doi: 10.1109/ITCOM.1999.781396

[6] Sharma A and De S November 2011 Exploiting fading dynamics along with amc for energy-efficient transmission over fading channels *IEEE Communications Letters* **15** pp 1218-20 doi: 10.1109/LCOMM.2011.090911.111472

IOP Conf. Series: Materials Science and Engineering **78** (2015) 012017 doi:10.1088/1757-899X/78/1/012017

Performance Analysis for LTE Wireless Communication

S Tholhath and T C Tiong

Department of Electrical and Computer Engineering, Faculty of Engineering and Science, Curtin University, Miri, Sarawak, Malaysia

E-mail : tiong.teck.chai@curtin.edu.my

Abstract. Long Term Evolution (LTE) is the new upgrade path for carrier with both GSM/UMTS networks and CDMA2000 networks. The LTE is targeting to become the first global mobile phone standard regardless of the different LTE frequencies and bands use in other countries barrier. Adaptive Modulation and Coding (AMC) is used to increase the network capacity or downlink data rates. Various modulation types are discussed such as Quadrature Phase Shift Keying (QPSK) and Quadrature Amplitude Modulation (QAM). Spatial multiplexing techniques for 4×4 MIMO antenna configurations are considered. This paper has outlined various estimation techniques to increase the throughput of the LTE network by simulating the estimation techniques with various parameters in the LTE downlink mode 4 (spatial multiplexing). Three techniques i.e. channel estimation technique, estimation of channel models and MIMO receiver algorithm are simulated to provide the ideal LTE wireless communication system.

1. Introduction

The increase in broad band multi-media services including voice over internet protocol (VoIP), mobile TV, audio and video messaging and internet access demand a larger wireless cellular network with better reliability. In addition, telecommunication operators are constrained with limited frequency spectrums purchased from the government. Thus the need arises to search for a feasible network which can increase the throughput without purchasing additional bandwidth and at providing good service quality.

The Long Term Evolution (LTE) standard, specified by the 3rd Generation Partnership Project (3GPP) in Release 8, defines the next evolutionary step in 3G technology. LTE offers significant improvements over previous technologies such as Universal Mobile Telecommunications System (UMTS) and High-Speed Packet Access (HSPA) by introducing a novel physical layer and reforming the core network. The main reasons for these changes in the Radio Access Network (RAN) system design are the need to provide higher spectral efficiency, lower delay, and more multi-user flexibility than the currently deployed networks [1].

In 2008 the International Telecommunication Union specified requirements for 4G standards and named it the International Mobile Telecommunication Advanced which was expected to provide transmission rates of up to 1 Gbps.

This paper analyzed LTE mode 4 (Closed-Loop Spatial Multiplexing) configurations of channel estimation techniques, channel models and multiple-input multiple-output (MIMO) receiver algorithms gives a lower bit error rate (BER) as to comprehend these needs that have arisen for the telecommunication operators [2]-[4].

2. MIMO

MIMO use multiple antennas for both transmitting and receiving antennas. The advantages of MIMO communication, which exploits the physical channel between many transmit and receive antennas, are currently receiving significant attention [5]-[6].

Figure 1 shows a MIMO system consists of 'n' receive antennas and 'm' transmit antennas. The connections from each antenna are designated as

$$H = \begin{bmatrix} h_{11} & h_{12} & .. & h_{1m} \\ h_{21} & h_{22} & .. & h_{2m} \\ .. & .. & .. & .. \\ h_{n1} & h_{n2} & .. & h_{nm} \end{bmatrix} \tag{1}$$

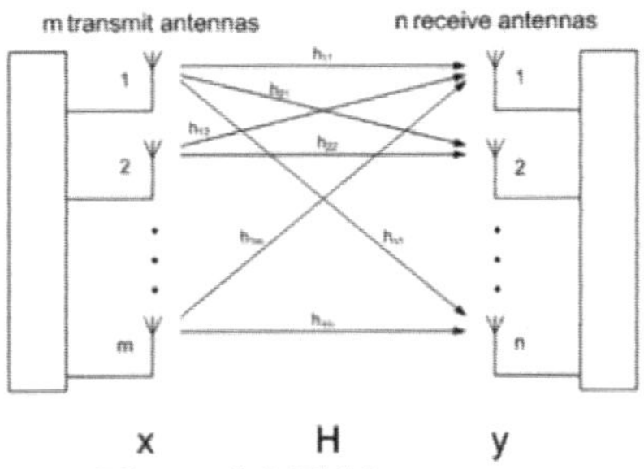

Figure 1. MIMO system.

Denote the receive vector '**y**', transmit vector '**x**' and noise '**n**'. We have

$$\mathbf{y} = \mathbf{Hx} + \mathbf{n} \tag{2}$$

2.1. Channel modes

2.1.1. Flat fading. A transmitted signal undergoes flat fading, or sometimes called as narrowband channels, when the signal bandwidth (BW) is less than the channel coherence bandwidth. In flat fading, though the spectrum of the transmitted signal is preserved, the received signal varies in amplitude and encounters deep fades of 20 to 30 dB. Instantaneous amplitudes of these received signals follow Rayleigh distribution.

2.1.2. Frequency selective. For Frequency selective fading or wide band channel, the signal bandwidth is greater than the channel coherence bandwidth. In frequency selective fading, time dispersion causes inters-symbol interference at the receiver making the multiple versions of the received signal attenuate and delay in time to a different degree.

2.1.3. LTE specific channels. There are three different multipath fading channels for LTE specific channels made by the Third Generation Partnership Project (3GPP). Namely,

- Extended Pedestrian A (EPA)

- Extended Vehicular A (EVA)
- Extended Typical Urban (ETA)

2.2. MIMO receivers
Receiver use equalizer to produce the best estimate of the transmitted symbols. Three algorithms are considered i.e.

- Zero Forcing (ZF) equalizer
- Minimum Mean Square Error (MMSE) equalizer
- Sphere Decoder (SD) equalizer

2.2.1. ZF equaliser. The ZF receiver inverses the MIMO channel matrix. The major drawback of the ZF receiver is noise enhancement. Although the ZF receiver eliminates the interference, performance becomes poor when the channel of the signal of interest is almost collinear to the interference subspace, or in other words when the channel matrix is almost rank deficient.

2.2.2. MMSE equaliser. The MMSE receiver minimises the average estimation error on the transmitted symbols. The average is taken over the transmitted symbols and the noise : the mean square error (MSE) is $E_{\mathbf{x},n}\|\hat{\mathbf{x}}-\mathbf{x}\|^2$, where $\hat{\mathbf{x}}$ is estimated symbol and $\mathbf{x}$ is the transmitted symbol. The ZF receiver also minimises the output MSE but under the constraint of complete inter symbol interference (ISI) elimination.

2.2.3. SD equaliser. SD receiver finds the transmitted signal vector with minimum maximum likelihood (ML) metric, that is, to find the ML solution vector. However it considers only a small set of vectors within a given sphere rather than all possible transmitted signal vectors. SD adjusts the sphere radius until there exists a single vector (ML solution vector) within a sphere. It increases the radius when there exists no vector within a sphere, and decreases the radius when there exists multiple vectors within the sphere.

3. OFDM
Orthogonal Frequency Division Multiplexing (OFDM) is a radical and revolutionary technology as it forms the basis for 4G technology. The basic concept of OFDM is to send samples concurrently using multiple orthogonal sub-channels rather than sending a sample signal using the entire bandwidth. The main advantage of OFDM is that because of the use of many sub-channels, only some of the sub-channels are affected by interference or multi-path effect [7]-[8].

3.1. OFDM
In OFDM a single channel utilizes multiple overlapping sub-carriers as shown in Figure 2. However, because these overlapping sub-carriers are orthogonal to each other, the sub-carriers will not interfere with each other. OFDM system can maximize spectral efficiencies.

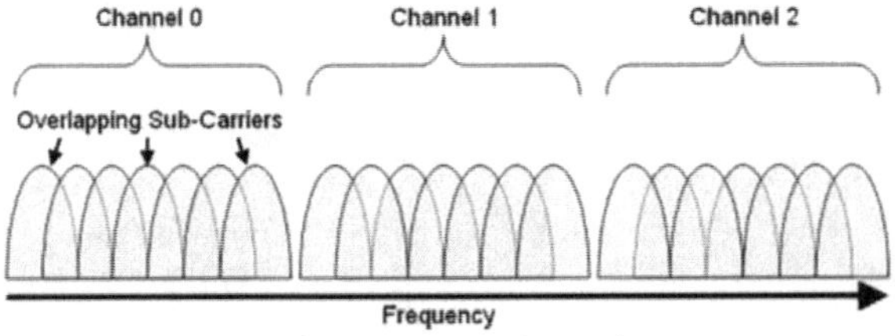

Figure 2. OFDM channel

3.2. Cyclic prefix
To completely eliminate the Inter Symbol Interference (ISI), cyclic prefix (CP) is introduced into the OFDM. CP is a repetition of the last samples of the data portion that is appended at the beginning of the data payload. As long as the CP duration is longer than the channel delay spread the ISI will be completely eliminated.

4. LTE physical layer
Designed to meet the present-day needs for high-speed media and data transfer, the 3GPP Long Term Evolution, or more commonly known LTE, symbolizes a major advance in cellular technology. It also includes multimedia unicast and broadcast services in addition to high-capacity voice support.

The LTE physical layer (PHY) acts as a highly efficient means for transmission of both data and control information between mobile user equipment (UE) and an enhanced base station (eNodeB). The technologies employed by the LTE PHY include OFDM and MIMO

Furthermore, it uses Orthogonal Frequency Division Multiple Access (OFDMA) on the downlink (DL) and Single Carrier – Frequency Division Multiple Access (SC-FDMA) on the uplink (UL). The function of OFDMA is to direct data to or from multiple users for a number of symbolic periods on a subcarrier-by-subcarrier basis.

4.1. Transmission Modes in LTE downlink

4.1.1. Mode 1 (SIMO). There is a single transmit antenna and multiple receiver antennas. Thus Mode 1 of LTE has only one transmit antenna.

4.1.2. Mode 2 (transmit diversity). The default MIMO mode, transmit diversity, sends the same data via various antennas. As Each antenna stream is using different frequency resources and different coding it provides a stronger transmission and causes the SNR to improve. Transmit diversity is used as a fall-back option in LTE for some transmission modes. For example, when spatial multiplexing (SM) cannot be applied, Control channels, such as physical broadcast channel (PBCH) and physical downlink control channel (PDCCH), is also transmitted using transmit diversity. A frequency-based version of the Alamouti codes space frequency block code (SFBC) is used for two antennas; whereas a combination of SFBC and frequency switched transmit diversity (FSTD) is used in the case of four antennas.

4.1.3. Mode 3 (open-loop spatial multiplexing). To attain higher date rates, this mode supports spatial multiplexing of two to four layers that are multiplexed to two to four antennas, respectively. Less user equipment (UE) feedback is required for it regarding the channel situation as precoding matrix indicator is excluded. It is used when channel rapidly changes or when there is missing channel information. An example of this can be UEs moving with high velocity. Furthermore, there is a specific delay in supplying the signal to every antenna known as the cyclic delay diversity (CDD). This in turn artificially creates a frequency diversity.

4.1.4. Mode 4 (closed-loop spatial multiplexing). This mode supports spatial multiplexing with up to four layers that are multiplexed to up to four antennas respectively. This is in order to attain higher data rates. To allow channel estimation at the receiver, cell-specific reference signals (RS) is transmitted by the base station and distributed over various resource elements (RE) and timeslots. A response regarding the channel situation is then sent by the UE. This includes information about which precoding is ideal from the defined codebook.

5. Performance Analysis

5.1. Examining effects for channel estimation techniques

The channel estimation techniques are

- Ideal Channel Estimation
- Channel Estimation on interpolation
- Channel estimation based on averaging over each slot
- Channel estimation based on averaging over each subframe

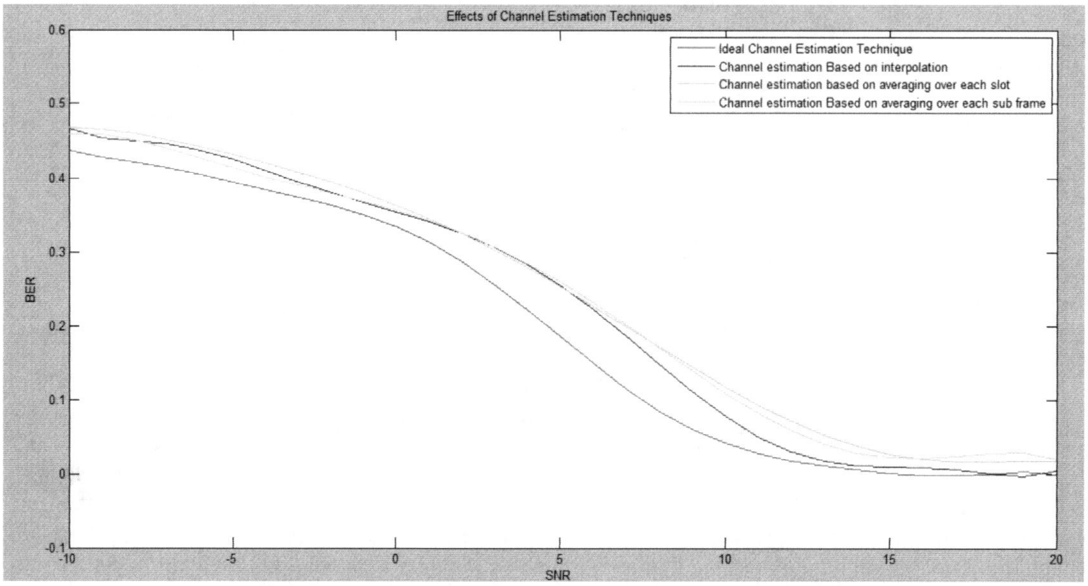

Figure 3. BER vs SNR for channel estimation techniques

Table 1. Channel estimation technique parameters.

Parameter	Value
Transmission Mode	4
Number of transmit antenna	2
Number of receiver antenna	2
Channel Bandwidth	10 MHz
Modulation Scheme	16QAM
Channel Model	frequency selective

Figure 3 shows the ideal channel estimation technique provides the best BER performance. As the variations within slots and subframes are allowed by interpolation it has a higher performance in an MMSE error-minimization context. Though the averaging methods don't improve the BER performance much it provides a continuity needed for better perceptual results.

The best channel estimation is Ideal channel Estimation, followed by Estimation based on Interpolation and then averaging over each slot and finally averaging over each sub frame.

5.2. Examining effects of channel models

Table 2. Effects of channel model parameters.

Parameter	Value
Transmission Mode	4
Number of transmit antenna	2
Number of receiver antenna	2
Channel Bandwidth	10 MHz
Modulation Scheme	16QAM
Channel estimation technique	Ideal channel estimation
MIMO Receiver	MMSE

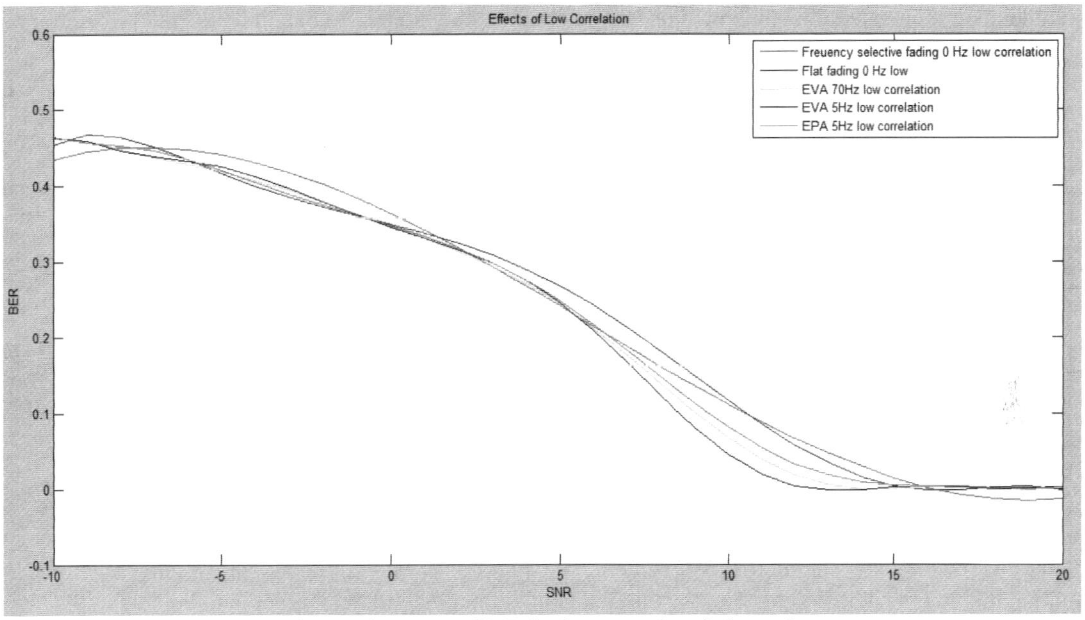

Figure 4. BER vs SNR for low correlated channels

Figure 4 shows there is not much difference in all the graphs until around SNR 7dB. At SNR 12 dB, As such the best to worst type of channel models can be listed as follows from the graph;

- EVA of a maximum Doppler shift of 5 Hz
- EVA of a maximum Doppler shift of 70 Hz
- EPA of a maximum Doppler shift of 5 Hz
- Flat Fading channel of a Maximum Doppler shift of 0 Hz
- Frequency Selective Fading channel of a maximum Doppler shift of 0 Hz

As noise profile increase, the performance becomes consistently reduced. Higher mobility as indicated by higher Doppler shift also reduce performance.

Above 15 dB, Flat Fading channel of a Maximum Doppler shift of 0 Hz outperform all others.

5.3. Examining effects of MIMO receiver algorithms

To examine the effects of MIMO receiver Algorithms, the MIMO receiver algorithms are compared i.e. ZF, MMSE and SD.

Table 3. MIMO receiver algorithm parameters.

Parameter	Value
Transmission Mode	4
Number of transmit antenna	2
Number of receiver antenna	2
Channel Bandwidth	10 MHz
Modulation Scheme	16QAM
Channel estimation technique	Ideal channel estimation

As the ideal channel estimation technique is the best estimation technique, Ideal channel estimation technique is used as a simulation parameter for channel estimation technique in this simulation. Figure 5 shows that SD has the best performance followed by MMSE and ZF.

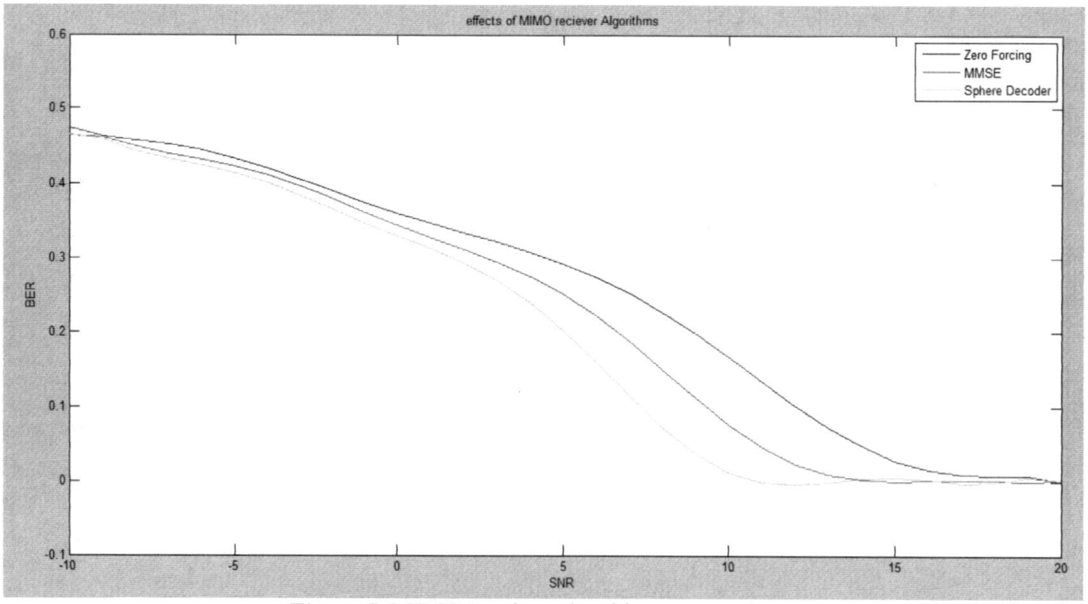

Figure 5. MIMO receiver algorithms comparison.

Though the SD outperforms MMSE and ZF, SD has higher computational complexity. Therefore Trade-off between performance and complexity decides the choice between SD and MMSE.

6. Conclusion

LTE was introduced to meet the demands of the users to download and upload packet data faster than its predecessor 3G. This paper has outlined various estimation techniques to increase the throughput of the LTE network by simulating the estimation techniques with various parameters in the LTE downlink Mode 4 (closed-loop spatial multiplexing).

Three techniques are used in this paper, channel estimation technique, estimation of channel models and MIMO receiver algorithm simulations. LTE physical layer is simulated to select ideal combination. The ideal combination is Ideal Channel Estimation Technique, Low correlated EVA with low Doppler shift and SD.

References

[1] Dahlman E , Parkvall S, Skold J and Beming P Jul. 2007 *3G Evolution: HSDPA and LTE for Mobile Broadband* (Academic Press)

[2] Foschini G J 1996 Layered space-time architecture for wire-less communication in a fading environment when using multi-element antennas *Bell Labs Tech. J.* **1(2)** pp 41–59

[3] Telatar I E 1999 Capacity of multi-antenna gaussian channels *Eur. Trans. Telecommun.* **10 (6)** pp 585–95

[4] Bliss D W, Forsythe K W, Hero A O and Swindle-hurst A L 29 Oct.–1 Nov. 2000 MIMO Environmental Capacity Sensitivity *Thirty-Fourth Asilomar Conf. on Signals, Systems & Computers* vol. 1 pp. 764–8.

[5] Bliss D W, Forsythe K W and Yegulalp A F 4–7 Nov. 2001 MIMO communication capacity using infinite dimension ran - dom matrix eigenvalue distributions *Thirty-Fifth Asilomar Conf. on Signals, Systems & Computers* vol 2 pp 969–74

[6] Ghosh A and Ratasuk R 2011 Essentials of LTE and LTE-A *Cambridge University Press*

[7] Sidrum B M, Mithun T P and Madhukar M 2013 BER, SNR and PAPR analysis for multiple accesses in LTE *Int. J. of Scientific & Engineering Research* vol 4 Issue 7 pp 773-7

[8] Kebede G M and Paul O O 2010 *Performance Evaluation of LTE Downlink with MIMO Techniques* Blekinge Institute of Technology

IOP Conf. Series: Materials Science and Engineering **78** (2015) 012018 doi:10.1088/1757-899X/78/1/012018

Optimizing electro-thermo fields for soot oxidation using microwave heating and metal

Haitham B Al-Wakeel, Z A Abdul Karim and Hussain H Al-Kayiem
Mechanical Engineering Department, University Technology PETRONAS, Bandar Seri Iskandar, 31750 Tronoh, Perak, Malaysia

haitham_alwakel@yahoo.com

Abstract. Soot is produced by incomplete combustion of various carbon-containing compounds. Soot is one of the main environmental pollutants and has become an important environmental and specific objective. To reduce soot from exhaust emission of diesel engine, a new technique is proposed and implemented by using metal inserted in the soot exposed to electromagnetic radiation. This paper presents a simulation to obtain optimum metal length and shape that give optimum electric field for attaining temperature enough for soot oxidation using microwave heating and a thin metal rod. Four cases were numerically examined to investigate the electric field and temperature distributions in a mono-mode TE_{10} microwave cavity having closed surfaces of perfect electric conductors. The operating frequency is 2.45 GHz, and power supply is 1500 W. The simulation methodology is coupling the absorbed electromagnetic energy with heat transfer energy. The absorbed electromagnetic energy is found from the electric field within the soot. The simulation was run using ANSYS based on finite element method. The results of the four simulation cases show that the optimum simulation is represented by case 2 where the value of electric field is 39000 V/m and heating time to arrive at the oxidation temperature (873 K) is 35 s using cylindrical metal rod of 8 mm length. It is revealed that the concept of achieving high temperature for soot oxidation by using thin metal rod inside a microwave cavity can be applied.

1. Introduction

Soot is elemental, amorphous, and black carbon. Soot consists of spherical particles with primary diameter between 10 to 500 nm [1]. It is agglomerated to form chainlike spherical particles with 100 to 800 nm [2]. Production of soot represents incomplete combustion; therefore it is usually undesirable because it causes air contamination and pollution, absorption of solar radiation and health hazard. After-treatment devices are used in diesel engines to filter and trap the soot particles in the exhaust gas with efficiency among 50% to 90% [3]. But the increasing of trapped soot in the filter leads to close the filter after a period of time and increases the back pressure. The higher back-pressure causes temperatures and pressures rise in the intake and exhaust lines, that resulting in some fuel penalty and changing the engine working conditions [4]. So periodically, a controlled regeneration process is required to oxidize the soot particulates at ignition temperature (600°C). In order to obtain a high temperature, a microwave heating was used and simulated in the literature to determine the electric field and temperature distribution on dielectric materials. Tada et al. [5] solved Maxwell equations employing two-dimensional finite deference time domain (FDTD) to investigate the electromagnetic field in a microwave applicator filled partially with a dielectric material, operating in the dominant TE10 mode at a frequency of 2.45 GHz. The electric field and power distributions were predicted for a lossy material and the results showed the electric field varies with dielectric material properties and position in a microwave cavity. Huang et al. [6] solved the coupled electromagnetic field equations, reaction equation and heat transport equation by using FDTD method to predict temperature rising and transmitted power through the electromagnetic cell and the findings indicated that electromagnetic energy is absorbed by the dielectric material and dissipated as heat energy leading to increase in the material temperature. Makul et al. [7] analyzed heat generation by microwave energy inside hardened cement paste. Electromagnetic distribution inside a rectangular wave guide and thermal models were taken into account. The compressive strength of the hardened cement paste increases and then slightly decreases from 14-18 days. Ciacci et al. [8] proposed two-dimensional mathematical model for the microwave-induced pyrolysis of wood block, based on a close combination of the propagation and absorption of electromagnetic waves into the material and heat/mass transfer with chemical reactions using a quasi-steady approximation for the electromagnetic field. The comparison with external heating showed that the differences are not only for the spatial temperature distribution, conversion rates and species densities but also for the weight loss characteristics. Salvi et al. [9] simulated numerically temperature profile in Newtonian fluids during continuous flow microwave heating by one way coupling electromagnetism, fluid flow, and heat transport in ANSYS and COMSOL Multiphysics softwares. They reported that the simulation of power loss distributions and temperature profiles are in close agreement by both softwares. Theoretical and experimental researches were conducted on materials mixed with metal powder for material sintering [10]. Experimentally proved there is no temperature rise for solid metal bar exposed to electric and magnetic fields in a microwave heating [11]. The numerical model of susceptor-assisted microwave heating reveals that the maximum temperature increase by using susceptor and the efficiency of the microwave heating depends on the sample size and thermal conductivity [12]. The theoretically and experimentally finding of thick metal objects post in a microwave oven reflect the microwave, while thin metal forms electric current on the metal surface [13].

In order to obtain high temperature in short time for soot microwave heating and oxidation with lower power consumption, a new technique is presented in this paper. This technique is inserting metal in the soot exposed to electromagnetic radiation. The formed electric field and temperature distribution were optimized numerically by conducting four simulation cases using ANSYS based on finite element method (FEM).

2. Problem formulation

The geometry of the analysis includes a microwave cavity WR430 [14] with dimensions (a= 0.10922 m on x-axis, b= 0.05461 m on y-axis and $\lambda g/2$ = 0.07386 m on z-axis). Where $\lambda g/2$ is half waveguide wavelength on z-axis. The cavity contains cylindrical sample of soot and cylindrical aluminum rod (the dimensions as mentioned in Table 1). The center of the soot is located in the center of x and y axes and 0.4λg in z-axis. The rod is posted vertically in the center of the sample as depicted in figure 1.

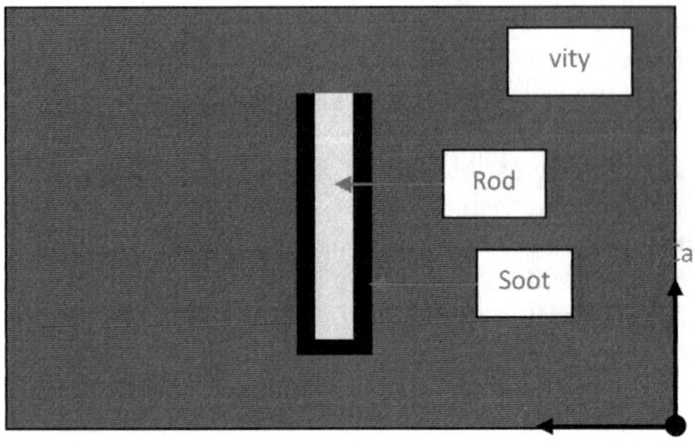

Figure 1. Position of analysis sample in microwave cavity

3. Methods

The governing equations of the electric field, absorbed electromagnetic energy and heat transfer equations in a microwave cavity are described in the following.

3.1. *Electric field in a microwave*

Maxwell's equations represent the coupling of unsteady electric, $\boldsymbol{E}$ and magnetic $\boldsymbol{H}$ field components. The wave has only one component of electric field and one component of magnetic field that are transverse to the direction of propagation. Therefore Maxwell's equations can be presented in terms of a transverse electric field (TE) or transverse magnetic field (TM) wave equation. To study the microwave heating, it is useful to present Maxwell's equations as one equation in terms of a steady transverse electric field equation called Helmholtz equation (1) for inhomogeneous vector wave [16].

$$\nabla X \nabla X \boldsymbol{E} = \left(\omega^2 \mu_0 \mu_r \varepsilon_0 \left(\varepsilon_r' - j\varepsilon_r''\right) - j\omega\mu_0\mu_r\sigma\right)\boldsymbol{E} \tag{1}$$

Complex permittivity ε_r is equal to:

$$\varepsilon_r = \varepsilon_r' - j\varepsilon_r'' \tag{2}$$

Let:

$$\gamma^2 = -\omega^2 \mu_0 \mu_r \varepsilon_0 \varepsilon_r' \left[1 - j\left(\frac{\varepsilon_r''}{\varepsilon_r'} + \frac{\sigma}{\omega\varepsilon_0\varepsilon_r'}\right)\right] \tag{3}$$

Replacing equation (3) into equation (1) gives:

$$\nabla X[(\nabla X \boldsymbol{E})] + \gamma\hat{\ }2.\boldsymbol{E} = 0 \tag{4}$$

The solution of Laplace's equation (4) is:

$$\boldsymbol{E} = E_0 e^{-\gamma^2 xyz} \tag{5}$$

3.2. *Electric field at the surfaces boundaries*

The boundary conditions (BC) of the cavity are perfect electric conductor (PEC), so the tangential component of the electric field is equal to zero on the cavity surfaces. At incident port, a continuous transverse electric field mode 10 (TE_{10}) is implemented in a rectangular waveguide WR430 and shorting port at end of the waveguide in propagation direction (z-axis). Then the components of $\boldsymbol{E}$ are:

$$E_x = 0\,,\ E_z = 0 \tag{6}$$

$$E_y = E_0 e^{-\gamma^2 xz} \tag{7}$$

3.3. *Electric field at the incident port*

The following mathematical model was formed to apply boundary condition at incident port of waveguide and represent the electric field in relation to source power.

$$E_y = \sqrt{2P_0/\beta_0.(a.b)}\ \ e^{-\gamma^2 xz} \tag{8}$$

Where P_0 is input power at waveguide incident port.

$$\beta_0 = \sqrt{(1-(f_c/f)^2)/(\mu/\varepsilon)} \tag{9}$$

Where;

$$f_c = \left(\frac{1}{2\pi\sqrt{\mu\varepsilon}}\right)\sqrt{(m\pi/a)^2 + (n\pi/b)^2} \tag{10}$$

m and n are integers numbers of sinusoidal half wavelength variation in x and y axes. a and b are dimensions of wave incident plane in x and y axes.

3.4. *Absorbed electromagnetic energy*

The absorbed electromagnetic energy is dissipated as thermal energy within the material inside a closed surfaces microwave cavity as in following equation

$$P_{abs} = JHEAT = \left(\frac{1}{2}\right)\omega\,\varepsilon_0\left(\varepsilon_r'' + \frac{\sigma}{\omega\varepsilon_0}\right)|E|^2 \tag{11}$$

(P_{abs}) is the average volumetric absorbed power dissipated into heat (*JHEAT*)

3.5. *General Transient Heat Transfer*

Heat transfer equation is given by [17]:

$$-\iiint\left[\frac{\partial}{\partial x}\left(k_x\frac{\partial T}{\partial x}\right)+\frac{\partial}{\partial y}\left(k_y\frac{\partial T}{\partial y}\right)+\frac{\partial}{\partial z}\left(k_z\frac{\partial T}{\partial z}\right)-JHEAT+\rho c_p\frac{\partial T}{\partial t}\right]dv = 0 \tag{12}$$

The boundary conditions (BC) are:

a. Heat convection BC

$$k_x\frac{\partial T}{\partial x}\cos\theta_x + k_y\frac{\partial T}{\partial y}\cos\theta_y + k_z\frac{\partial T}{\partial z}\cos\theta_z = -h(T-T_a) \tag{13}$$

b. Heat flux BC

$$k_x\frac{\partial T}{\partial x}\cos\theta_x + k_y\frac{\partial T}{\partial y}\cos\theta_y + k_z\frac{\partial T}{\partial z}\cos\theta_z = -q \tag{14}$$

Where, ($\cos\theta$) is the direction normal to the surface of heat transfer (*ds*)
The initial condition is:

$$T = T_i \ at\ time = 0 \tag{15}$$

After the substitution of equation (13) and equation (14) in equation (12), the general heat transfer equation in FEM will be obtained [18]:

$$-\iiint\left[\frac{\partial}{\partial x}\left(k_x\frac{\partial T}{\partial x}\right)+\frac{\partial}{\partial y}\left(k_y\frac{\partial T}{\partial y}\right)+\frac{\partial}{\partial z}\left(k_z\frac{\partial T}{\partial z}\right)-JHEAT+\rho c_p\frac{\partial T}{\partial t}\right]dv - \iint q\,ds - \iint h(T-Ta)\,ds = 0 \tag{16}$$

4. Numerical analysis

The simulation was conducted by using finite element method of ANSYS software. The simulation couples the harmonic high frequency electromagnetic analysis with transient thermal analysis to find the electric field distribution in the microwave cavity and dissipated heat and temperature distribution in the soot contains metal object and posts inside a mono-mode TE_{10} microwave cavity WR430, working under a frequency of 2.45 GHz and 1500 W power supply. Investigation of the temperature distribution during a microwave heating process is obtained by coupling the equations of absorbed electromagnetic energy (dissipated heat) with heat transfer energy. The absorbed energy is found from the electric field formed from the interaction of electromagnetic wave with metal and dielectric material. The dissipated heat was applied as elements body load in thermal analysis.

The inputs of High Frequency Electromagnetic code are dielectric properties of air (μ_r= 1 and $\varepsilon_r' =$ 1), soot (μ_r= 1, $\varepsilon_r' = 8.6$ and $\varepsilon_r'' = 7.4$) [18] and metal rods ($\mu_r = 1$, $\varepsilon_r' = 1$ and $\tan\delta = \sigma/2\pi f \varepsilon_0 \varepsilon_r'$). The boundary conditions for the cavity are perfect electric conductor surfaces on x and y axes and mono-mode incident port (TE_{10}) and shorter port with distances equal to $+\lambda_g/2$ and $-\lambda_g/2$ on z-axis respectively. Element type for high frequency electromagnetic code is tetrahedral solid (HF119).

The inputs of Transient Thermal code are thermal properties (density, specific heat and thermal conductivity) of the soot and the metal object. The thermal boundary condition is 9 W/m^{2}K heat transfer coefficient for soot. Conduction heat transfer was used between the soot and metal object (aluminum). The initial condition is 25°C. The soot reduction was presented by "killing" each element upon attaining the oxidation temperature of 600°C. The analysis is nonlinear so the solution process requires some intermediate equilibrium steps to correct the heat flow. This is accomplished by step-by-step incremental analysis to reach applied heat at time n and performing Newton-Raphson iteration at each step. Auto time step was selected by program according to Newton-Raphson method for nonlinear equations controlled between (0.001-0.25) s. Element type is tetrahedral (Solid87).

The simulation is based on mesh independency test using 10 elements per each dimension for the cavity, 20 elements per each dimension for soot, and 0.0002 m element size for metal.

In order to apply the simulation, some reasonable assumptions were made, these are: The simulation work did not deal with chemical reaction. KILL command was used to remove soot elements of temperature more than 600°C. The working substance is soot powder which was treated as a packed bed in the simulation work.

5. Results and discussion

For the purpose of soot oxidation strategy many simulation cases were conducted to obtain optimum electric field to control the temperature distribution by using high frequency electromagnetic radiation and metal objects inside a microwave cavity. Table 1 explains the results of four simulation cases that were conducted to compare and discuss the effect of length and shape of metal rod on the formed electric field and temperature inside the soot.

TABLE 1. Electric field and temperature exist by an Aluminum rod (of multi lengths) covered with soot (2 mm diameter) located inside a microwave cavity.

Case No.	Metal Rod			Max. *E*sum inside the soot (v/m)	Time of Oxidation (873 K) (s)
	Length (mm)	Diameter (mm)	Shape		
1	2	1	cylinder	15,000	78.5
2	8	1	cylinder	39,000	35
3	2	1	cone	61,000	48
4	8	1	cone	610,000	not limited

5.1. Effect of metal rod length on electric field and temperature distribution
Case 1 reveals the maximum value of E formed at rod tips (2 mm length) posted inside a soot sample located in a microwave cavity is 15 kV/m as depicted in figure 2. Where the rod tips represent the positive and negative terminals and the rod middle is neutral. The difference in the temperatures are about 2 K due to small sample dimensions and the temperature at sample tips are higher than the middle section (as shown in figure 6) due to the effect of ***E*** on dissipated heat according to equation 11. The heating time for whole soot oxidation is 78.5 s as reported in Table 1.

Case 2 implies the value of maximum ***E*** is 39 kV/m as displayed in figure 3. The difference in the temperatures between sample middle and tips are more than 10 K as modeled in figure 7. The increase in the value of maximum E resulted from the increase in rod length from 2 mm to 8mm according to equation 5 and the increased ***E*** leaded to increase the temperature close to rod terminals while the temperature at the sample middle stay close to initial temperature as shown in figure 7a. The time of heating for whole soot oxidation is 35 s as recorded in Table 1. The comparison between cases 1 and 2 shows there is a reduction in oxidation time in case 2 due to the higher temperature resulted from higher formed E in case 2.

5.1. Effect of metal rod shape on electric field and temperature distribution
Case 3 shows the maximum amount of ***E*** is 61 kV/m formed at the head of the metal cone where the cone have a 2 mm length and a point head as presented in figure 4. This means the energy is focused in one direction by tapered rod (cone), so it is increased. Also as in case 1 there is small difference in temperature between tips and middle as shown in figure 8a. The time for whole soot oxidation is 48s heating. The comparison between cases 1 and 3 show about 40% of the oxidation time is reduced due to the energy focusing by using the short conical rod.

Case 4 indicates that the higher ***E*** is obtained by using conical shape metal with 8 mm length, where the value of ***E*** attained 610 kV/m at cone head as simulated in figure 5. The increasing in cone length leads to increase the focusing of ***E*** as mentioned in case 2. The predicted difference in the temperatures between the sample bottom tip and middle are more than 100 K (as shown in figure. 9a) due to the very high ***E*** formed at cone tip. The soot oxidation at sample tips achieved after 18s heating as shown in figure 9b, but the noticed there is no oxidation for the whole soot where the same quantity of soot still not oxidized during the time from the 20th second to 60th second due to the reduction in temperature in spite of continuous microwave heating as shown in figures 9c and 9d. The reason of this unexpected phenomenon belongs to the rapid oxidation of elements at the high temperature (the sample tips). The rapid oxidation did not give enough time to increase the temperature of middle elements due to the heat flow by conduction from high to low temperature. Consequently the remaining soot temperature was decreased with time progress due to the generated heat at middle of the sample is less than the lost heat by convection.

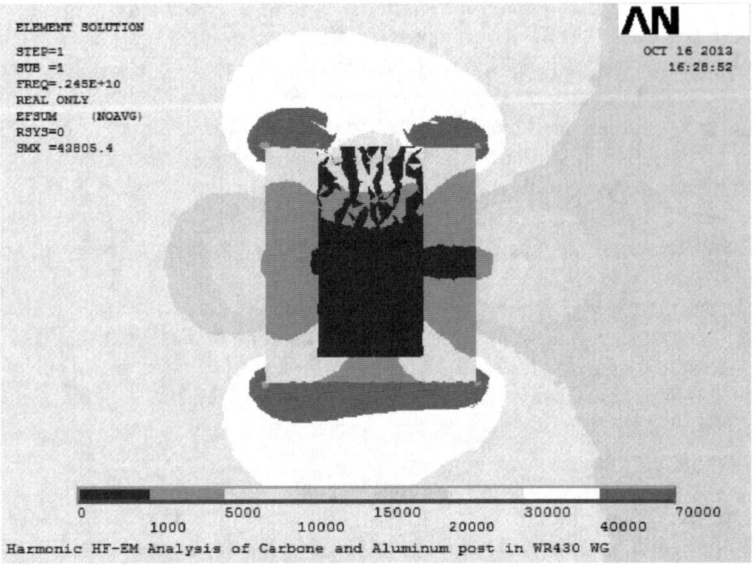

Figure 2. Electric field distribution of case 1

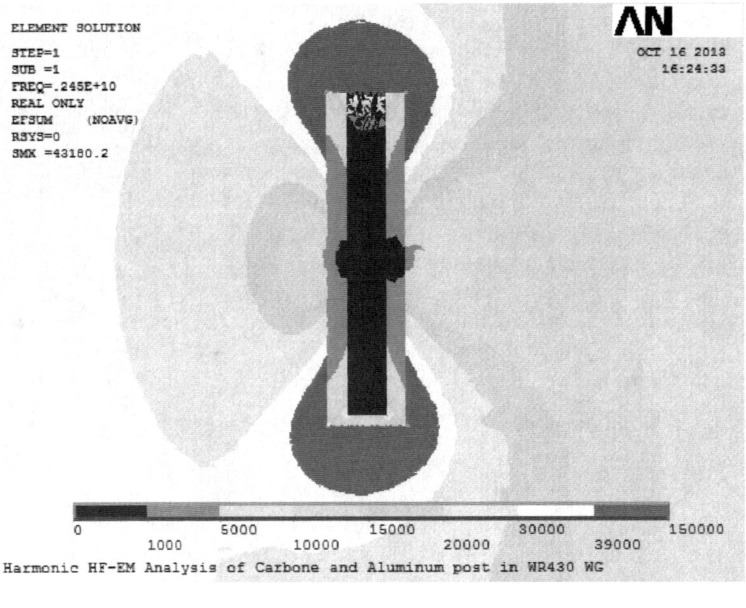

Figure 3. Electric field distribution of case 2

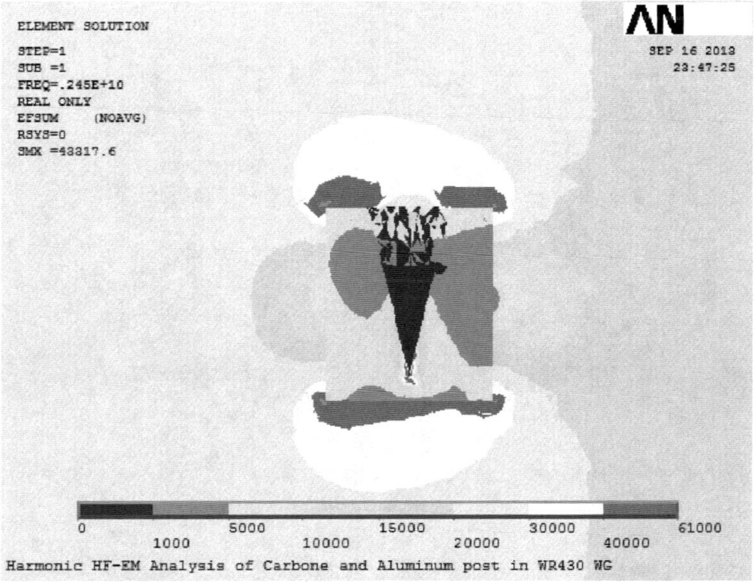

Figure 4. Electric field distribution of case 3

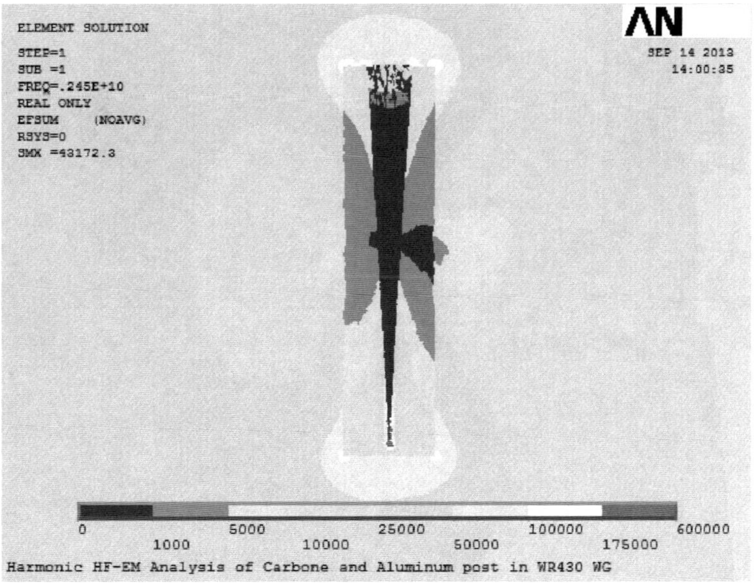

Figure 5. Electric field distribution of case 4

IOP Conf. Series: Materials Science and Engineering **78** (2015) 012018 doi:10.1088/1757-899X/78/1/012018

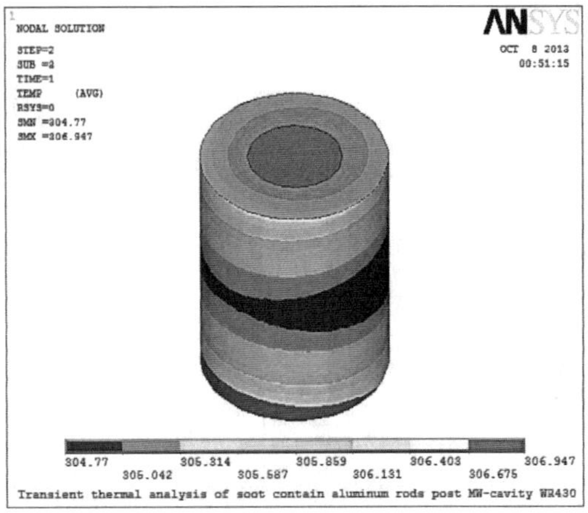

Figure 6a. Temperature distribution of the sample at 1s, case 1

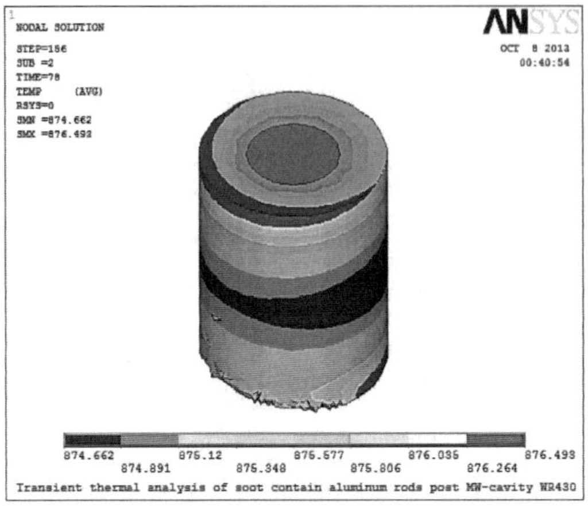

Figure 6b. Temperature distribution of the sample at 78s, case 1

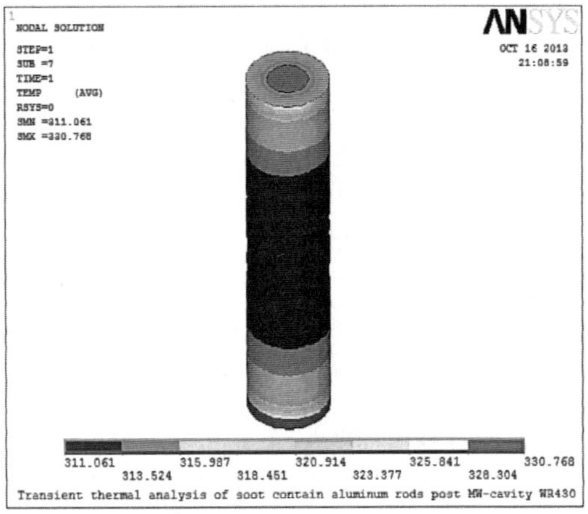

Figure 7a. Temperature distribution of the sample at 1s, case 2

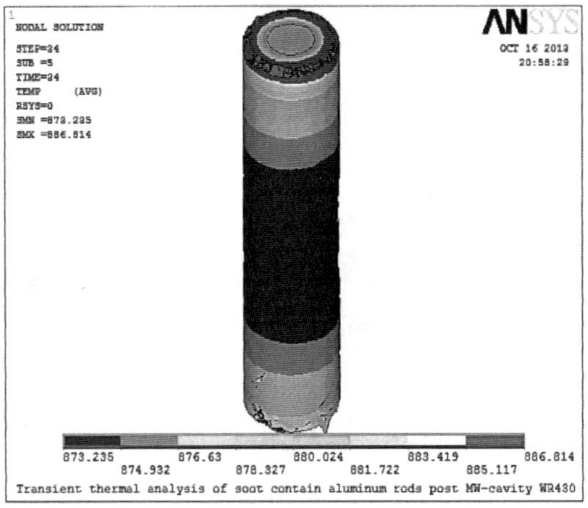

Figure 7b. Temperature distribution of the sample at 34s, case 2

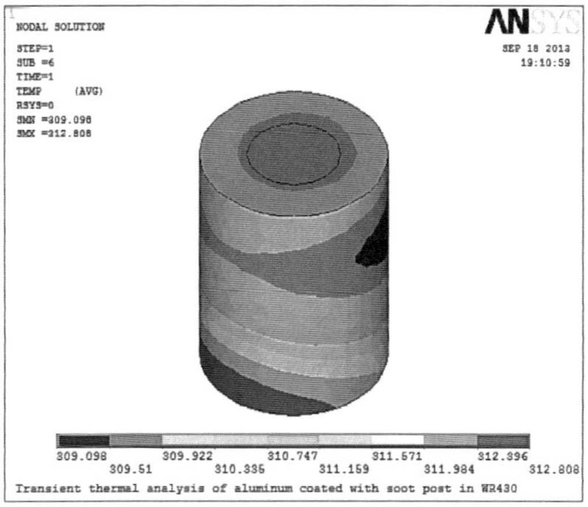

Figure 8a. Temperature distribution of the sample at 1s, case 3

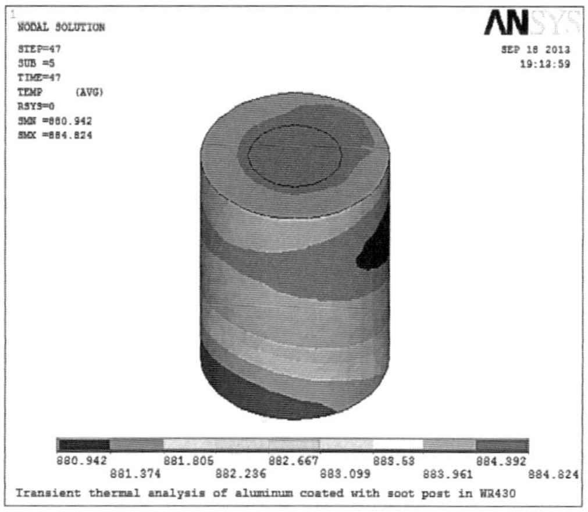

Figure 8b. Temperature distribution of the sample at 47s, case 3

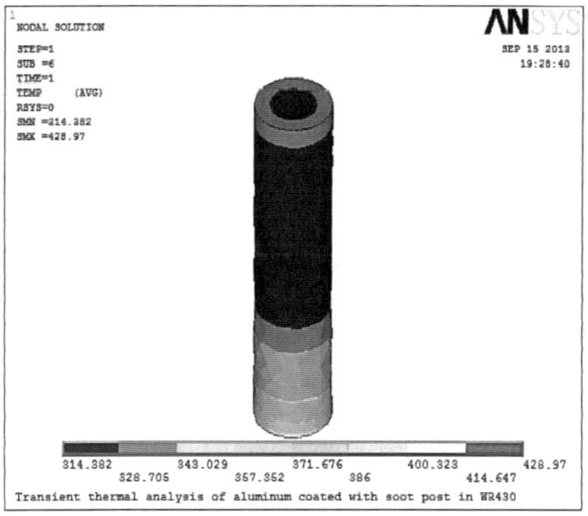

Figure 9a. Temperature distribution of the sample at 1s, case 4

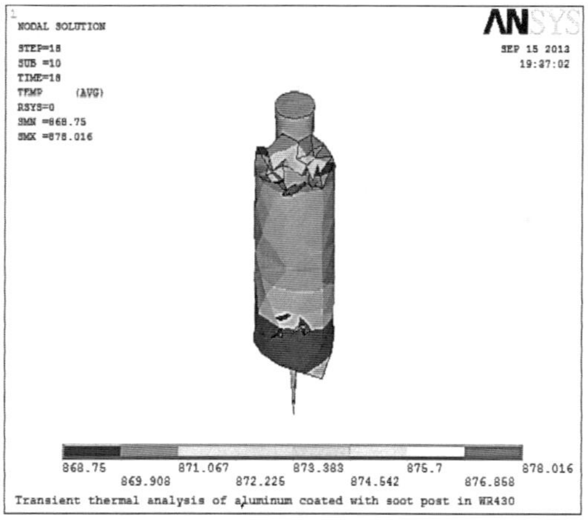

Figure 9b. Temperature distribution of the sample at 18s, case 4

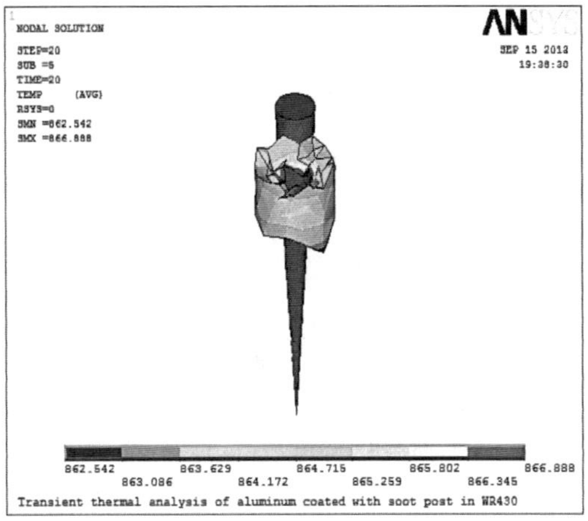

Figure 9c. Temperature distribution of the sample at 20s, case 4

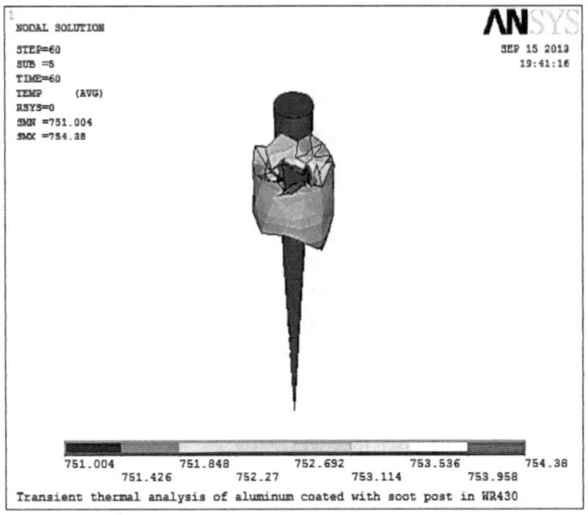

Figure 9d. Temperature distribution of the sample at 60s, case 4

6. Conclusions

This work presents a new technique to get locally high electric field to obtain rapidly high temperature for soot oxidation by using microwave heating and thin metal objects inside a microwave cavity. Four cases were examined numerically to simulate the electric field and temperature distribution. The methodology of simulation was implemented using ANSYS based on FEM. The findings from the results are:

- The increasing of electric field leads to increase dissipated heat and consequently increase material temperature.
- The metals inside a sample redistribute the electric fields with high and low values.
- Case 2 represent the optimum simulation for the value of electric field is 39000 V/m, and heating time to arrive oxidation temperature (873 K) is 35 s using cylindrical metal rod 8 mm length.
- The technique of achieving high temperature for soot oxidation by using metal object inside a microwave cavity can be applied. The temperature distribution can be controlled by size and shape of metal object.

Nomenclature

C_p	Specific heat
$\mathbf{D}$	electric flux vector
ds	Surface integral
dv	Volume integral
E_0	Amplitude of electric field intensity
f	Operating frequency
f_c	Cutoff frequency
h	Heat transfer coefficient
k	Thermal conductivity
I	current

j	Imaginary part of a complex number
T	Temperature
t_a	Ambient temperature
t_i	Initial temperature
Z_0	Intrinsic impedance for medium
Z_{xy}	Transverse wave impudence
γ	Complex propagation constant
ε	Complex Permittivity
ε_0	Permittivity of vacuum or air (8.854x10^{-12} F/m)
ε_r'	Relative dielectric constant
ε_r''	Relative dielectric loss factor
λ_c	cutoff wavelength
λ_g	waveguide wavelength on z-axis
λ_f	Free space wavelength
μ	Magnetic permeability
μ_0	permeability of vacuum or air (1.257x10^{-6} H/m)
μ_r	Relative permeability
ρ	Density
σ	Electric conductivity
ω	Angular frequency

Acknowledgement
The authors would like to acknowledge Universiti Teknologi PETRONAS (UTP) for the research grand (FRGS) from Ministry of Higher Education, Malaysia to conduct this research work.

References
[1] Information: http://www.cancarb.com/thermax1.html.
[2] Information: http://www.inchem.org/documents/iarc/vol65/carbon.html.
[3] Michel R P, Baican R and Schubert E 1993 Soot particle properties in the microwave range *Proc. Conf. on Microwave (Madrid, Spain)* vol 23 pp 959-60
[4] Lapuerta M, Rodríguez-Fernández J and Oliva F 2012 Effect of soot accumulation in a diesel particle filter on the combustion process and gaseous emissions *Energy* **47** 543-52
[5] Tada S, Echigo R and Yoshid H 1998 Numerical analysis of electromagnetic wave in a partially loaded microwave applicator *J. Heat and Mass Transfer*, **41** 709-18
[6] Huang K M, Lin Z and Yang X Q 2004 Numerical simulation of microwave heating on chemical reaction in dilute solution *Progress In Electromagnetics Research* **49** 273-89
[7] Makul N, Keangin P, Rattanadecho P, Chatveera B and Agrawal D K 2010 Microwave-assisted heating of cementitious materials *J. of Int. Communications in Heat and Mass Transfer* **37** 1096-105
[8] Ciacci T, Galgano A and Di Blasi C 2010 Numerical simulation of the electromagnetic field and the heat and mass transfer processes during microwave-induced pyrolysis of a wood block *J. of Chemical Engineering Science* **65** 4117-33

[9] Salvi D, Bolder D, Ortgo J, Aita G M and Sabliov C 2010 Numerical modeling of continuous flow microwave heating J. *of Microwave Power and Electromagnetic Energy* **44** (4) 187-97

[10] Gupta M and Wong Wai Leong E 2007 *Microwaves and Metals* (Singapore: Wiley)

[11] Cheng J, Roy R and Agrawal D 2001 Experimental proof of major role of magnetic field losses in microwave heating of metal and metallic composites *J. of Materials Science Letters* **20** 1561-63

[12] Shukla A K, Mondal A and Upadhyaya A 2010 Numerical modeling of microwave heating *Science of Sintering* **42** 99-124

[13] Shayeganrad G and Mashhadi L 2009 Theoretically and experimentally investigation of sparking of metal objects inside a microwave oven *Proc. of Progress in Electromagnetics Research Symp.* (Beijing, China, 23-27 March 2009) pp 632-41

[14] Chow Ting Chan T V and Reader H C 2000 *Understanding Microwave Heating Cavities* (London: Artech House Inc.)

[15] Jin J 2002 *The Finite Element Method in Electromagnetics*, 2nd ed (New York: Wiley)

[16] Kraus J D and Fleisch D A 1999 *Electomagnetics with Applications*, 5th ed (Singapore: McGraw Hill)

[17] Liu G R and Quek S S 2003 *The Finite Element Method-A Practical Course*, 1st ed (Singapore: Butterworth Heinmann)

[18] Lewis R W, Nithiarasu P and Seetharamu K N 2004 *Fundamental of the Finite Element Method for Heat and Fluid Flow* (England: Wiley)

[19] Henrichsen M and Popuri S 2001 Development of a microwave assisted particulate filter regeneration system *Cummins Inc.* www.osti.gov/energycitations/seirlets/purl/827991.../827991.pdf

IOP Conf. Series: Materials Science and Engineering **78** (2015) 012019 doi:10.1088/1757-899X/78/1/012019

Comparative performance analysis of a dual-solenoid mechanical oscillator

V C C Lee[1,3], **H V Lee**[1], **H G Harno**[1] **and K C Woo**[2]

[1] Department of Mechanical Engineering, Faculty of Engineering and Science, Curtin University Sarawak, Miri, Sarawak, Malaysia.
[2] Department of Mechanical, Materials and Manufacturing Engineering, Faculty of Engineering, University of Nottingham Malaysia Campus, Semenyih, Selangor Darul Ehsan, Malaysia.

Email: [1] vincent@curtin.edu.my, [2] woo.ko-choong@nottingham.edu.my

Abstract. An innovative dual-solenoid electro-mechanical-vibro-impact system has been constructed and experimentally studied. Comparative studies against a mechanical spring system and a permanent magnet system have been performed, where it is shown that the dual-solenoid system is able to produce oscillations better than the permanent magnet system and more energy efficiently. Comparison with a higher-powered dual solenoid system has also been conducted where a stationary solenoid has shown to be a more dominant parameter. In addition, it is also discovered that a mechanical oscillator in the dual-solenoid system is independent of the angular frequency.

1. Introduction

Impact vibration technology was first introduced in the Soviet Union in 1940 for industrial applications. Tsaplin [1], in 1940, introduced the first impact vibration hammer for use in piling. The system consisted of a vibrating element, an eccentric mass and springs oriented in a vertical position for the piling work. Soil penetration rate by pure vibration and that produced by a vibro-impact process was compared by Tsaplin, and the rate of penetration by vibro-impact process was shown to be eight times more effective than that of a pure vibration. This phenomenon in the context of cohesion-less soil was later examined by Rodger and Littlejohn [2] in 1980 and showed the phenomenon of reduction in shear strength. The usefulness of such vibration was later demonstrated by Preobrazhenskaya [3], in 1955 and showed that the dynamic resistance of clay soil was similarly reduced as well. This impact vibration technology was later adapted in performing other earth boring works such as moling by Woo et al. [4] in 2000.

An electro-mechanical-vibro-impact system was constructed and studied experimentally by Nguyen et al. [5, 6] in 2008 and through numerical analysis by Ho et al. [7] in 2011. The system involves the use of a solenoid driven by an electrical circuit, coupled with a solid state relay, to generate large electro-magnetic forces acting on a conductor, which oscillates within the solenoid. Impacts are generated by means of a stop in the path of bar oscillations and forward progression of the device can

[3]Author to whom any correspondence should be addressed.

then be generated, where its performance is closely related to the control frequency of the solid state relay. Lee et al. [8] in 2011 performed experimental analyses on an optimized electro-mechanical-vibro-impact system which generates impact forces using a permanent magnet as replacement to a mechanical spring, in which it is effectively acting as a highly non-linear magnetic spring. Consequently, impact forces have been generated to occur at high frequencies, which have proven to be beneficial to a downward progression rate in soil. Its magnetism is however a variable with respect to time and does not ensure longevity of the system.

This paper describes an innovative electro-mechanical-vibro-impact system with use of two solenoids at both ends of the mechanical oscillator. Three comparative studies are presented in this work to evaluate the mechanical oscillator performance of the new system. The aim of this study is to evaluate the proposed alternative to the permanent magnet system, in particular the issue on decaying magnetism in the system. Outcomes of the comparative studies will be beneficial to the future design of energy efficient and fully magnetized electro-mechanical-vibro-impact system.

2. Experimental methods and materials

Experimental approach is chosen in this study such that the responses of the mechanical oscillator can be explicitly measured and analyzed. The analysis of the system response will emphasize on oscillatory displacement made and in comparison to the system by Lee et al. [8] and Ho et al. [9]. For consistency in comparison, Ho et al.'s work is repeated in this study to serve as controlled data as well as the benchmark for comparison, denoting its maximum displacement as one. Comparison against Lee et al.'s work will be emphasized to evaluate the performance of both systems, in which both are dual-magnetized but in different configuration.

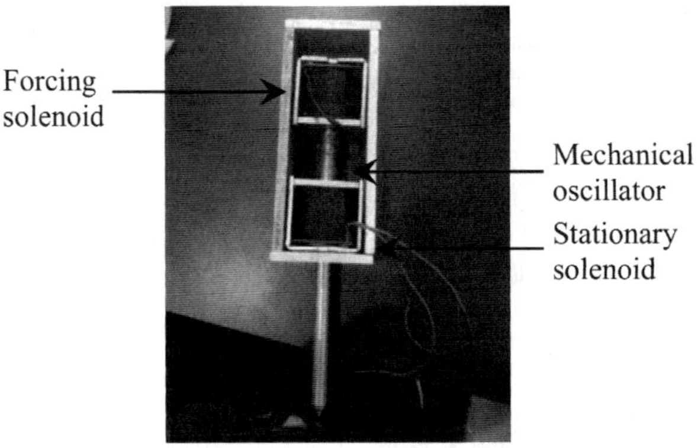

Figure 1. Proposed new dual-solenoid system.

A new electro-mechanical-vibro-impact system of dual solenoid as illustrated in Figure 1 is constructed and described. The solenoids are connected vertically with a gap of 50 mm in between and supported with aluminium plates at both vertical sides. The mechanical oscillator is placed within the two solenoids, in which the solenoid positioned on top (forcing solenoid) generates the oscillatory magnetic force and bottom (stationary solenoid) is supplied with a constant magnetic force. The new system undergoes a major change in terms of its configuration such that, previously the forcing solenoid is place at the bottom while in the new system it is installed at the top. This is due to the challenges faced by the stationary solenoid in attracting the oscillator upwards due to the gravitational pull. As such, the placement of solenoid is hence inverted so that the oscillation is made possible. The forcing solenoid is connected to power supply with 15 Volts AC and the stationary solenoid is supplied with 2.5 Volts DC. In order to eliminate the uncertainties in the design of system, similar driving circuit as [8] and [9] will be used in this study. Overall, all the environmental settings in this experiment are similar to [8] and [9] with exception on the use of an additional solenoid. For repeatability of the experiment, each data point is repeated six (6) times with uncertainty measured at 0.2 at resonance.

3. Mathematical formulations

Mathematical formulations for all three systems are derived in accordance to classical equation of motion. The motion of a mechanical spring system has been widely discuss in the literature [5-9] as expressed by (1), where m is the mass of the oscillator, F is the force from the forcing solenoid and k is the stiffness of the spring. ω is the angular frequency, x and $\ddot{x}$ is the displacement and acceleration of the oscillator, respectively and t is the time.

$$m\ddot{x} = F\sin\omega t - kx \quad (1)$$

Similarly, for a permanent magnet system, the motion of the oscillator can be expressed as,

$$m\ddot{x} = F\sin\omega t - k_{mag}x \quad (2)$$

where k_{mag} is the non-linear magnetic spring. For a dual-solenoid system, its motion is subjected to the oscillating force induced from the forcing solenoid and restoring magnetic force from the stationary solenoid. The motion of its oscillator is expressed in (3) where the negative sign shows that the placement of the solenoids is inverted as compared to the previous two and that k_{stat} is the non-linear magnetic spring induced by the stationary solenoid.

$$m\ddot{x} = -F\sin\omega t + k_{stat}x \quad (3)$$

With reference to (1)-(3), it is shown that the motion of the mechanical oscillator is dependent on two parameters. The first term on the right hand side denotes the sinusoidal force exerted from the forcing solenoid. The second term denotes the different types of restoring force as represented in the three comparative studies, a mechanical spring, a permanent magnet and a stationary solenoid. It is shown in the second term that it is a function of displacement, which is the main parameter of interest in this study. The forcing term is the controlled parameter and an independent variable throughout the study.

4. Results and discussion

Displacement of the mechanical oscillator in a dual-solenoid system is measured and compared its performance against those of [8] and [9]. Three comparative studies are presented in this paper, the first being comparison against a mechanical spring system, second, against a permanent magnet system, and finally against a higher-powered dual solenoid system. As the study is emphasizing on the innovative dual-solenoid system, the foci of discussion will revolve with the permanent magnet system and the higher-powered dual-solenoid system.

4.1. Comparison against mechanical spring system

Typically, the mechanical spring system generates a displacement of approximate 3-4 mm as reported by Ho et al in [9]. Figure 2(a) shows the displacement generated as presented by Ho et al. in [9] at angular frequency of 18 rad/s and Figure 2(b) shows the displacement generated in a repeated experiment with Ho et al.'s system under influence of different frequencies.

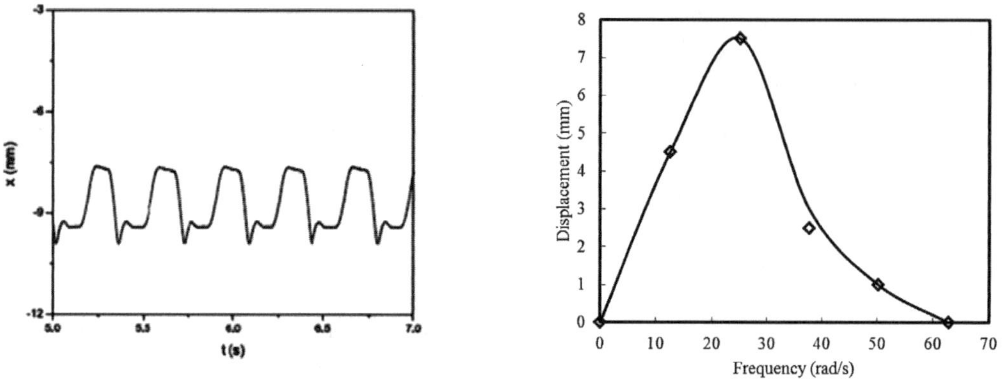

Figure 2(a). Displacement generated in a mechanical spring system at 18 rad/s by Ho et al. [9] **(b).** Displacement of mechanical oscillator under different frequencies for a mechanical spring system.

The power supplied for the mechanical spring system is rated at 15 Watts. Taking the mechanical spring system as benchmark of one, the displacement generated is compared in the dual-solenoid system. The new dual-solenoid system, on the other hand, drew only a total of 2 Watts for the forcing solenoid and 0.01 Watts for the stationary solenoid. The displacement induced for the dual-solenoid system has shown to be at 2 mm top, giving a normalized displacement of 0.38. It is, however, noted that the working range for the dual-solenoid system has shown to be rather consistent over a span of frequencies as shown in Figure 3. In addition, the power consumption is approximated a magnitude lower as compared to the mechanical spring system.

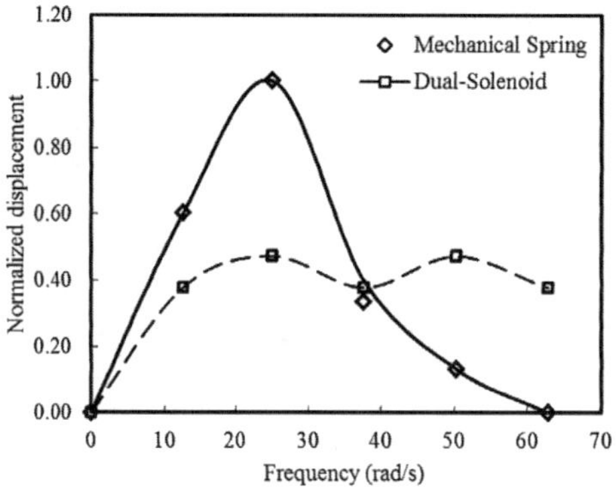

Figure 3. Comparison of mechanical spring and dual-solenoid system.

4.2. Comparison against permanent magnet system
For a permanent magnet system, the power supplied is similar to a mechanical spring system, at 15 Watts. The displacement induced by the oscillator is however smaller with only 1 mm top as presented in [8], giving a normalized displacement of 0.2 against the mechanical spring. Figure 4 shows the displacement plot of the oscillator in a permanent magnet system generated at 213.5 rad/s. It is also discovered that for a permanent magnet system, its operating window is rather narrow with angular frequency ranging from 207 to 220 rad/s, which is also shown to be one magnitude higher than the mechanical spring system.

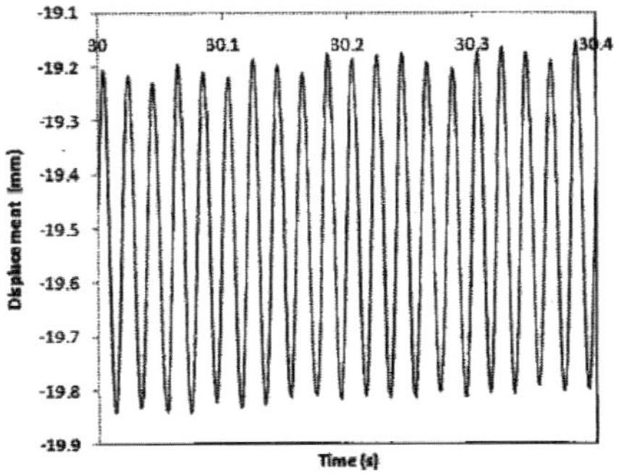

Figure 4. Displacement generated in a permanent magnet system at 213.5 rad/s by Lee et al. [8].

Comparing its performance with a dual-solenoid system, the displacement generated from a dual-solenoid system is approximately 100% higher than the permanent magnet system. In addition, despite lower power consumption by 86%, the angular frequency for a dual-solenoid system has shown to be one magnitude lower with advantage in a wider working range. With reference to Figure 3 for the plot of dual-solenoid system, it is also shown that the displacement of the mechanical oscillator is rather independent of the driving frequency.

4.3. Comparison against higher-powered dual-solenoid system

For the study of different power consumption for a dual-solenoid system, the stationary solenoid is maintained at a power rating of 0.01 Watts, for the purpose of consistency on the system environment. The variable change is the forcing solenoid, given as 2 Watts and 4 Watts. All other operating conditions remain the same throughout the experiment.

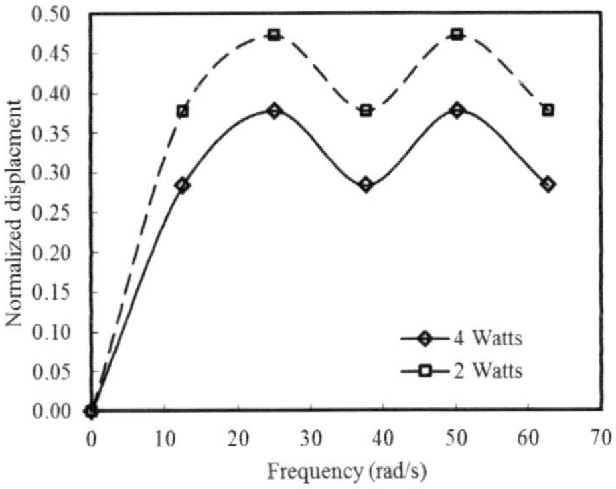

Figure 5. Comparison of a 2 Watts and 4 Watts powered dual-solenoid system.

Figure 5 shows the comparison of a higher-powered dual-solenoid system with rating of 4-Watt against a 2-Watt system. While the trend of the displacement is shown to be similar for both systems, the displacement of the 4-Watt system is shown to be relatively lower than the 2-Watt system by 10%. From Figure3-5, it may be deduced that the dynamics of the mechanical oscillator in dual-solenoid system is independent of the angular frequency. Increasing its power input to the system, while conserving the similar trend, has shown to reduce its performance by 10%. With respect to (3), it is shown that the forcing solenoid term is in negative notation hence increasing that term will eventually deteriorate the motion of the mechanical oscillator. It can then be deduced that in order to increase the performance of a dual-solenoid system, the key parameter to be altered would be the magnetic force of the stationary solenoid.A comparative performance analysis of all three systems is tabulated in Table 1. All comparative studies are made with respect to the mechanical spring system. Based on the analysis, the 2-Watt dual-solenoid has shown to have generated higher oscillation over the permanent magnet system and the 4-Watt system with significantly lower power consumption. Comparing against the mechanical spring system, although the oscillation has reported to be lesser by 62%, the power consumption, on the other hand, has shown to be more efficient by 86%.

Table 1. Performance analysis on mechanical oscillator of different systems.

Parameters	Mechanical spring (benchmark)	Permanent magnet	2-Watt dual-solenoid	4-Watt dual-solenoid
Displacement (mm)	5.3	1.0 (81% lower)	2.0 (62% lower)	1.8 (66% lower)
Power Consumption (Watts)	15	15	2 (86% lower)	4 (73% lower)

5. Conclusion

An innovative dual-solenoid electro-mechanical-vibro-impact system has been constructed and experimentally studied on the performance of its mechanical oscillator. It is observed that a dual-solenoid system generates less displacement as compared to the mechanical spring system by 62%. The dual-solenoid system, however, tops the permanent magnet system by 100% higher. The 4-Watt dual-solenoid system, on the hand, induced a lower displacement by 10% against the 2-Watt system, but topped against the permanent magnet system by 80%. The power consumption of the dual-solenoid system has proven to be more energy efficient by 86% and 73% for the 2-Watt and 4-Watt system, respectively. It is also shown that the displacement of the mechanical oscillator is independent of the angular frequency and that the stationary solenoid is the dominant parameter for the system

References

[1] Tsaplin S 1953 Vibratory impact mechanism for road and bridge construction *Autotranzidat: Moscow*(translated by the National Engineering Laboratory)
[2] Rodger A A and Littlejohn G S 1980A study of vibratory driving in granular soils *Geotechnique***30** 3 pp 269-293
[3] Preobrazhenskayaa N A 1955 Experimental investigations of vibration sinking of piles and channels *No. 27. NII osnovaniy, Dinamika Gruntov, Gosstroizdat*
[4] Woo K C, Rodger A A, Neilson R D and Wiercigroch M 2000 Application of the harmonic balance method to ground moling machines operating in periodic regimes *Chaos, Solitons & Fractals***11**15 pp 2515-2525
[5] Nguyen V D and Woo K C 2008 New electro-vibro-impact system *Proc. Inst.Mech. Eng. Part C J. Mech. Eng. Sci.***222** C4 pp 629-642
[6] Nguyen V D, Woo K C and Pavlovskaiia E 2008 Experimental study and mathematical modelling of a new vibro-impact moling device *Int. J. Non-linear Mech.***43** 6 pp 542-550
[7] Ho J H, Nguyen V D and Woo K C 2011 Nonlinear dynamics and chaos of new electro-vibro-impact system *Nonlinear Dyn.***63** pp 35-49
[8] Lee V C C, Woo K C and Abakr Y A 2011 Optimised progression rates in soil by means of vibro-impact motion, Poster Presentation *IUTAM Symposium on 50 Years of Chaos: Applied and Theoretical* pp 128-129
[9] Ho J H, Woo K C, Lee V C C and Abakr Y A 2013 Mechanical oscillator in a magnetic field *Proc. IUTAM Symposium "Nonlinear dynamics for advacned technologies and engineering design***32** pp 347-356

CUTSE2014 IOP Publishing
IOP Conf. Series: Materials Science and Engineering **78** (2015) 012020 doi:10.1088/1757-899X/78/1/012020

Improving of the photovoltaic / thermal system performance using water cooling technique

Hashim A. Hussien , Ali H. Numan, and Abdulmunem R. Abdulmunem
Eletromechanical Engineering Department, University of Technology, Baghdad-Iraq

E-mail: doctorhashim2004@yahoo.com , and ahnuman@uotechnology.edu.iq

Abstract. This work is devoted to improving the electrical efficiency by reducing the rate of thermal energy of a photovoltaic/thermal system (PV/T).This is achieved by design cooling technique which consists of a heat exchanger and water circulating pipes placed at PV module rear surface to solve the problem of the high heat stored inside the PV cells during the operation. An experimental rig is designed to investigate and evaluate PV module performance with the proposed cooling technique. This cooling technique is the first work in Iraq to dissipate the heat from PV module. The experimental results indicated that due to the heat loss by convection between water and the PV panel's upper surface, an increase of output power is achieved. It was found that without active cooling, the temperature of the PV module was high and solar cells could only achieve a conversion efficiency of about 8%. However, when the PV module was operated under active water cooling condition, the temperature was dropped from 76.8oC without cooling to 70.1oC with active cooling. This temperature dropping led to increase in the electrical efficiency of solar panel to 9.8% at optimum mass flow rate (0.2L/s) and thermal efficiency to (12.3%).

1. Introduction

Developing alternative energy resources with high efficiency and low emission has become of great importance with increasing concerns about fossil fuel deficit, high oil prices, global warming, and damage to environment and ecosystem [1]. In this concern, photovoltaic solar energy is a clean, renewable, energy with a long service life and high reliability [2]. A photovoltaic system converts sunlight into electricity, where, the basic device of a photovoltaic system is the photovoltaic cell. Cells may be grouped to form panels or modules [3].

Performance of a solar-photovoltaic (PV) system not only depends on its basic electrical characteristics; maximum power, tolerance rated value %, maximum power voltage, maximum power current, open-circuit voltage (Voc), short-circuit current (Isc), maximum system voltage, but also is negatively influenced by several obstacles such as ambient temperature, relative humidity, dust storms and suspension in air, shading, global solar radiation intensity, spectrum and angle of irradiance [3-4].

There are several reasons which motivate the development of the PV/T system. One of the main reasons is that PV/T system can provide higher efficiency than individual PV and thermal collector system. With increased the efficiency, the payback period of the system can also be shorten [5].

The electrical efficiency reduction of PV modules due to their temperature increase can be partially avoided by water or air heat extraction. PV heating is mainly the result of the absorbed solar radiation that is not converted into electricity and PV cooling is considered necessary to keep electrical efficiency at a satisfactory level. Natural or forced air circulation is a simple and low cost method to remove heat from PV modules, but it is less effective if ambient air temperature is over 20oC, as it is usual for many months in low latitude countries.The operating temperature plays an essential role in the PV energy conversion process [6-7].

The performance of PV panels reduces with increasing temperatures. Therefore, most panels' will not operate under ideal conditions due to different weather conditions or real ones. Since PV panels are more efficient at lower temperatures, PV systems have to design with active and passive cooling [8-9]. The water heat extraction is more expensive than air heat extraction, but it is considered practical for the above case, as the water temperature from mains is under 20oC almost all year. The usual mode of PV cooling by water is the circulation of it through a heat exchanger in thermal contact with the PV module rear surface, to avoid pressure and electrical problems.

The theoretical cell electrical efficiency (ηe) and this parameter are functioned of the cell temperature [10].

$$\eta_e = \eta_o[1 - \beta(T_c - T_o)] \tag{1}$$

$$\eta_e = \frac{\int VI\,dt}{A_c \int G(t)dt} \tag{2}$$

The electrical efficiency of the PV module can be described as following equation:

$$\eta_o = \frac{V_{mp}\,I_{mp}}{GA} \tag{3}$$

The thermal efficiency can be computed with the following equation [11]:

$$\eta_{th} = \frac{m \cdot C_p \int (T_{out} - T_{in})\,dt}{A_c \int G(t)dt} \tag{4}$$

The total efficiency of the hybrid PV/T system is:

$$\eta_{total} = \eta_{th} + \eta_e = \frac{m \cdot C_p \int (T_{out} - T_{in})dt + \int VI\,dt}{A_c \int G(t)dt} \tag{5}$$

The electrical and thermal efficiencies are presented in Eqs. (2) and (4). It can be seen that the solar irradiation is a function of time and those parameters which are affected by the solar irradiation, such as inlet and outlet temperatures, PV voltage and PV current, are also functions of time. That is the reason to integrate the equation with time.

2. Experimental methodology

In this experimental work a PV-array were cooled by made a cooling technique which is consist of heat exchanger and water circulating pipes are placed at PV module rear surface to reduce heat stored inside the PV cells during operation.

The aim of the suggested technique is to increase the electric current flow. Moreover, water circulating pipes should further decrease the panel surface temperature and increase power output, therefore, gaining better electrical efficiency.

2.1 System configuration

Figs. (1) and (2) presents the photograph and schematic diagram of the PV/T cooling system used in this paper. This system consists of seven pipes connected in parallel and placed in PV module rear surface. Solar irradiation was measured by the pyranometer, which was put at the same level as the solar panels. In this experiment, the mass flow rate was measured and the temperature of working fluid and PV module was obtained by using the k-type thermocouple directly connected to the data logger. The voltage and current of the solar panels were directly recorded by the data logger. The capacities of cooling system is 4 liters and the material of piping system is cupper with inner diameter 21mm and thickness 1mm and 7 parallel rows because cupper have high thermal conductivities to transfer the temperature from PV panels to working fluid.

Fig.1 Photograph of the PV/T cooling system.

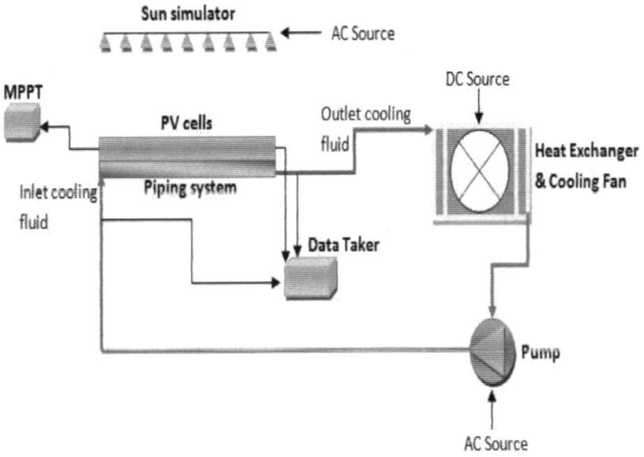

Fig. 2 Schematic diagram of the PV/T cooling system.

2.2 Measurements

The measurements recorded during this work. The flow rate of water is measured by an ULTRASONIC FLOMETER-TDS-100H flow meter. AC to AC transformer is used to control the speed of motor to change the mass flow rate of working fluid. A 90 bulbs of (12V 50W) connected in series are used to simulate the sun light and controled by 3 AC to AC transformers to give different irradiation arrive up to 1000 W/m2 . The temperatures of inlet and water outlet, were measured to measure the temperature difference (ΔTw=heating rate). The maximum power point tracer is used to seek this point in order to maximize the power output of the panels under the different irradiation. PV temperature was measured using a k-type thermocouple. k-type thermocouples are suitable for measurements in the range at -180 to +1300 °C. K-type thermocouples comprise 2 wires, copper and constantan (copper-nickel alloy).

3. Results and discussion

The proposed water cooling technique contributes to considerably decrease the PV panel temperature and then leads to increase the output power from the PV system.

The thermal performance of the PV system using water as working fluid with difference mass flow rate (0.1, 0.2, 0.3) L/s is shown in Figs. 3&4. It should be noted that under lowest value of mass flow rate (0.1L/s) a better thermal gain is obtained compared with increase of water mass flow rate due to at lowest mass flow rate the working fluid will take more time in absorbing heat from PV panel than that at highest mass flow rate.

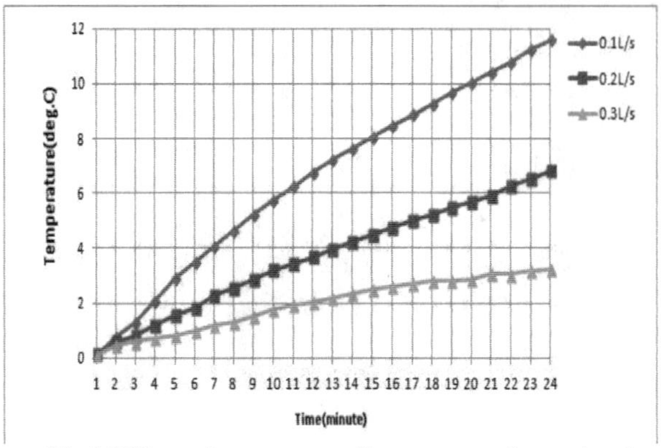

Fig.3 Effect of water mass flow rate on thermal gain.

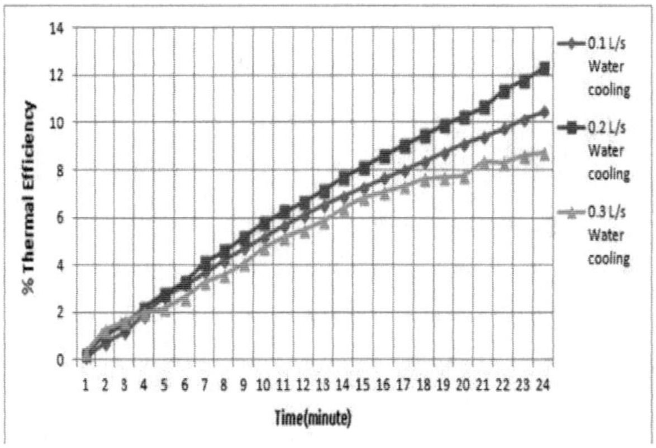

Fig.4 Effect of water mass flow rate on thermal efficiency.

Fig.5 show that the mass flow rate of water working fluid at (0.3L/s) give a better cooling for PV panel than others mass flow rates used in this test. This cooling process for all mass flow rate give an enhancement on PV power generated when increased from (0.1L/s) to (0.3L/s) because more heat dissipate in cooling system (radiator) with increasing mass flow rate of circulated of working fluid.

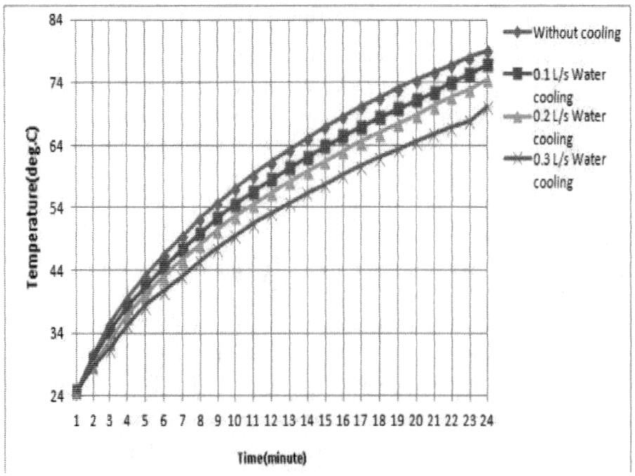

Fig.5 Effect of water mass flow rate on PV panel temperature.

Fig.6 demonstrate that the mass flow rate of water working fluid at (0.3L/s) led to produce more maximum power and improve the PV panel electrical efficiency as shown in Fig.7. The performance enhancements using cooling process in Impp & Vmpp is shown in Figs. 8&9. Compared with the process without cooling treatment because decreasing in PV panel temperature will increase in power generated.

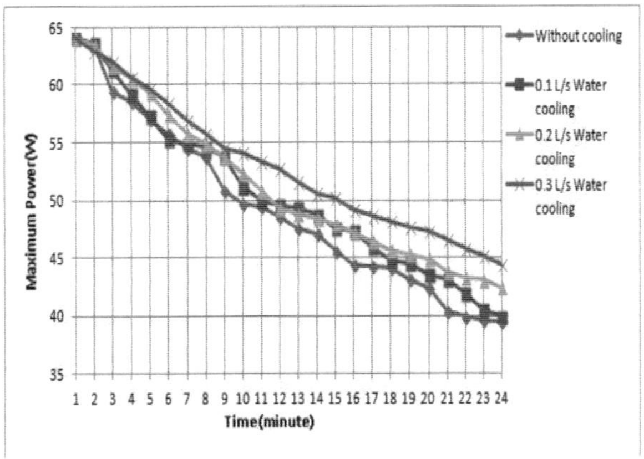

Fig.6 Effect of water mass flow rate on PV panel maximum power generated.

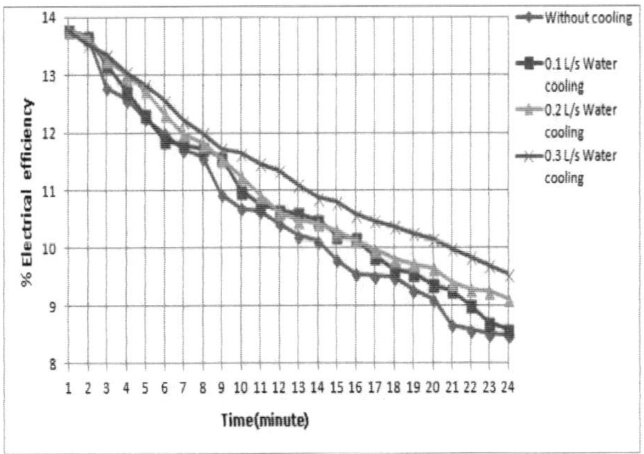

Fig.7 Effects of water mass flow rate on electrical efficiency of the PV panel.

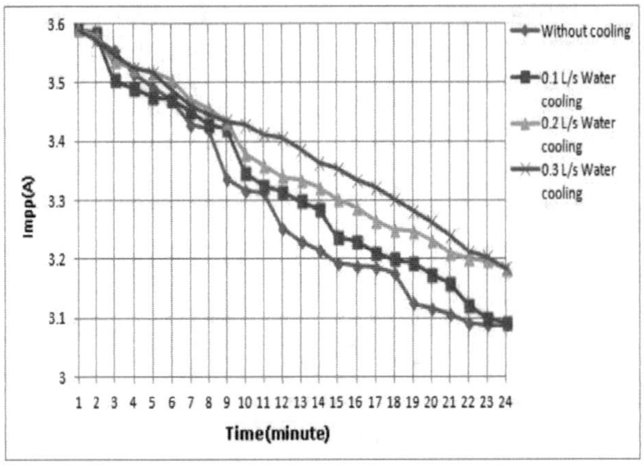

Fig.8 Effect of water mass flow rate on Impp of PV panel.

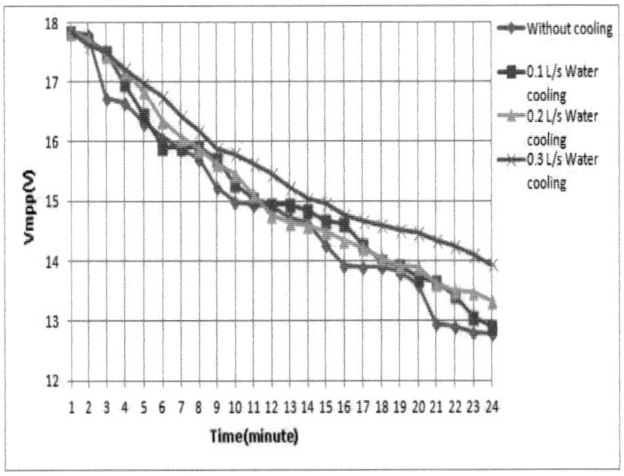

Fig.9 Effect of water mass flow rate on Vmpp of PV panel.

The MPPT have shown enhancement in electrical performance of PV panel under the cooling process with different water mass flow rate as shown in Figs. 10, 11 and 12 respectively at the end of each test.

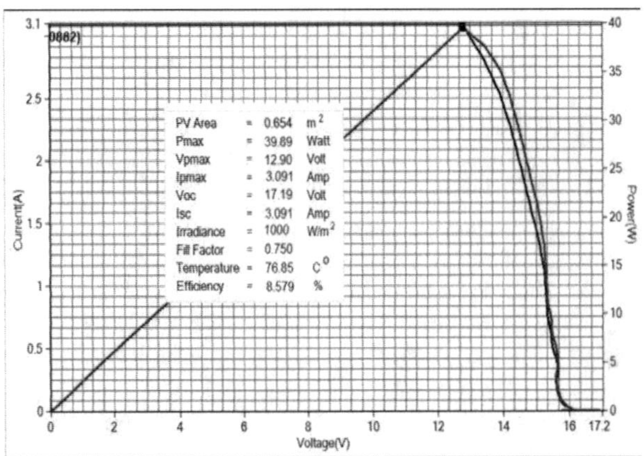

Fig.10 MPPT trace at end of testing the effect of water cooling at (0.1L/s) on PV panel performance.

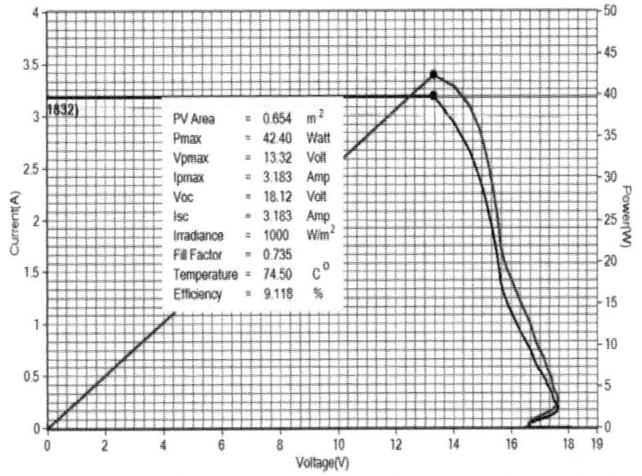

Fig.11 MPPT trace at end of testing the effect of water cooling at(0.2L/s) on PV panel performance.

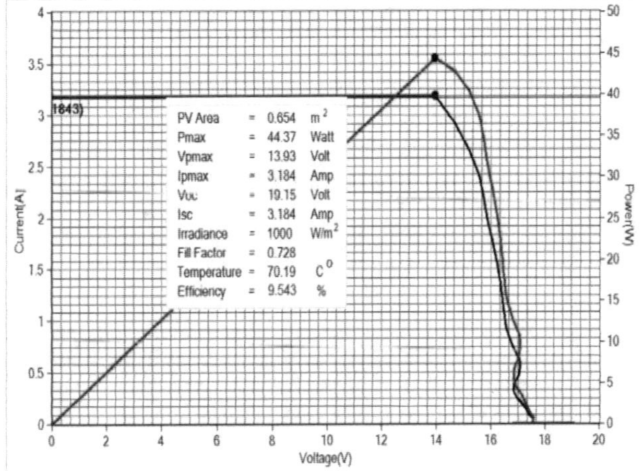

Fig.12 MPPT trace at end of testing the effect of water cooling at(0.3L/s) on PV panel performance.

Figs.13, 14, and 15 are theoretical relationships of water physical properties with time at different mass flow rate (0.1, 0.2, 0.3) L/s obtained by using MATLAB simulation program, we saw that no effect of mass flow rate on water physical properties as seen on start of these tests at the same temperature but rising of temperature with time led to changing of these properties.

Figs.13 and 14 have shown that the density and specific heat, these properties are decreasing with increasing of temperature and the important role of mass flow rate in the change of absorbed temperature with changing of the mass flow rate, with increasing of water mass flow rate the range of absorbed temperature is dropped and that led to increase in both water density and specific heat and dynamic viscosity respectively as shown in Fig.14 and Fig.15 the rising of temperature led to increasing on thermal conductivities and thermal diffusivity respectively and the dropping on absorbed heat with increasing of mass flow rate led to decreasing of these ranges.

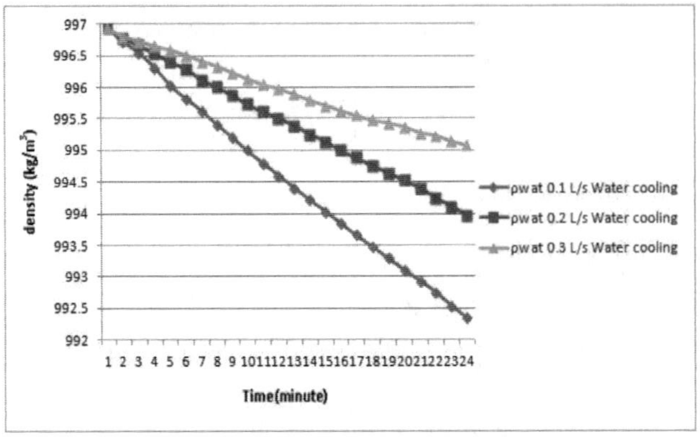

Fig.13 Effect of water mass flow rate on water density.

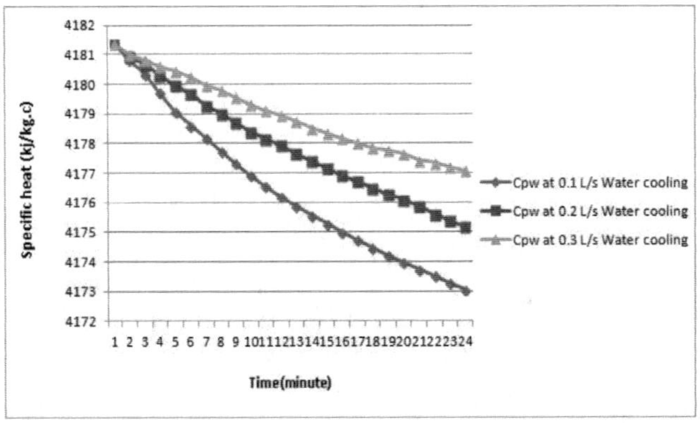

Fig.14 Effect of water mass flow rate on water specific heat.

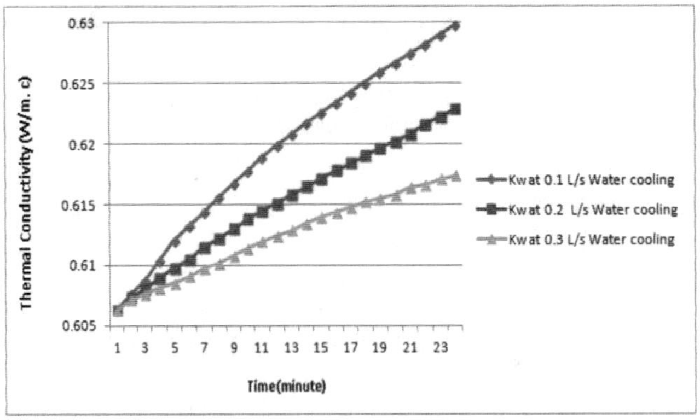

Fig.15 Effect of water mass flow rate on water thermal conductivity.

4. Conclusions

The increase in efficiency and the power output depends largely on reducing the temperature limit, which allows the greatest benefit from the whole system. Using cooling technique of water on the rear surface of PV/T system as a coolant cools it down enough to cause it to generate more power than one without water. Experimental results showed that the PV cells power is increased due to cooling water for the photovoltaic cells. This can significantly increase the system efficiency. It was found that without active cooling, the temperature of the PV module was high and solar cells could only achieve a conversion efficiency of about 8%. However, when the PV module was operated under active water cooling condition, the temperature was dropped to (76.8°C at 0.1L/s), (74.5oC at 0.2L/s) and (70.1°C at 0.3L/s). This temperature dropping led to increase in the efficiency of solar panel to (8.6%, 9.1% and 9.6%) respectively, depending on water mass flow rate, and the thermal efficiency to (12.3%).

5. References

[1] Bari, S., Optimum slope angle and orientation of solar collectors for different periods of possible utilization, Energy Conversion. Management. Vol.41, (2000),pp.855-860.

[2] Yakup, M.A., Malik, A.Q., Optimum tilt angle and orientation for solar collector in Brunei Darussalam, Renew. Energ , Vol.24, (2001),pp.223-234.

[3] A.S.Joshi, A.Tiwari, G.N Tiwari, I. Dincer B.V.Reddy, Performance evaluation of a hybrid photovoltaic thermal (PV/T) (glass-to-glass) system, International Journal of Thermal Science, Vol.48, (2009), pp.154-164.

[4] Swapnil Dubey, G.S.Sandhu, G.N.Tiwari, Analytical expression for electrical efficiency of PV/T hybrid air collector, Applied Energy ,Vol. 86, (2009),pp.697-705.

[5] T.T. Chow, A review on photovoltaic/thermal hybrid solar technology, Applied Energy, Vol.87, (2010),pp.365-379.

[6] B.J.Huang, T.H.Lin, W.C.Hung and F.S.Sun. , Performance Evaluation if Solar Photovoltaic /Thermal Systems, Solar Energy, Vol.70, (2001),pp.443-448.

[7] Jie Ji, Jian-Ping Lu, Tin-Tai Chow, Wei He, Gang Pei., A sensitivity study of a hybrid photovoltaic/thermal water-heating system with natural circulation, Applied Energy, Vol.84, (2007),pp.222-237.

[8] C. H. Cox, III and P. Raghuraman, Design Consideration for Flat Plat Photovoltaic/Thermal Collectors, Solar Energy, Vol.35, (1985),pp.227-241.

[9] Fedele L, Colla L, Bobbo S, et al., Experimental stability analysis of different water-based nanofluids, Journal Article,Nanoscale Res Lett; Vol.6, (2011),pp.289-300

[10] Rott N. Note on the history of the Reynolds number, Annual Review of Fluid Mechanics, Vol.22, (1990), pp.1–11.

[11] Gnielinski V., New equations for heat and mass transfer in turbulent pipe and channel flow, International Chemical Engineering , Vol.16, (1996),pp.359–368.

IOP Conf. Series: Materials Science and Engineering **78** (2015) 012021 doi:10.1088/1757-899X/78/1/012021

On radial flow between parallel disks

AYL Wee[1] and A Gorin

Faculty of Engineering, Computing and Science
Swinburne University of Technology Sarawak Campus
Kuching, Malaysia

E-mail: awylin@swinburne.edu.my

Abstract. Approximate analytical solutions are presented for converging flow in between two parallel non rotating disks. The static pressure distribution and radial component of the velocity are developed by averaging the inertial term across the gap in between parallel disks. The predicted results from the first approximation are favourable to experimental results as well as results presented by other authors. The second approximation shows that as the fluid approaches the center, the velocity at the mid channel slows down which is due to the struggle between the inertial term and the flowrate.

1. Introduction

Converging flow is defined by fluid flowing from the periphery towards the center. As the fluid flows radially towards the center, the velocity increases in order to satisfy continuity equation. Analytical, numerical and experimental studies have been carried out by others in order to characterize the phenomena [1]-[5]. Lee and Lin [1] presented a simplified solution by linearizing the Navier-Stokes equation by replacing radial velocity with mean radial velocity which resulted in dimensionless pressure gradient equation. According to Savage [2], Livesey J.L. considered the inertial term and used integral approach and the assumption of parabolic velocity profile in order to solve the problem. Savage [2] obtained a series solution by perturbing the creeping-flow solution. Vatistas [3] also included the inertial term and linearized the inertial term in the momentum equation by taking radial velocity from the continuity equation averaged over the gap. Then, the linearized differential equation was solved using method of separation variables. The author derived the static pressure distribution, radial component of the velocity as well as the friction coefficient [3]. In paper [4], the authors presented a numerical solution which resulted in the characterization of the static pressure distribution and the radial velocity. Zitouni and Vatistas [5] solved the problem by using the power series solution. This paper focuses on an approximate analytical solution to a nonlinear problem for radial flow in between parallel disks. Equations for static pressure distribution and velocity profiles are derived by averaging the inertial term over the space between the disks.

[1] To whom any correspondence should be addressed

2. Analysis

2.1. First Approximation

The flow is assumed to be laminar, incompressible, non-swirling and purely radial. The continuity equation and Navier-Stokes momentum equation in cylindrical coordinates are transformed into nondimensional form by using the following parameters:

$v_r = \frac{v_r^*}{v_0^*}$ $\quad v_z = \frac{v_z^*}{v_0^*}$ $\quad r = \frac{r^*}{R_0^*}$

$z = \frac{z^*}{h^*}$ $\quad Re = \frac{\rho v_0^* h^*}{\mu}$ $\quad \xi = \frac{h^*}{R_0^*}$

$P = \frac{P^*}{\rho^* v_0^{*2}}$; $P^* = P_r - P_{inlet}$

where ρ = density $\quad \mu$ = dynamic viscosity $\quad P_r$ = current pressure

Note: (*) symbolizes dimensional parameters for the velocity component and pressure distribution in r and z direction

The equations are simplified and can be written as:
Continuity equation

$$\frac{1}{r}\frac{\partial v_r r}{\partial r} + \frac{\partial v_z}{\partial z} = 0 \qquad (1)$$

r momentum equation

$$-v_r \frac{dv_r}{dr}\xi - v_z \frac{dv_r}{dz} = +\frac{dP}{dr}\xi - \frac{1}{Re}\frac{d^2 v_r}{dz^2} \qquad (2)$$

where
v_r = radial velocity $\quad v_z$ = axial velocity $\quad P$ = dimensionless pressure

Re = Reynolds number $\quad r$ = dimensionless radial coordinate

z, ξ = dimensionless axial coordinate

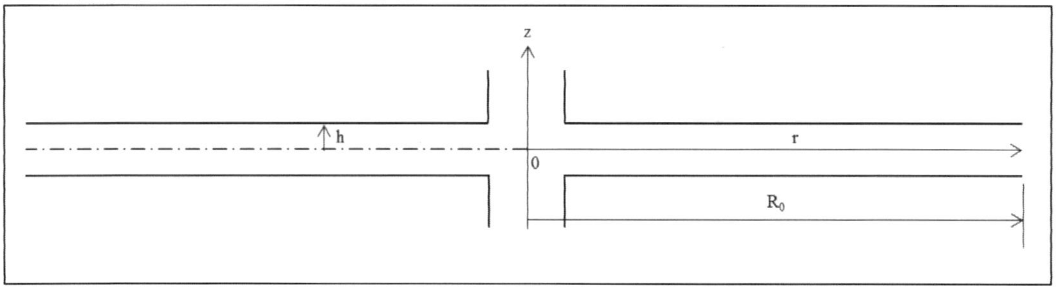

Figure 1. Cross section of flow between parallel disks.

For the first approximation, the nonlinear convective acceleration term is linearized by averaging across the gap between the parallel disks,

$$B(r) = \int_0^1 \left(-v_r \frac{dv_r}{dr}\xi - v_z \frac{dv_r}{dz}\right) dz \qquad (3)$$

Then (2) can be written as:

$$\frac{1}{Re}\frac{d^2 v_r}{dz^2} = \frac{dP}{dr}\xi - B(r) = A(r) \qquad (4)$$

The velocity function can be calculated by integrating (4) twice with respect to the distance between the parallel disks as shown:

$$\frac{1}{Re} v_r = \frac{A(r)z^2}{2} + c_1 z + c_2 \tag{5}$$

where c_1 and c_2 are constants of integration. The evaluation of the constants of integration is made using the following boundary conditions:

$$BC1 : z = 0, \frac{dv_r}{dz} = 0$$

$$BC2 : z = 1, v_r = 0$$

From continuity equation,

$$rv_r = D = \text{constant} \tag{6}$$

Integration of (6) with respect to z (from 0 to 1) yields,

$$\therefore A(r) = \frac{3D}{rRe} \tag{7}$$

Thus,

$$v_{r(1)} = \frac{3D}{2r}(z^2 - 1) \tag{8}$$

Based on (8), it is a parabolic velocity profile. Thus, (3) is reduced to:

$$B(r) = \int_0^1 -v_r \frac{dv_r}{dr} \xi \, dz$$

$$\therefore B(r) = \left(-\frac{8}{15}\right) \frac{\xi Re^2 A(r) A'(r)}{4} \tag{9}$$

Then from (7) and (9) yield

$$\frac{dP}{dr} = \frac{3D}{rRe\xi} + \frac{6D^2}{5r^3}$$

$$\Delta P = P_r - P_i \, ; \; P_i = \text{inlet static pressure}$$

Thus, the pressure distribution is given by:

$$P_r - P_{i=1} = \ln r \frac{3D}{Re\xi} - \frac{3D^2}{5r^2} + \frac{3D^2}{5(1)^2} \tag{10}$$

where $Re\xi = \text{Reduced Reynolds number}, \overline{Re}$

2.2. Second Approximation

Second approximation is carried out in order to check the accuracy of the first approximation.
r momentum equation can be written as:

$$-v_r \frac{dv_r}{dr} - \frac{v_z}{\xi} \frac{dv_r}{dz} = + \frac{dP}{dr} - \frac{1}{\overline{Re}} \frac{d^2 v_r}{dz^2} \tag{11}$$

Substituting the inertial term with results obtained from the first approximation and integrating (11) twice together with boundary condition BC1 and BC2 results in:

$$\therefore v_{r(2)} = \overline{Re}\left(\frac{dP}{dr} \times \frac{z^2}{2} - \frac{9z^6}{120r^3} + \frac{18z^4}{48r^3} - \frac{9z^2}{8r^3} + \frac{33}{40r^3} - \frac{1}{2} \times \frac{dP}{dr}\right) \tag{12}$$

From Continuity Equation,

$$r\int_0^1 -v_r \, dz = \int_0^1 D \, dz$$

$$\frac{dP}{dr} = \frac{3D}{r\overline{Re}} + \frac{54}{35r^3} \tag{13}$$

Integrating with respect to r, the pressure distribution for the second approximation is given by:

$$P_r - P_{i=1} = \frac{-27}{35r^2} + \frac{3D}{\overline{Re}} \ln r + \frac{27}{35} \tag{14}$$

Substitute pressure gradient (13) into the velocity profile (12),

$$v_{r(2)} = \frac{-3D}{2r} + \frac{3\overline{Re}}{56r^3} + \frac{3Dz^2}{2r} - \frac{99\overline{Re}z^2}{280r^3} + \frac{3\overline{Re}z^4}{8r^3} - \frac{3\overline{Re}z^6}{40r^3} \tag{15}$$

where D=1 after normalizing in terms of flow rate as shown:

$$rv_r = D = 1 \tag{16}$$

3. Results and Discussion

The analytical solutions are validated by comparing the equation with equations developed by other authors and experimental data. The first approximation results in a parabolic velocity profile as shown in figure 2.

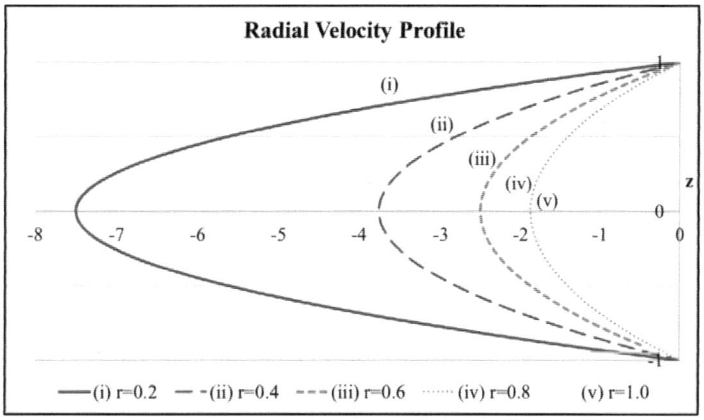

Figure 2. 1st Approximation radial velocity profile.

However, the second approximation shows a different flow kinematics as shown in figure 3. Based on the calculations, the profile flattens as the fluid flows towards the center. At radial distance (r) less than 0.3, it can be seen the maximum velocity profile is no longer at the mid channel. This phenomenon is caused by the struggle between inertial term and flowrate and is verified as the continuity equation holds.

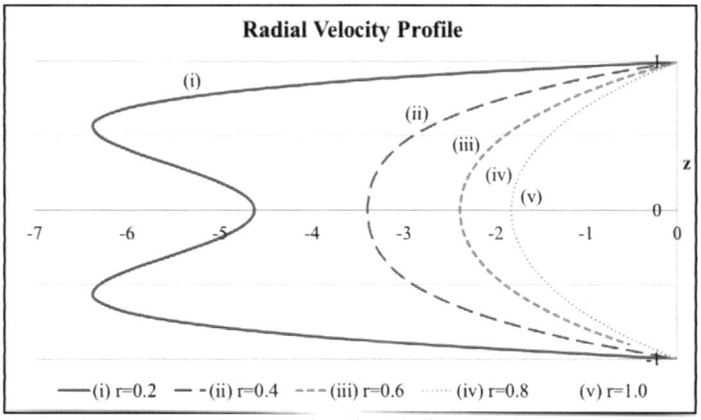

Figure 3. 2nd Approximation radial velocity profile.

The static pressure distribution obtained in this analysis is compared with results from Vatistas [3], Lee &Lin [1], Livesey [2] and Savage and Kwok & Lee which were presented in Vatistas, Ghila &

Zitouni [4] and are in favor with their solution as shown in figure 4 and figure 5. The first approximation has the exact same solution as obtained by Kwok and Lee [4] and Livesey [2]. The second approximation is very close to the results obtained by Savage [4] and shows a larger change in pressure as compared to the first approximation. This is due to the higher wall shear stress experienced by the second profile as compared to the parabolic profile obtained from the first approximation. Equation (17) and (18) shows shear stress for first approximation and second approximation respectively.

$$\tau = \frac{3z}{rRe} \tag{17}$$

$$\tau = \frac{1}{Re}\left(\frac{3z}{r} - \frac{99\overline{Re}z}{140r^3} + \frac{3\overline{Re}z^3}{2r^3} - \frac{9\overline{Re}z^5}{20r^3}\right) \tag{18}$$

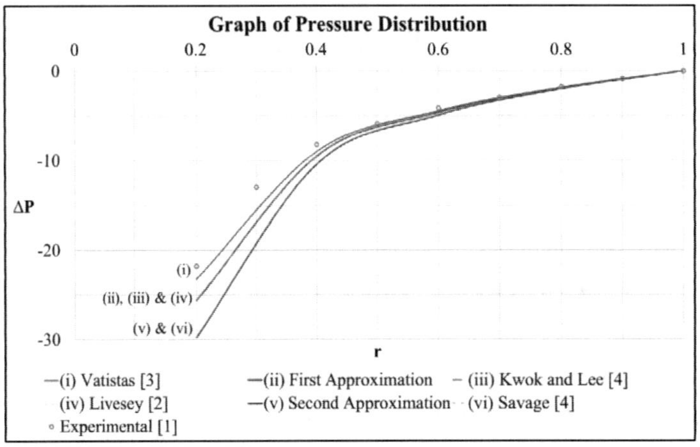

Figure 4. Pressure distribution of water along the radius ($\overline{Re} = 0.43$).

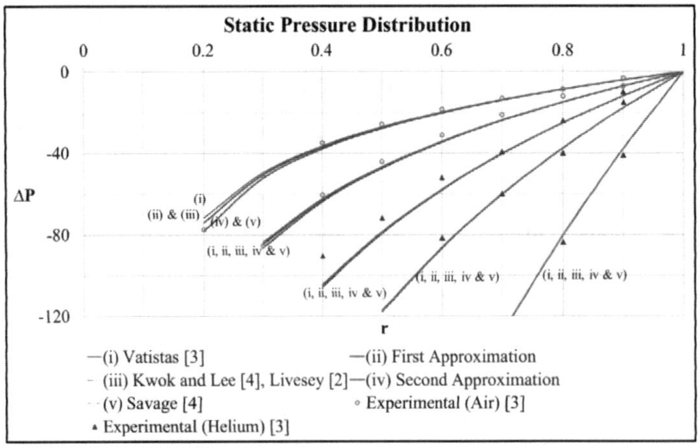

Figure 5. Static pressure distribution using different reduced Reynold's number.

Table 1 shows pressure distribution comparison presented by Vatistas, Ghila & Zitouni [4] and complimented by our results. However, there are differences between Savage and Livesey's results as cited by Savage [2] and not exactly the same as presented by Vatistas, Ghila & Zitouni [4].

Table 1. Pressure distributions.

	ΔP
Vatistas [3]	$\frac{6\bar{r}^2 \ln r - \overline{Re}(1-r^2)}{2r^2\overline{Re}}$
Kwok and Lee [4] Livesey [2]	$0.6 - \frac{0.6}{r^2} - \frac{3\ln r}{\overline{Re}(r^2-1)} + \frac{3r^2 \ln r}{\overline{Re}(r^2-1)}$
Savage [4]	$0.77 - \frac{0.77}{r^2} - \frac{3\ln r}{\overline{Re}(r^2-1)} + \frac{3r^2 \ln r}{\overline{Re}(r^2-1)}$
1st Approximation	$\frac{15r^2 \ln r - 3\overline{Re}(1-r^2)}{5\overline{Re}r^2}$
2nd Approximation	$\frac{-27}{35r^2} + \frac{3}{\overline{Re}}\ln r + \frac{27}{35}$

4. Conclusion

The approximate analytical solution for Navier-Stokes on radial flow between parallel disks is developed. The pressure and radial velocity distribution are obtained by averaging the inertial term across the gap between the parallel disks. This approach allows one to develop the second approximation for the velocity profile and pressure gradient. The static pressure distributions are in agreement with experimental results and results obtained by other authors. Thus, the analytical method is verified. This study also shows that in contrary to the other authors, the velocity profile is not constantly in a parabolic shape as the fluid flow towards the center and it is verified as the continuity equation holds.

References

[1] Lee PM and Lin S 1985, 'Pressure distribution for radial inflow between narrowly spaced disks', *Journal of Fluids Engineering*, **107**, pp. 338-341.

[2] Savage SB 1964, 'Laminar radial flow between parallel plates', *Journal of Applied Mechanics*, pp.594-596.

[3] Vatistas GH 1988, 'Radial flow between two closely placed flat disks', *AIAA Journal*, **26**, pp. 887-889.

[4] Vatistas GH, Ghila, A and Zitouni, G 1995, 'Radial flow between two flat disks', *Acta Mechanica* **113**, pp. 109-118.

[5] Zitouni G and Vatistas GH 1997, 'Purely accelerating and decelerating flows within two flat disks', *Acta Mechanica* **123**, pp. 151-161.

Temperature and voltage responses of a molten carbonate fuel cell in the presence of a hydrogen fuel leakage

M C Law[1], G V Y Liang, V C C Lee, S K Wee[2]

[1]Department of Mechanical Engineering, Faculty of Engineering and Science, Curtin University, Miri, Sarawak, Malaysia
[2]Department of Chemical Engineering, Faculty of Engineering, Universiti Malaysia Sabah, Kota Kinabalu, Sabah, Malaysia

E-mail: [1]m.c.law@curtin.edu.my, [2]weesiawkhur@ums.edu.my

Abstract. A two dimensional (2-D), dynamic model of a molten carbonate fuel cell (MCFC) was developed using COMSOL Multi-physics. The model was used to investigate the dynamic behaviour of the MCFC in the presence of hydrogen fuel leakage. A leakage was modelled as a known outflow velocity at the anode gas channel. The effects of leakage velocity and the leakage location were investigated. The simulations show that anode electrode temperature increases as the leakage velocity increases. The voltage generated is shown to decrease at the start of the leakage occurrence due to loss of hydrogen gas. Later the voltage increases as the anode temperature increases. The results also show that the changes of temperature and voltage are more significant if a leakage occurs nearer to the inlet compared to that at the outlet of anode gas channel.

1. Introduction

Fossil fuels have been the main energy source for centuries. However, due to the issues of environmental pollution and scarcity of the fossil fuels, the demand for greener energy sources is increasing. A molten carbonate fuel cell (MCFC) appears to be one of the strong energy alternatives in this respect.

A MCFC stack consists of a number of anode and cathode electrodes which are separated by an electrolyte as shown in the figure 1.

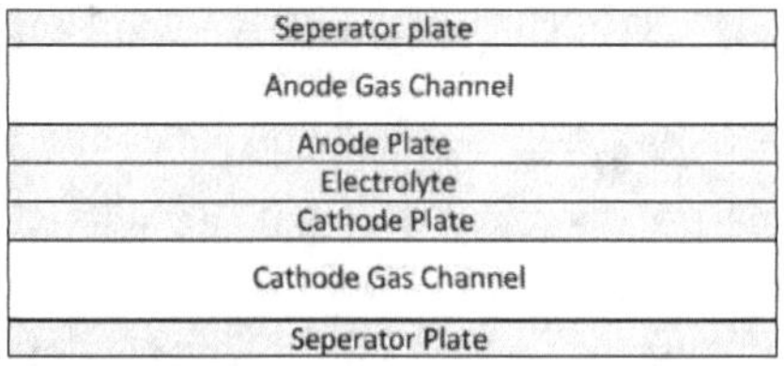

Figure 1. Schematic of a typical MCFC.

[1] To whom any correspondence should be addressed.

At the cathode electrode, the following reaction occurs,

$1/2O_2 + CO_2 + 2e^- \rightarrow CO_3^-$ (R1)

At the anode electrode, hydrogen gas is used as a fuel,

$H_2 + CO_3^- \rightarrow H_2O + CO_2 + 2e^-$ (R2)

The overall reaction can be written as,

$H_2 + 1/2O_2 \rightarrow H_2O$ (R3)

which produces water as a by-product.

As hydrogen is a flammable gas which could cause explosion, it is important to take precautionary measures to prevent the hydrogen leakage. If the leakage does occur, it is important to ensure that the leakage can be detected as early as possible.

From the literature conducted, there is a limited amount of work which has been done on the hydrogen leakage detection. From the study of Ashraf Khorasani et al. [1], the location of the leakage in the fuel cell could be determined based on the data received from the voltage-time graphs of the stack under hydrogen sudden stop condition. The authors suggested an approximate expression for the cells voltage loss due to a decrease in partial pressure of anode and cathode reactants. Tian et al.[2] presented data of a failed fuel stack due to the internal gas leakage. With the help of statistical analysis tools, it enables the detection of failed cells inside the fuel cell stack in two different cases, namely anode/cathode crossover and anode/cooling compartment leak. Voltage and the pressure readings of the cell were analysed to detect the cell with leakages. Husar, Serra and Kunusch [3] reported that cross-over in polymer membrane fuel cell resulted in unexpected changes in system variables such as temperature, pressure and voltage. Kim et al.[4] predicted the hydrogen dispersion in a residential fuel cell systems by using computational fluid dynamics (CFD) simulations. A three dimensional geometric model of the system was constructed and the flow pattern and hydrogen concentration in the residential fuel cell system was studied.

The objective of the current work is to investigate the performance changes of a MCFC in the presence of hydrogen leakages. The result is helpful in developing a model-based hydrogen leakage detection method in future.

2. Model development

2.1. Governing equations

The governing equations of the MCFC model are as follows:
For the gases in the channels, Navier- Stokes equation is used,

$$\frac{\partial \rho}{\partial t} + \nabla(\rho \mathbf{u}) = 0 \tag{1}$$

$$\rho \frac{\partial \mathbf{u}}{\partial t} + \rho \mathbf{u} \nabla \mathbf{u} = -\nabla p + \mu\left(\nabla \mathbf{u} + (\nabla \mathbf{u})^T\right) - \frac{2}{3}\left(\mu(\nabla \mathbf{u})\right) \tag{2}$$

where ρ, μ and $\mathbf{u}$ are the gas density, dynamic viscosity and velocity vector respectively.

For the gas phase momentum conservation of porous anode and cathode electrodes is described by Darcy equation,

$$\frac{\rho}{\varepsilon_p}\left(\frac{\partial \mathbf{u}}{\partial t} + \frac{(\mathbf{u}\cdot\nabla)\mathbf{u}}{\varepsilon_p}\right) = \nabla \cdot \left\{-p\mathbf{I} + \left[\frac{\mu}{\varepsilon_p}\left(\nabla \mathbf{u} + (\nabla \mathbf{u})^{\mathrm{T}}\right) - \frac{2\mu}{3\varepsilon_{\mathrm{p}}}(\nabla \cdot \mathbf{u})\mathbf{I}\right]\right\} - \mu\mathbf{u}/\kappa \tag{3}$$

where ε_p and κ are the porosity and permeability of the porous media.

For the transport of concentrated gas species ω_i, it can be described as,

$$\rho\frac{\partial\omega_i}{\partial t}+\nabla\cdot\mathbf{J_i}+\rho(\mathbf{u}\cdot\nabla)\omega_i=\dot{R}_i \tag{4}$$

$\mathbf{J}_i$ is the mass flux, i relative to the average mass velocity. For the water gas shift (WGS) reaction, the equation that takes place in the anode is as follows [5],

$$\dot{R}_{WGS}=k_{WGS,f}\left(p_{\mathrm{H_2O}}p_{\mathrm{CO}}-p_{H_2}p_{CO_2}/K_{p,WGS}\right) \tag{5}$$

$$k_{WGS,f}=0.0171\exp(-\frac{103191}{R_uT_g}) \tag{6}$$

$$K_{p,WGS}=\exp(-0.2935Z^3+0.635Z^2+4.1788Z+0.3169) \tag{7}$$

and

$$Z=\frac{1000}{T_g}-1 \tag{8}$$

T_g is the gas phase temperature and p_i is the partial pressure of a gas species, i.

For the cathode electrode, oxygen and carbon dioxide are consumed. Thus, the rates of consumption of the gas species are as follows,

$$\dot{R}_{\mathrm{O_2}}=-i_{ct}M_{\mathrm{O_2}}/(4F\delta_{cat}) \tag{9}$$

$$\dot{R}_{\mathrm{CO_2}}=-i_{ct}M_{\mathrm{CO_2}}/(2F\delta_{cat}) \tag{10}$$

where the supplied current density is denoted as i_{ct}. The molecular weight of a gas species i is denoted as M_i and F and δ are Faraday's constant and thickness of an electrode respectively.

For the anode electrode, carbon dioxide and water vapour are produced and hydrogen gas is consumed. Thus, the rates of consumption of the gas species are as follows [6],

$$\dot{R}_{\mathrm{H_2}}=M_{\mathrm{H_2}}\left[\dot{R}_{WGS}-i_{ct}/(2F\delta_{an})\right] \tag{11}$$

$$\dot{R}_{\mathrm{H_2O}}=M_{\mathrm{H_2O}}\left[i_{ct}/(2F\delta_{an})-\dot{R}_{WGS}\right] \tag{12}$$

$$\dot{R}_{\mathrm{CO_2}}=M_{\mathrm{CO_2}}\left[\dot{R}_{WGS}+i_{ct}/(2F\delta_{an})\right] \tag{13}$$

It is assumed that the electrodes of the MCFC are porous structures and thermal equilibrium. Hence, the heat transfer of the MCFC is governed by convection, conduction and radiation mechanisms. Although the effect of the thermal radiation is found to be negligible by Koh et al [7], it is still be consider in the present work with case study. Hence, the temperature of the fuel cell, T_p is calculated according to,

$$\rho_{eff}c_{p,eff}\partial T_p/\partial t+\rho_{eff}c_{p,eff}\mathbf{u}\cdot\nabla T_p=\nabla\cdot(k_{eff}\nabla T_p)+\dot{Q}_{\mathrm{Ohm}}+\dot{Q}_r+Q_{\mathrm{rad}} \tag{14}$$

where $c_{p,eff}$ and k_{eff} are the specific heat capacity and the thermal conductivity respectively.

The reaction heat source for the anode electrode,

$$\dot{Q}_r=-\left[\Delta H_{WGS}\dot{R}_{WGS}+\Delta H_{f,\mathrm{H_2O}}M_{H_2O}i_{ct}/(2F\delta_{an})\right] \tag{15}$$

whilst it is zero at the cathode electrode. The enthalpies of reaction for water-gas shift and water formation are denoted as ΔH_{WGS} and $\Delta H_{f,\mathrm{H_2O}}$ respectively.

Heat transfer by thermal radiation (Q_{rad}) is described as,

$$Q_{rad}=\sigma_B\varepsilon_s(T_S{}^4-T_{sur}{}^4) \tag{16}$$

where σ_B is the Stefan-Boltzmann constant. T_S, T_{sur} and ε_s are cell separator temperature, surrounding temperature and emissivity of the separator plates, respectively.

The Ohmic heat source $\dot{Q}_{\text{Ohm}}$ is defined as,

$$\dot{Q}_{\text{Ohm}} = V_{cell} i_{ct} / \delta_k \tag{17}$$

where δ_k is the thickness of anode, cathode electrodes or current collector.

The temperature of the gas domain in the gas phase is governed by the following equations,

$$\rho c_{p,g} \partial T_g / \partial t + \rho c_{p,g} \mathbf{u} \cdot \nabla T_g = \nabla \cdot (k_g \nabla T_g) + \dot{Q}_r \tag{18}$$

where $\dot{Q}_r = -\left(\Delta H_{WGS} \dot{R}_{WGS}\right)$ for the anode gas phase and $\dot{Q}_r = 0$ for the cathode gas phase.

The impact of electrochemical dynamics of cell voltage is accounted for in the MCFC model. The voltage of the MCFC can be calculated via [8]:

$$V_{\text{cell}} = E_{\text{eq}} - i_{ct}\left(\eta_{\text{act}} + \eta_{\text{conc}} + \eta_{\text{Ohm}}\right) \tag{19}$$

where E_{eq} is calculated by Nernst equation and the voltage losses due to activation, concentration and Ohmic overpotentials are calculated as follows,

$$\eta_{act} = 2.27 \times 10^{-9} \exp(\frac{6435}{T_p}) p_{H_2}^{-0.42} p_{CO_2}^{-0.17} p_{H_2O}^{-1} \tag{20}$$

$$\eta_{conc} = 7.505 \times 10^{-10} \exp\left(\frac{9298}{T_p}\right) p_{O_2}^{-0.43} p_{CO_2}^{-0.09} \tag{21}$$

$$\eta_{\text{Ohm}} = 0.5 \times 10^{-4} \exp\left[3016\left(1/T_p - 1/923\right)\right] \tag{22}$$

Detailed discussion on the modeling can be found in [9].

2.2. Computational domain

A number of leakages are introduced at the separator of the anode gas channels as shown in figure 2.

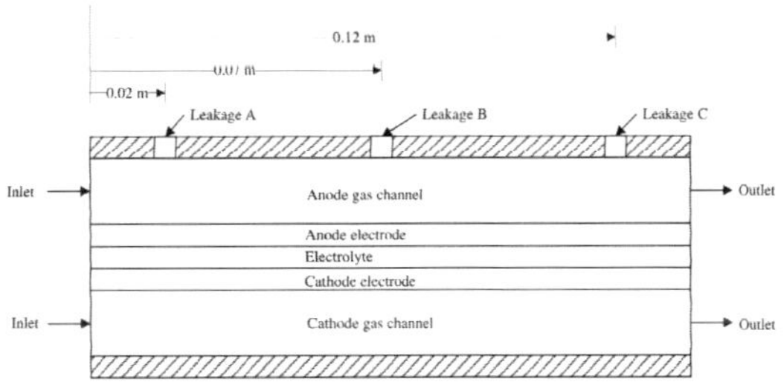

Figure 2. Computational domain of the MCFC.

2.3. Leakage modelling and boundary / operating conditions

The leakage modelling is carried in two stages. The first stage is to assume no leakage in the fuel cell. The above-mentioned governing equations are solved to obtain the steady-state solution. The leakage domains A, B and C are assumed to be a porous media with a low porosity ($\varepsilon_p \approx 10^{-5}$) and the boundary conditions of the leakages are described as walls. The second stage involves solving the transient governing equations (1) to (22). The initial conditions are taken from the previously obtained stead-state solution. This is important because it improves the convergence of the transient solution. The leakage domains A, B and C are now treated as part of the gas channel ($\varepsilon_p = 1$). The boundary condition of the leakage is prescribed by a known flow velocity. The leakage is not prescribed by atmospheric pressure because it avoids the need to generate a high number of meshes around a small leakage region.

The inputs of the model [10] are summarized in table 1.

Table 1. Operating conditions and geometry details of the model.

Input	Values
Anode gas velocity (m/s)	1.24
Cathode gas velocity (m/s)	3.73
Molar fraction in the anode gas channel, H_2:CO_2:H_2O	0.4:0.4:0.2
Molar fraction in the cathode gas channel, O_2:CO_2:N_2	0.148:0.296:0.556
Inlet gas temperature at the anode and cathode gas channel (K)	873
Current density demand (A/m^2)	1400
MCFC length (m)	0.14
Anode / cathode gas channel width (m)	0.0016
Emissivity of the separator plate	0.286

3. Results and discussions

Two case studies are conducted using the current model. These include: The effects of leakage flow rates and the locations of leakage.

3.1. The effects of leakage flow rates

In the first case study, a leakage is assumed to occur at the middle of the MCFC, i.e. leakage B as in figure 2. Figure 3 shows the velocity magnitude of the anode gas channel for the case when the leakage velocity is 1.5m/s. It can be observed that upon entering the gas channel, the velocity increases due to the entrance effect. The anode gas velocity remains constant before it encounters the leakage. The velocity is reduced after it passes by leakage as a portion of the gas is lost to the environment through the leakage.

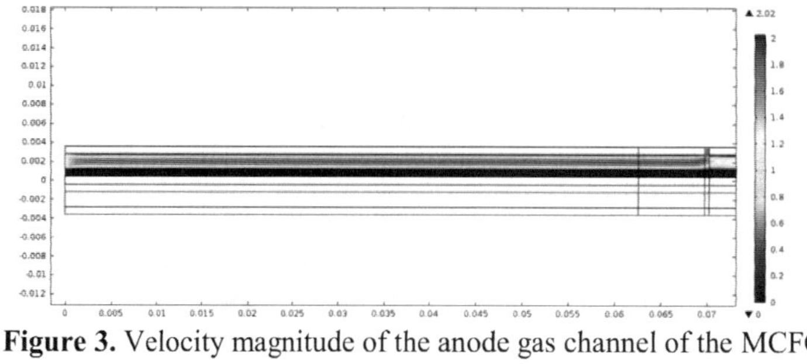

Figure 3. Velocity magnitude of the anode gas channel of the MCFC

Figure 4 shows the temperature of anode electrode at the outlet. It shows that as the anode temperature increases leakage velocity increases. This is because a larger leakage velocity causes the in-channel gas flow to decrease. A lower channel gas temperature is unable to dissipate heat quickly which causes an increase of the anode temperature.

Figure 5 shows the changes of the voltage of MCFC when the leakage occurs. It can be observed that there is a voltage drop initially due to the lower electrochemical reaction owing to the smaller convective mass flux at the electrodes. After a while, the voltages increase. The graph shows that the voltage becomes higher if the leakage velocity increases. This observation is due to the increasing electrode temperature which causes the reduction in the polarization voltage, i.e. activation, concentration and Ohmic polarizations. The changes of these polarizations are shown in figure 6 to figure 8. These graphs show that as the electrode temperature increases, the overpotentials decrease. The overpotentials drop more noticeably as the leakage velocity increases from 0.5 m/s to 1.5 m/s. From equation (19), the voltage generated increases as the overpotentials decrease. Among the three overpotentials, activation overpotential is relatively small compared to the other two. Figure 6 and figure 7 show that the change of overpotential decreases gradually as the temperature increases. On the other hand, a drastic drop can be observed for the activation overpotential at the onset of leakage at t = 0 s. The decrease is more pronounced for a higher leakage which is due to the decrease of gas pressure at the anode.

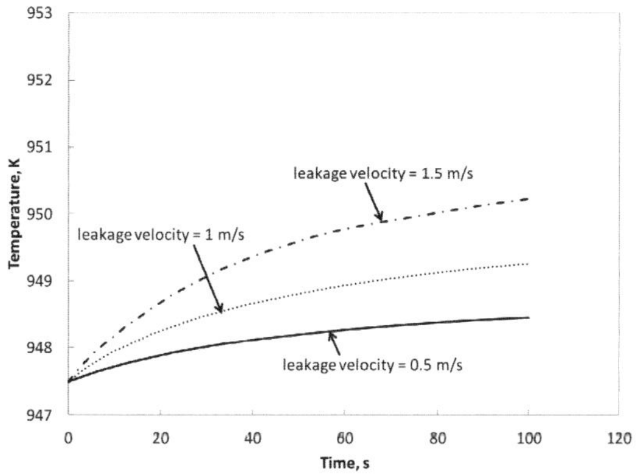

Figure 4. The temperature of the anode electrode at the outlet of the gas channel for the case of middle leakage

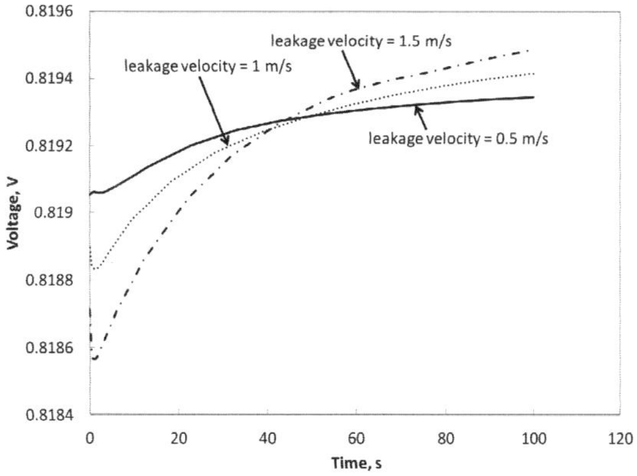

Figure 5. The voltage generated by the MCFC during various leakage conditions for the case of middle leakage

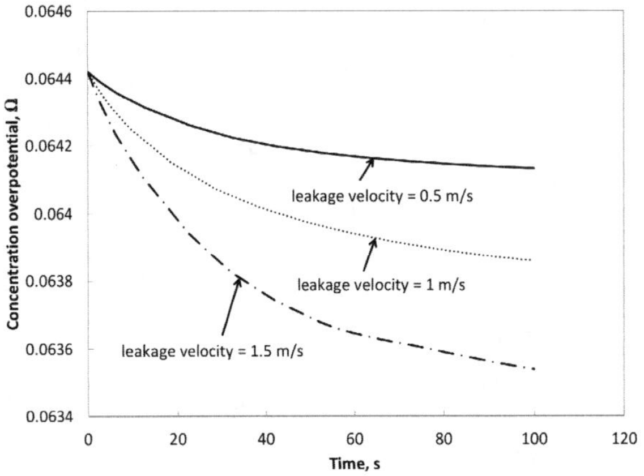

Figure 6. The concentration overpotential of the MCFC during various leakage conditions for the case of middle leakage

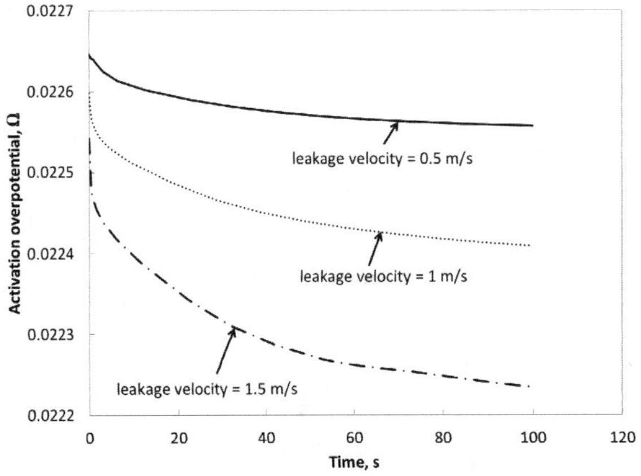

Figure 7. The activation overpotential of the MCFC during various leakage conditions for the case of middle leakage

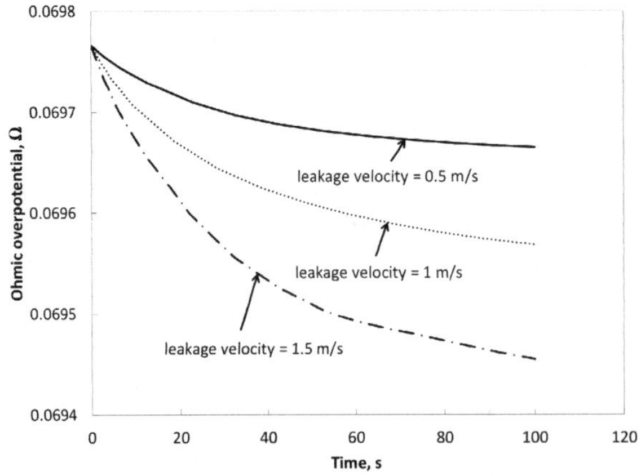

Figure 8. The Ohmic overpotential of the MCFC during various leakage conditions for the case of middle leakage

3.2. The effects of leakage locations

The model was used to study the effects of leakage locations on the performances of the MCFC. Figures 9 and 10 show the anode electrode temperatures at the outlet for the cases of front and end leakages. By comparing figures 4, 9 and 10, it can be seen that the temperature rise is larger as the leakage is nearer to the inlet. This is because gas loss from the front leakage will result in smaller gas velocity in majority of the gas channel. On the other hand, much smaller temperature rise is observed as it is nearer to the outlet as lower gas velocity is available only at a small section of the gas channel after the leakage.

The voltages generated by both front and end leakages are also compared as shown in figure 11. It can be observed that the changes of voltage by end leakage are insignificant. By comparing figure 5 and figure 11, it can be concluded that front leakage will cause a noticeable change in voltage reading compared to the mid and end leakage conditions. The results can be used to develop a model-based hydrogen leakage detection system. Data collected from sensors will be used to compare with the simulation data. If the difference of both data is within a predefined tolerance, alarm will be triggered to inform the presence of hydrogen leakage from a MCFC.

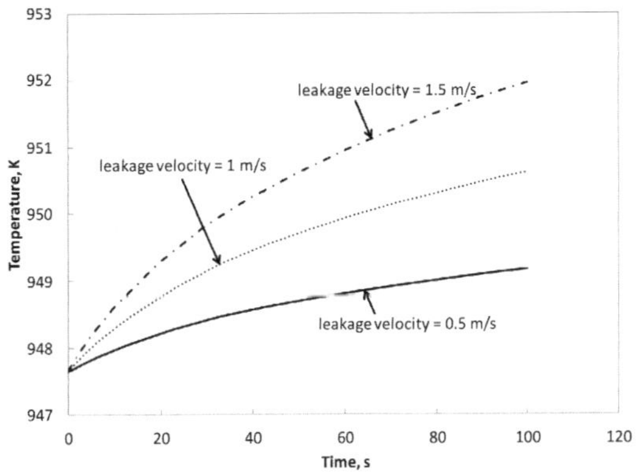

Figure 9. The temperature of the anode electrode at the outlet of the gas channel for the cases of front leakage

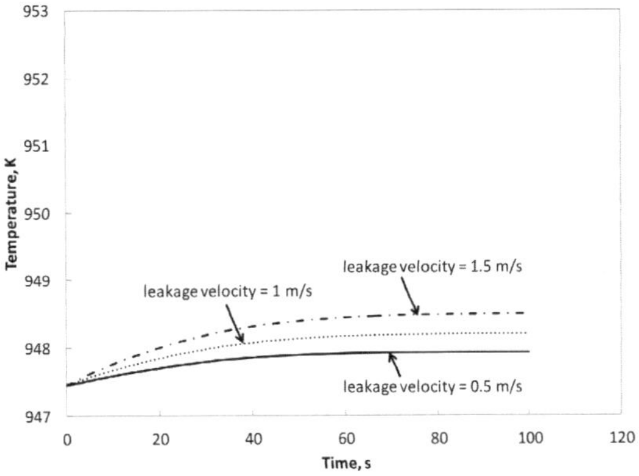

Figure 10. The temperature of the anode electrode at the outlet of the gas channel for the case of end leakage

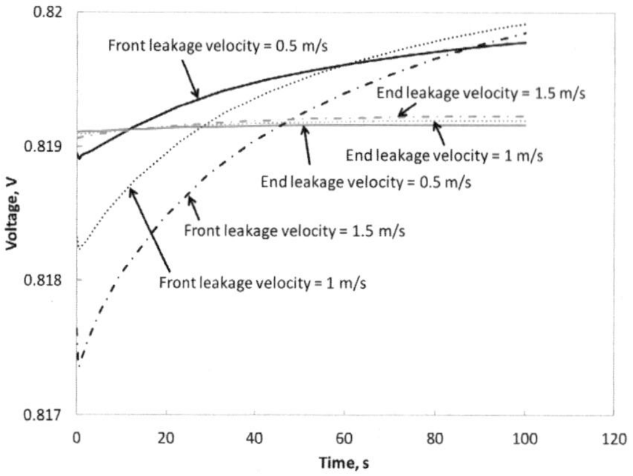

Figure 11. The voltage generated by the MCFC under front and end leakage conditions

4. Conclusions

In the current study, a dynamic model of a MCFC was developed. The model was used to investigate the dynamic performance of the MCFC in the presence of hydrogen gas leakage. The study indicates that when a leakage occurs, the in-channel anode gas velocity will decrease. This in turn causes the anode electrode temperature to increase. The increase of anode electrode temperature is proportional to the leakage velocity. The simulation also shows that when a leakage occurs, the MCFC voltage decreases initially due to a lower electrochemical reaction rate. After that, the voltage increases due to the electrode temperature rise which encourages a higher reaction rate. In addition, it has been shown that a leakage location has a noticeable effect on the temperature and voltage of a MCFC. For the same leakage flow rate, a leakage nearer to the inlet of the gas channel results in larger temperature / voltage increase compared to that at the end of the gas channel. This is because the gas velocity in the majority section of the gas channel becomes smaller after it passes by the leakage in the former case. The current result is useful to develop a model-based hydrogen leakage detection system.

5. References

[1] Ashraf Khorasani M R, Asghari S, Mokmeli A, Shahsamandi M H and Faghih Imani B 2010 *Int. J. Hydrogen Energ.* **35** pp 9269-9275
[2] Tian G, Wasterlain S, Candusso D, Harel F, Hissel D and François X 2010 *Int. J. Hydrogen Energ.*, **35** pp 2772-2776
[3] Husar A, Serra M, and Kunusch C 2007 *J. Power Sources*, **169** pp 85-91
[4] Kim Y, Nam J H, Shin D, Chung T Y, Kim Y G 2010 *Current Applied Physics*, **10** pp S81-S85
[5] Haberman B A, Young J B 2004 *Int. J. Heat Mass Tran*, **47** pp 3617-3629
[6] Akhtar N, Decent S P, Loghin D, Kendall K 2009 *Int. J. Hydrogen Energ.* **34** pp 8645-8663
[7] Koh J H, Kang B S, Lim H C 2000 *J. Power Sources* **91** pp 161-171
[8] Brouwer J, Jabbari F, Leal E M, Orr T 2006 *J. Power Sources* **158** pp 213-224
[9] Law M C, Lee V C C, Tay C L 2015 *Applied Thermal Engineering* (accepted for publication)
[10] Yu L J, Ren G P, Jiang X M 2008 *Energ. Convers. Manage.* **49** pp 873-879

Influence of Carbon Nano Tubes on the Thermo-Mechanical Properties of Unsaturated Polyester Nanocomposite

A K M Moshiul Alam, M D H Beg, Rosli Mohd Yunus

Faculty of Chemical and natural Resources Engineering, University Malaysia Pahang, 26300, Gambang, Kuantan, Pahang, Malaysia

akmmalam@gmail.com

Abstract. To date nano fillers are renowned reinforcing agent for polymer materials. In this work, unsaturated polyester (UPR) nanocomposites were fabricated by 0.1, 0.3 and 0.5 wt% multi walled carbon nanotubes (MWCNTs) through solution dispersion and casting method. The influence of MWCNT content was investigated by thermo-mechanical properties. Dispersion of nanotubes was observed by fracture morphology. The strength of nanocomposites rose with raising the CNT content. Moreover, DSC thermograms of nanocomposites represent noticeable improvement of glass transition temperature (T_g), melting temperature (T_m) and enthalpy (ΔH_m). Micro-crystallinity of nanocomposites increased with increasing the CNT content. Moreover, the stiffness increased with increasing the CNT content.

Key word: Solvent dispersion, Nanocomposites, Characterization

1. Introduction

The unsaturated polyester resins (UPRs) are common thermosetting resins and steadily increasing their applications for several purposes because of their sound properties, cost effectiveness as well as simple handling. However, cross-linked UPRs have limited structural reliability, therefore , before cross-linking they are often mixed with reinforcing materials such as natural fibers, synthetic fibers, nanofillers as well as mineral fillers and so on [1,2,3,4] . The reinforced composites are devoted for construction, marine and automotive industries due to their light weight and durability.

The carbon nanotubes (CNTs) are more attractive fillers in the vicinity of polymer composites because of their outstanding properties therefore they are substitute of conventional macro and micro fillers [5,6,7]. CNTs are geometrically distinctive to their surface area, provide an immense resources of interaction with any continuous phase giving rise to great opportunity for effective load transfer [8].Moreover, small amount of CNT with efficient dispersion in matrix exhibit considerable enhancement of different properties [9].The significant improvement of properties are determined by the degree of CNTs dispersion and interfacial adhesion into the composite system [10]. Conversely, several phenomena restrict carbon nanotube dispersions, such as nanotube morphology and Van der Waal's forces between nanotube surfaces. Not only that but also the high aspect ratios together with the high flexibilities noticeably increase the possibilities for entanglements. These entangled aggregates are very complicated to separate into individual nanotubes [11].The interaction of CNTs reveal high potential energy which naturally making them more difficult to separate as individuals nanotubes [12,13].Therefore, different physical and chemical methods are demonstrated for control dispersion of individual nano tubes in matrix. Physical methods consider for direct mixing through mechanical force. Chemical methods are carried out by surfactants action, functionalization of carbon nanotube, surface modification, and polymer wrapping technology [14,15]. For instance, studies related to thermoset nanocomposite systems have shown that share mixing strategies for dispersion of CNT. It has carried out at room temperature with different non hydrogen bonding Lewis base solvents to take away the nanotube surface interaction [16,17,18].

Corresponding author: Faculty of Chemical and Natural Resources Engineering, Universiti Malaysia Pahang, Gambang 26300, Kuantan, Malaysia. akmmalam@gmail.com (A.K.M.Moshiul Alam)

Moreover, UPR has huge market compare to other resins, there have only been limited investigations of MWCNT dispersions in UPR [19,20]. Apparently, the small number of efforts has been paid to disperse MWCNTs in UPR and characterization of thermo-mechanical properties. Consequently, first of all we have emphasis on the dispersion of MWCNTs in UPR matrix in association of solution and sonication techniques. We report the effect of CNT content in UPR, as well as focus on the interaction between matrix and MWCNTs in the nanocomposite networking system. Furthermore, compare the properties of neat resin and nanocomposites for determining efficiency of MWCNT.

2. Materials and methods

2.1 Materials

The orthopthalic unsaturated polyester resin (Polymal) was used as received from Luxchem Polymer Industries *Sdn. Bhd.*, Malaysia. The viscosity of resin is 700-800mPa.s at 25°C.volatile content 30-35%, Gel time 8-15min. Multi Wall Carbon Nanotubes (MWCNT) produced by moving-bed catalysis technique, diameter <8nm, length between 10 - 30 μm, and the carbon purity 95% received from Timesnano, China. Tetra-hydro Furan (THF), received from Merk, Germany. Methyl Ethyl keton peroxide (MEKP) purchased from Sigma Aldrich, USA.

2.2 Composite fabrication

First of all, 0.1, 0.3 and 0.5 wt% MWCNT mixed separately with THF solvent, the MWCNT: THF ratio was maintained as 1:25. The suspension stirred by magnetic stirrer for 15 minutes followed by sonication in ultra sound bath for 1hour. After that, it mixed with resin and stirred again for 15 minutes, subsequently sonicated for 2 hours. Secondly the resin/CNT suspension heated at the boiling temperature of THF (66°C) for 5 minutes to evaporate the solvent. The warm suspension placed in a cold water bath to cool at room temperature. 1 wt% MEKP added in this suspension and gently stirred for 3 minutes, then placed in vacuum to remove the bubbles. Finally, the bubble free mixture poured on the specimen mold and cured at room temperature. In this manner, the samples prepared for subsequent analysis were neat unsaturated polyester resin (UPR), 0.1wt% MWCNT reinforced UPR nanocomposite (0.1MWCNT- UPR), 0.3wt% MWCNT reinforced UPR nanocomposite (0.3MWCNT-UPR) and 0.5wt% MWCNT reinforced UPR nanocomposite (0.5MWCNT- UPR).

2.3 Field emission scanning electron microscopy (FESEM)

Composites fracture surfaces investigated by using a (JOEL, JSM-7800F, Japan) field emission scanning electron microscope. Samples mounted on aluminium stubs with carbon tape followed by sputter coated with platinum to make them conductive prior to FESEM observation.

2.4 Tensile testing of composites

Tensile testing was conducted according to ASTM 638-08, using a Shimadzu (Model: AG-1) Universal tensile testing machine fitted with a 5 kN load cell operated at a cross-head speed of 1mm/min and keeping 65 mm gauge length. Tensile strength (TS), tensile modulus (TM) and elongation at break (EB) are obtained by this testing method. Five samples of each category were tested for analysis the data.

2.5 Differential scanning calorimetry

Differential scanning calorimetry (DSC) was performed to determine transition states of material, for instance, the glass transition temperature (T_g), crystallization temperature (T_c) and melting temperature (T_m), using a TA/Q1000 apparatus under nitrogen atmosphere. During DSC analysis, the samples were initially heated at 30 – 400°C with a heating rate 10°C min^{-1} in nitrogen atmosphere.

3. Result and discussion

3.1 Mechanical properties

Figure 1 represents the (a) TS and TM and (b) EB of nanocomposites as a function of MWCNTs content. The results represent those MWCNTs able to improve the mechanical properties of UPR. Furthermore, 0.1, 0.3 and 0.5 wt% MWCNTs loaded MWCNT-UPR nanocomposites exhibited enhancement of TS, TM. The increased TS of 0.1MWCNT-UPR, 0.3MWCNT-UPR and 0.5MWCNT-

UPR nanocomposites are 10%, 29% and 21% correspondingly, similarly the TM rose of 51%, 88%, and 80% respectively and the EB% decreased as well.

It is obvious, that the 0.3MWCNT-UPR exhibits better mechanical performance than 0.1MWCNT-UPR and 0.5MWCNT-UPR. When the CNT content increased more than 0.3 wt% MWCNT undergo partial agglomeration represents in FESEM micrograph, which limits the efficiency of reinforcement. Such a phenomenon also observed in other reports [21,22]. It is true that, individual nano tubes enjoy both large surface area and strong interfacial interactions with matrix these inherent advantages lead to efficient stress transfer between the MWCNTS and the UPR, and avoid the deformation or fracture of the composites under external force [21]. In addition, the lowest elongation of 0.3MWCNT-UPR nanocomposite credibly MWCNTs are complimentary of the non elastic deformation.

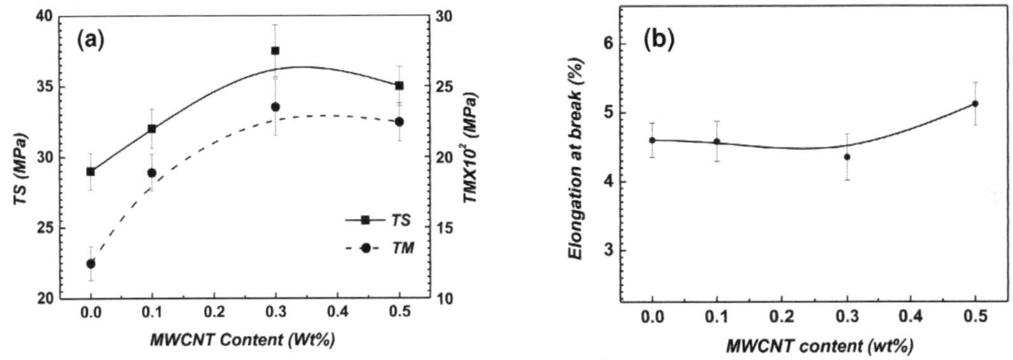

Figure 1. (a) Tensile strength and tensile modulus (b) Elongation at break of nanocomposites as a function of MWCNT content

3.2 Fracture Morphology

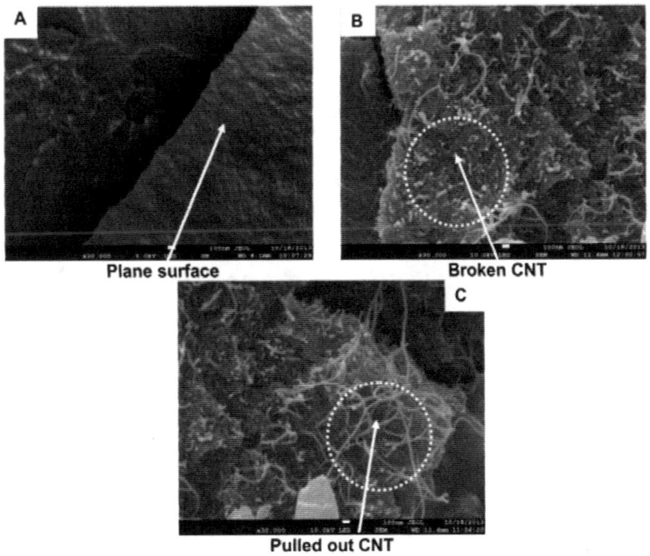

Figure 2. Fracture morphology of (a) UPR, (b) 0.3MWCNT- UPR and (c) 0.5MWCNT-UPR

Figure 2 illustrates the fracture morphology of (a) UPR, (b) 0.3MWCNT- UPR and (c) 0.5MWCNT-UPR nanocomposites. The surface of neat resin in figure 2(a) is smooth and distinctly cracking whereas nanocomposites surfaces are rough and CNTs tend to bridge the cracking. Several bright fractured tips of nanotubes represent by doted circle remained in the matrix and they were not pulled out during stretching of 0.3MWCNT-UPR nanocomposite shown in figure2b. In this case, matrix tends to stick to the CNTs by well wetting and Van der Waals force, as a result, increased the composite TS and TM. Furthermore, the nanotubes tend to nucleate cracking and propagate the rough surface of nanocomposites which implies that the CNT–UPR interface is strong as well as providing for an additional reinforcement effect thus more energy is needed to break the specimens. On the other hand, in case of 0.5MWCNT-UPR nanocomposite in figure 2(c) CNT pull out region evidently mentioned by doted circle. This is because high concentration of CNT formed agglomeration, thus declined the resultant composite's reinforcement.

3.3 DSC Analysis

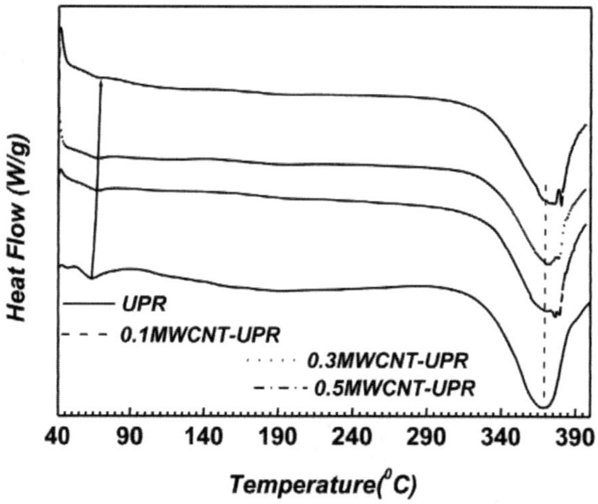

Figure 3. DSC thermogram of (i) neat UPR, (ii) 0.1MWCNT-UPR, (iii) 0.3MWCNT-UPR and (iv) 0.5MWCNT-UPR

Figure 3 illustrates the DSC thermograms of (i) neat UPR, (ii) 0.1MWCNT-UPR, (iii) 0.3MWCNT-UPR and (iv) 0.5 MWCNT-UPR nanocomposites. They notice the information of chain intercalation and thermal transitions of nanocomposites. The endotherm in the lower temperature region around 60-67°C ascribe to the glass transition (T_g). Furthermore, the endothermic transition around 366-382°C are related to the melting temperature(T_m).

Table1. The T_g , T_m and ΔH_m values obtained from DSC thermograms.

Samples	**T_g°C**	**T_m°C**		**ΔH_m Jg^{-1}**
		T_{m1}°C	**T_{m2}°C**	
UPR	64	369	---------	0.56
0.1MWCNT-UPR	67	371	376	0.64
0.3MWCNT-UPR	67	371	382	0.81
0.5MWCNT-UPR	65	371	382	0.61

The nanocomposites exhibit a split melting endotherm into two peaks (T_{m1} and T_{m2}), instead of one endothermic peak as shown by pure matrix. The presence of double-melting peaks for the nanocomposites may indicate the bond formation between MWCNT and UPR. The melting peak at T_{m2} is a precursor to indentify the binding energy among nanotubes and UPR molecules. The T_g , T_m and ΔH_m values of these samples represent in table 1.

MWCNT content significantly influenced the glass transition temperature of nanocomposites, which can be explained by the formation of microcrystalline segments, they are relatively well-ordered as well as enhance the degree of curing, moreover which restrict the resin chain mobility in the surrounding area of CNT [23,24,25].Taking into account, amorphousness of UPR matrix, some nucleating effect provided by MWCNTs is to be assumed. In fact, DSC thermogram of neat UPR confirms the amorphous behaviour. In presence of MWCNT nanofillers, nucleation of micro crystallization takes place in the host matrix. In case of nanocomposites the apparent crystallization takes place thus the corresponding melting enthalpies are higher than neat matrix. The greater degree of micro crystallinity in 0.3MWCNT-UPR composite is reflected in the higher enthalpy value. It is also interesting that the heat of transition increased significantly in case of composite specimens. It is obvious that MWCNT acting as a key nucleating agent.

4. Conclusion

0.3MWCNT-UPR shows the highest mechanical properties compared to neat UPR and other nanocomposites. These results indicate a considerable enhancement of materials properties by the incorporation of MWCNT in matrix. The fracture morphology confirmed 0.3 wt% MWCNT is preferable for well dispersion and interaction between UPR and MWCNT. The glass transition temperature and melting temperature and micro crystallinity were noticeably increased in nanocomposites.

Acknowledgement

Authors would like to acknowledge for providing financial support through the projects GRS 130356 and RDU 140319.

References

[1] Albdiry M T and Yousif B F 2012 Morphological structures and tribological performance of unsaturated polyester based untreated/silane-treated halloysite nanotubes. *Materials and Design* **48**, 68-76.

[2] Alam A K M M, Beg M D H, Mina M F, Khan M R and Prasad D R M 2012 Structures and Performances of Simultaneous Ultrasound and Alkali Treated Oil Palm Empty Fruit Bunch Fiber Reinforced Poly(Lactic Acid) Composites Compo. Part A **43** 1921-1929 .

[3] A K M M Alam, Mina M F, Beg M D H, Mamun A A, Bledzki A K and Shubhra Q T H 2014 Thermo-Mechanical and Morphological Properties of Short Natural Fiber Reinforced Poly (Lactic Acid) Biocomposite: Effect of Fiber Treatment *Fibers and Polymers*, **15**(6) 1303-9,.

[4] Shubhra Q T H, Alam A K M M and Quaiyyum M A 2011 Mechanical properties of polypropylene composites: A review *J Therm. Comp. Mat.* **26**(3) 362-91.

[5] Kayatin M J and Davis V A 2009 Viscoelasticity and Shear Stability of Single-Walled Carbon Nanotube/ Unsaturated Polyester Resin Dispersions *Macromolecules* **42** 6624–32.

[6] Mina M F, Beg M D H, Islam M R, Nizam A, Alam A K M M and Yunus R M 2014 Structures and Properties of Injection Molded Biodegradable Poly(Lactic Acid) Nanocomposites Prepared with Untreated and Treated Multi-Walled Carbon Nanotubes *Pol. Eng. & Sci.* **54**(2) 317-26.

[7] Haque M A, Mina M F, Alam A K M M, Rahman M J, Bhuiyan M A H, and Asano T 2012 Multi-Walled Carbon Nanotubes Reinforced Isotactic Polypropylene Nanocomposites: Enhancement of Crystallization and Mechanical,Thermal and Electrical Properties Poly. Comp. **33**(7) 1094–1104.

[8] Gojny F H, Wichmann M H G, Köpke U B, Fiedler and Schulte K 2004 Carbon nanotubereinforced epoxy-composites: Enhanced stiffness and fracture toughness at low nanotube content *Compos Sci Technol.* **64,** 2363–71.

[9] Kota A K, Cipriano B H, Duesterberg M K, Gershon A L, Powell D S R, Raghavan and Bruck H A 2007 Electrical and rheological percolation in Polystyrene/MWCNT composites *Macromolecules* **40** (20) 7400–06.

[10] Montazeri A, Javadpour J, Khavandi A, Tcharkhtchi A, Mohajeri A 2010 Mechanical properties of multi-walled carbon nanotube/epoxy composites *Mater Des* **31** 4202–8.

[11] Breton Y, D'esarmot G, Salvetat J P 2004 Mechanical properties of multiwall carbon nanotube/epoxy composite: influence of network morphology *Carbon* **42** 1027–30.

[12] Sun C H, Yin L C, Li F, Lu G Q, and Cheng H M 2005 Van der Waals interactions between two parallel infinitely long single-walled nanotubes *Chem. Phys. Lett.* **403**(4-6) 343–346.

[13] Girifalco L A, Hodak M and Lee R S 2000 Carbon nanotubes, buckyballs, ropes, and a universal graphitic potential *Phys. Rev. B* **62** (19) 13104–10.

[14] Hilding J, Grulke E A, Zhang Z G and Lockwood F 2003 Dispersion of Carbon Nanotubes in Liquids J Disp. Sci. Tech. **24**(1) 1–41.

[15] Mackay M E, Tuteja A, Duxbury P M, Hawker C J, Van Horn B, Guan Z, Chen G and Krishnan R S 2006 General strategies for nanoparticle dispersion *Science* **311**(5768) 1740 – 43.

[16] Ausman K D, Piner R, O Lourie, and Ruoff R S 2000 Organic Solvent Dispersions of Single-Walled Carbon Nanotubes: Toward Solutions of Pristine Nanotubes *J. Phys. Chem. B*, **104**(38).

[17] Gojny F H, Wichmann M H G, Fiedler B and Schulte K 2005 Influence of different carbon nanotubes on the mechanical properties of epoxy matrix composites – A comparative study Comp. Sci. Tech. **65** 2300–13.

[18] Sabba Y and Thomas E L 2004 High-Concentration Dispersion of Single-Wall Carbon Nanotubes Macromolecules **37** (13) 4815–20.

[19] Battisti A, Skordos A A, Partridge I K 2009 Monitoring dispersion of carbon nanotubes in a thermosetting polyester resin *Compos. Sci. Technol.* **69** (10) 1516–20.

[20] Seyhan A T, Gojny F H, Tanoglu M and Schulte K 2007 Critical Aspects Related to Processing of Carbon Nano-tube/unsaturated Thermoset Polyester Nanocomposites *Eur. Polym. J.* **43** (2) 374–79.

[21] Jialiang Wang, Zixing Shi, Yu Ge, Yan Wang, Jinchen Fan and Jie Yin 2012 Functionalization of unzipped carbon nanotube via in situ polymerization for mechanical reinforcement of polymer *J. Mater. Chem.* **22** 17663-70.

[22] Hansang Kwon and Marc Leparoux 2012 Hot extruded carbon nanotube reinforced aluminum matrix composite materials *Nanotechnology* **23** 415701.

[23] Pavlidou S, Papaspyrides C D 2008 A review on polymer–layered silicate nanocomposites *Prog. Poly. Sci.* **33**1119–98.

[24] Allaoui A and Bounia N E 2009 How carbon nanotubes affect the cure kinetics and glass transition temperature of their epoxy composites? – A review *eXPRESS Polymer Letters*, **3**(9) 588–94.

[25] Seymour R W and Cooper S L 1973 Thermal analysis of polyurethane block polymers *Macromolecules* **6**(I).

Thin-walled composite tubes using fillers subjected to quasi-static axial compression

Haidar F. AL-Qrimli*[1]
[1]Department of Mechanical Engineering, Curtin University,
CDT 250 98009, Miri, Malaysia

Fadhil A. Mahdi [2]
[2]Department of Electrical Engineering, University of Baghdad,
Jadriyah, Baghdad, Iraq

Firas B. Ismail [3]
[3]Deptment of Mechanical Engineering, Universiti Tenaga Nasional
43000 Kajang, Selangor, Malaysia

Ibrahim S. Alzorqi [4]
[4]Department of Chemical & Environmental Engineering
The University of Nottingham, 43500 Semenyih, Malaysia

*Corresponding author. E-mail: haidar.fa@curtin.edu.my, halqrimli@yahoo.com

Abstract. It has been demonstrated that composites are lightweight, fatigue resistant and easily melded, a seemingly attractive alternative to metals. However, there has been no widespread switch from metals to composites in the automotive sector. This is because there are a number of technical issues relating to the use of composite materials that still need to be resolved including accurate material characterization, manufacturing and joining process. The total of 36 specimens have been fabricated using the fibre-glass and resin (epoxy) with a two different geometries (circular and corrugated) each one will be filled with five types of filler (Rice Husk, Wood Chips, Aluminium Chips, Coconut Fibre, Palm Oil Fibre) all these type will be compared with empty Tubes for circular and corrugated in order to comprehend the crashworthiness parameters (initial failure load, average load, maximum crushing load, load ratio, energy absorption, specific energy absorption, volumetric energy absorption, crushing force efficiency and crush strain relation) which are considered very sufficient parameters in the design of automotive industry parts. All the tests have been done using the "INSTRON Universal machine" which is computerized in order to simply give a high precision to the collection of the results, along with the use of quasi-static load to test and observe the behaviour of the fabricated specimens.

[1] Haidar F. AL-Qrimli, Department of Mechanical Engineering, Curtin University Sarawak

Keywords: Crashworthiness; Energy absorption, Natural fibre; Axial compression.

1. Introduction

These Today, fibre reinforced composites used in a variety of structures, ranging from space-craft to buildings and bridges. This wide use of composites has been facilitated by the introduction of new materials, improvements in manufacturing processes and developments of new analytical and testing methods. Their unique integration of mechanical properties along with their customized fabrication provides the building blocks of our modern infrastructure. From aerospace to the automotive industry, from household products to optical devices, composite materials with unique characteristics are moving aggressively [1-5]. Although the fibre reinforced materials primarily comprise a reason system plus a fibrous reinforcement, other components are frequently added to produce decorative effects, aid handling, and improve moulding characteristics and cured properties. Fillers are used to modify the properties of resin systems either during processing or in the cured state or perhaps both. Often, the primary reason for adding the filler is to reduce the cost of the system; however, invariably, other properties are modified as well. To maximize the benefits of using filler, the correct type and grade needs to be chosen with care, taking into account the modelling process and end use of the moulding. Almost any powdered material may be used as filler; indeed, such materials as crushed egg shells and ground coconut shells have been used. The most common fillers, however, are those obtained from the natural deposits such as chalk slate, quartz or clay. These may be wet or dry ground to produce suitable powders, or subjected to chemical treatment to purify them. Other fillers used include metal powders, and glass spheres and flakes [6-10]. . Mallik, 2008 [11] give Examples of natural fibres are jute, flax, hemp, remi, sisal, coconut fibre (coir), and banana fibre (abaca). All these fibres are grown as agricultural plants in various parts of the world and are commonly used for making ropes, carpet backing, bags, and so on. The components of natural fibres are cellulose microfibrils dispersed in an amorphous matrix of lignin and hemicelluloses. Depending on the type of the natural fibre, the cellulose content is in the range of 60–80 wt% and the lignin content is in the range of 5–20 wt%. In addition, the moisture content in natural fibres can be up to 20 wt%. Menachem, 2007 [12] said that the initial structure of a cotton fibre is determined by biosynthesis, a series of processes that are subject to substantial influence during fibre growth. After the boll opens, there are many factors that affect the structure, from the weather before the fibre is harvested to the industrial processes such as mercerization. Cotton fibres are composed mostly (e.g., 95%) of the long-chain carbohydrate molecule, cellulose (the sugar of cell walls). A large number of impact tests have been carried out on woven-roving and chopped-strand-mat glass reinforced; hand laid-up polyester laminates of different thicknesses and weights. Deflection, force, absorbed energy and damaged area results for these tests have been presented and compared, and the failure modes described. A simple analysis assuming shear-dominated defection gave good correlations with the experimental data. Prediction of the damaged area proved to be more difficult, although some correlation was seen. Further investigation into the nature of the damage is required (Sutherland et al. 1999) [13]. Collapse behaviour of aluminium thin conical frusta with shallow spherical caps (shells of combined geometry) is studied both experimentally and numerically. These shells were of four different thicknesses and were subjected to axial compression between two rigid platens under both quasi-static and impact loading (Gupta et al. 2008) [14]. The bending fatigue behaviours were investigated in glass fibre-reinforced polyester composite plates, made from woven-roving with four different weights, 800, 500, 300, and 200 g/m2, random distributed glass-mat with two different weights 225, and 450 g/m2 and polyester resin (Raif et al. 2007) [15]. Experimental investigations were carried out in order to study the effects of segmentation on the crushing behaviour of quasi static laterally compressed composite tubes. Load–deformation curves and failure mechanism histories of typical specimens are presented and discussed (Abosbaia et al. 2004) [16]. In this paper specimens made of glass fibre/epoxy have been experimentally fabricated with two different geometries (circular and corrugated). Five types of fillers have been used namely; Rice Husk, Wood Chips, Aluminium Chips, Coconut Fibre and Palm Oil Fibre.

2. Fabrication of Tubes

Woven roving glass fibre E-600 has been used mixed with epoxy (EP-A215C1) and hardener (EP-B215) to fabrication the specimens, by mixing the epoxy and resin by weight 100 for the epoxy and 20 for the resin (see Figure 1). All the specimens have been fabricated in the same way by using hand lay-up method with 0/90° fibre orientation angle. Four types of fillers used (Rice husk, Wood Chips, Aluminium Chips and Coconut fibre); Figure 2 shows the type of filler that used in this paper. Wood has been chosen as mandrel. There are two types of mandrels. The first one has a circular cross sectional area with a 10 cm diameter and length 30 cm, while the second mandrel has a corrugated cross sectional area with a 10 cm diameter and 30 cm length as shown in Figure 3. The procedure of Hand lay-up method is shown in Figure 4. After fabricating the specimens, than cut them to the desired dimension. The next step will be the filling process. The cross sectional surfaces (top and bottom) of the specimen flattened to distribute the load on the whole top and bottom surfaces. Firstly, one side of the specimens covered by very light material to reduce the weight of the specimen, sticker tape used to cover the specimen by sticking it from one side to another as shown in Figure 3. For each type three specimens fabricated to insure superior results. By subtracting the weight of the specimen before the filling process and after the filling process, the weight of the filler can be found. The density of the filler calculated by dividing the mass of the filler to the volume of the specimen, Table 1 shows the volume and mass for the tubes.

The crushing test had been performed on all the specimens by using an electrical computer-controlled servo-hydraulic INSTRON machine type 4469. The speed was constant with 15 mm/min and the maximum crushing displacement at 90% from the original length which is 100 mm. The specimens were axially crushed between two steel flat platens; one is static and the other one moving with constant speed. The type of failure modes has been observed such as debonding, delamination, fibre pullout and matrix cracking.

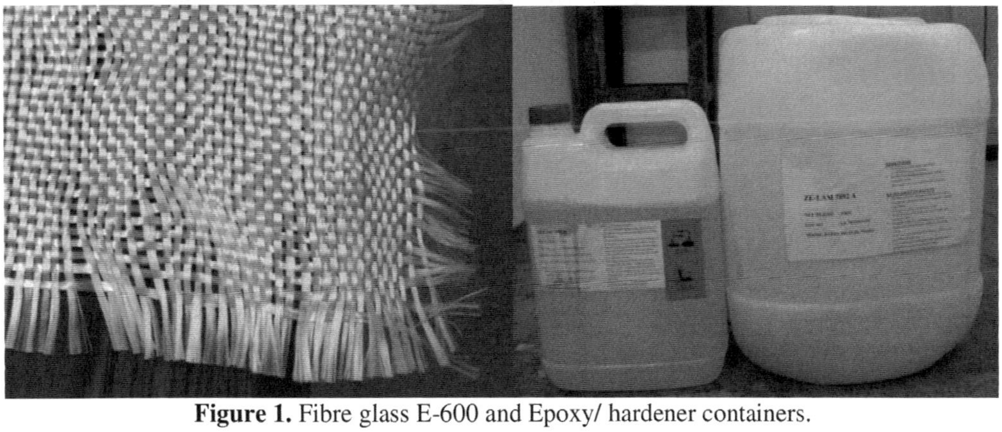

Figure 1. Fibre glass E-600 and Epoxy/ hardener containers.

Figure 2. Different fillers used.

Figure 3. Circular and corrugated wooden mandrels. Preparing specimens to add the fillers.

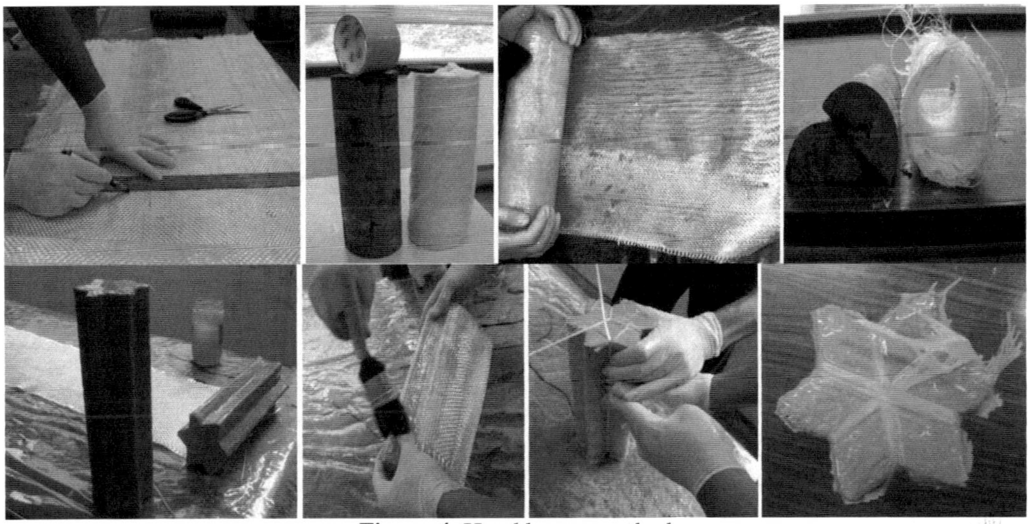

Figure 4. Hand lay-up method.

Table 1. Mass and Volume for the Tubes

Specimen ID	M(Kg)	V (m^3)	DI	DF	Thickness(m)	Length (m)
CUE	0.127	0.000749	0.001730	0.09	0.00251	0.1
CUR	0.201	0.000749	0.001608	0.07415	0.00251	0.1
CUW	0.197	0.000749	0.001857	0.08338	0.00251	0.1
CUC	0.202	0.000749	0.003220	0.09	0.00251	0.1
CUAL	0.295	0.000749	0.001224	0.07188	0.00251	0.1
CUP	0.213	0.000749	0.003220	0.08804	0.00251	0.1
COE	0.141	0.000785	0.002228	0.09	0.00285	0.1
COR	0.2	0.000785	0.001848	0.09	0.00285	0.1
COW	0.203	0.000785	0.002229	0.08995	0.00285	0.1
COC	0.191	0.000785	0.002256	0.09	0.00285	0.1
COAL	0.313	0.000785	0.001730	0.09	0.00285	0.1
COP	0.227	0.000785	0.001983	0.09	0.00285	0.1

IOP Conf. Series: Materials Science and Engineering **78** (2015) 012024 doi:10.1088/1757-899X/78/1/012024

3. Results and Discussion

3.1. Load Displacement Relations

Load-displacement curve can be classified into two main zones the first zone represent the elastic behaviour stage which is called the pre-crushing stage of the specimen then the initial failure will observed (Pi), after the initial failure the plastic zone starts and it's the second zone of the load displacement curve and it's called the post-crushing stage. The load displacement curve shows the maximum load (P_{max}), average load (P_{av}) and the displacement of the failure. From here, the other crashworthiness parameters can be also calculated. Table 2 shows a comparison between the specimens. The empty corrugated tube fluctuated in the same range of the crushing load. And for the corrugated tubes which are filled with filler material can recognize that the load displacement curve increase gradually to the end (See Figure 5 and 6).

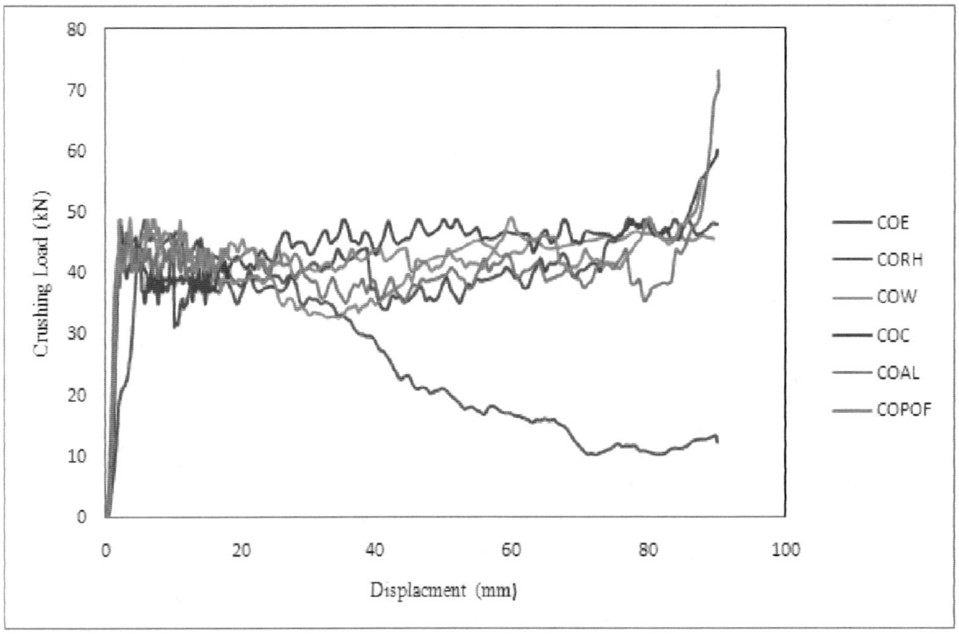

Figure 5. Load displacement curves of the circular specimens with different fillers.

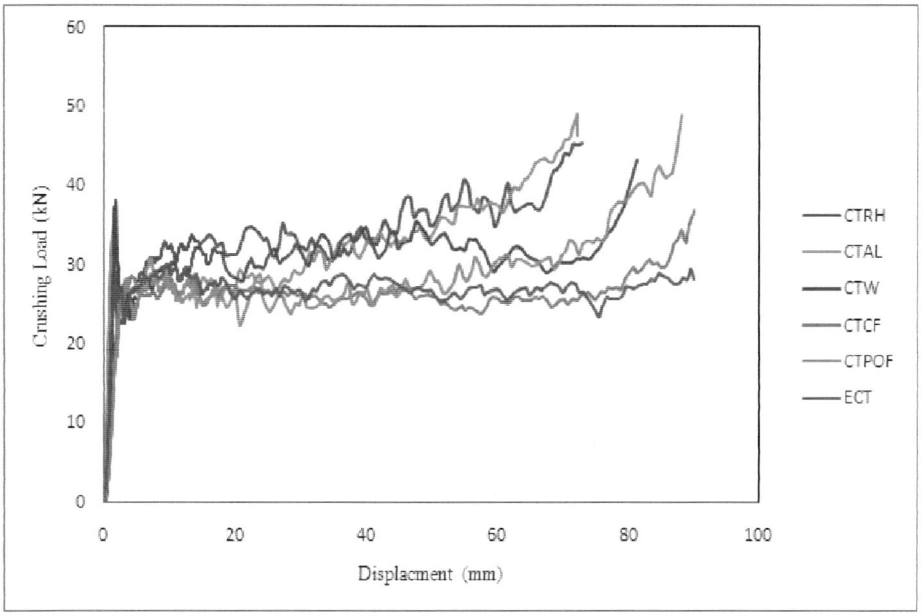

Figure 6. Load displacement curves of the corrugated specimens with different fillers.

3.2. Crashworthiness parameters
The crush worthiness parameters indicate how well the specimen absorbs the energy.

3.2.1 Initial Failure Load (P_i)
Initial failure load is the first peak in the curve which represents a first crush failure. This point lies between the pre-crushing load (elastic) zone and post-crushing load (plastic) zone. In the elastic zone, the structure has the ability to return to its original length without any deformation, but after this point (Pi) the specimen enters the plastic zone and the specimen will be deformed permanently. The corrugated tubes which are filled with Rice Husk (RH) and aluminium chips (AL) have the same range of Pi which is about 48 kN. The corrugated tube with wood chips has the maximum value. There are very slight differences between corrugated tube with rice husk, the circular tubes with palm oil fibre, coconut fibre and the empty circular tube which in turn has the minimum value of initial failure as shown in Table 2.

3.2.2 Maximum Crushing Load (Pmax)
The maximum crushing load is the maximum load that the specimen can endure during the crushing process. The maximum crushing load should be obtained for all specimens, it is very important in the design because it's the maximum amount of the load that can be carried by the composite tube. Figure 7 shows the maximum load for the corrugated tubes which are filled with five different fillers and the empty corrugated tube, totally 6 different types. slight difference in Pmax can be recognized clearly especially the difference between the corrugated tubes which are filled with rice husk (RH), wood chips (W), aluminium chips (AL) and the corrugated tube without filler. The corrugated tube which is filled with coconut fibre produced the maximum value of Pmax as shown in Table 2. Figure 8 shows the difference in the maximum load for the circular tubes which are filled with five different types of fillers and the tube without filler. The circular tube which is filled with coconut and the empty circular tube have the lowest maximum load endured. It is clear that the circular tubes which are filled with rice husk, aluminium chips and palm-oil fibre have the maximum value for maximum load, and the

circular tubes which are filled with rice husk have the maximum load between the bigger one and smaller one.

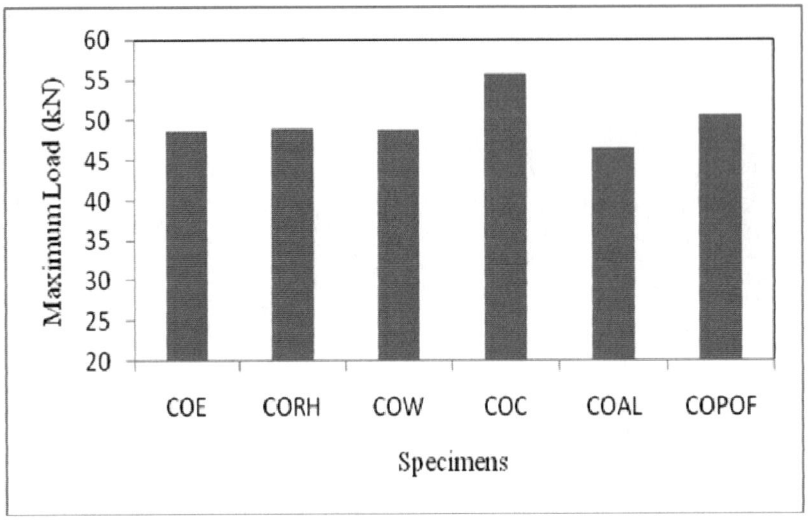

Figure 7. Maximum Loads for Corrugated Tubes

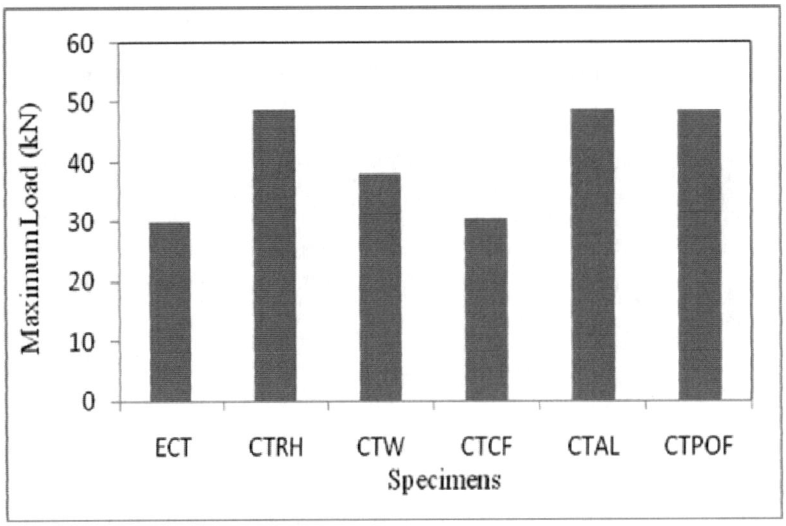

Figure 8. Maximum Loads for Circular Tubes

3.2.3 Average Crushing Load (P_{av})

The average crushing load is the mean of all crushing loads during the crushing process in the plastic zone. The empty tube has the lowest value of average crushing load P_{av}. A very large difference of P_{av} is found between the corrugated tubes which are filled with filler and the empty corrugated tube. The corrugated tube which is filled with rice husk has the lowest value of P_{av} of the corrugated tubes with fillers, the maximum P_{av} for the corrugated tubes which are filled with coconut fibre and palm-oil fibre. The circular tube which is filled with coconut fibre has the lowest value of P_{av}, the corrugated tubes which are filled with palm oil fibre and the empty tube have smaller values of P_{av} but they are greater than the circular tube which is filled with coconut fibre. The circular tubes which are filled with aluminium chips and wood chips have large values of P_{av}. The maximum value of P_{av} is circular tube filled with rice husk.

3.2.3 Load Ratio (LR)

Load ratio (LR) can be defined as the ratio between the initial crushing and the maximum crushing during the process. This parameter is very important to study the failure modes during the crushing process. If the load ratio is less than 1 it means the structure is initially crushed by a matrix failure, if the load ratio equals 1, it means that the structure is initially crushed in ultimate failure mode. The results show that the empty corrugated tube has the lowest LR. This value is increased for the corrugated tube with coconut filling.

3.2.3 Total Energy Absorbed (TEA)

Total energy absorbed is the energy that can be absorbed from the structure and this value can be obtained from the area under the load displacement curve. The empty corrugated tube has the lowest energy absorption; very large differences can be recognized between the empty corrugated tube and the corrugated tubes which are filled by filler material. The corrugated tube which is filled with wood chips has the maximum energy absorption, and there are very slight differences between the filled corrugated tubes as shown in Table 2.

3.2.3 Specific Energy Absorption (SEA)

Specific energy absorption (SEA) means the energy absorption divided by the weight. It is one of the most important parameters because weight is an important factor in reducing the load and this leads to reduced fuel consumption for cars, motorcycles and airplanes. The corrugated tube which is filled with aluminium chips has the lowest value of specific energy absorption (SEA) because it has a relatively heavy weight. When it is compared with the other fillers, for corrugated tube with coconut fibre has the maximum specific energy when compared with other filler materials.

3.2.3 Volumetric Energy Absorption (VEA)

Volumetric energy absorption (VEA) is the ratio between the energy absorption and the volume of the structure. The corrugated tube filled with palm-oil fibre has the largest value due to high efficiency in absorbing the energy, while the empty corrugated tube had the lowest value of (VEA) because it is the least efficient type in absorbing energy. The other filled corrugated tubes had varied values but close to one another. The values of the (VEA) for all types of circular tubes varied slightly, but also it shows that the circular tube filled with wood chips had the largest value of (VEA), while the circular tube filled with aluminium chips had the smallest value of (VEA). The other filled circular tubes along with the empty tube had similar values of (VEA) between one another.

3.2.3 Crush Force Efficiency (CFE)

Crushing Force Efficiency is the ratio between the average load and the maximum load (P_{av}), (p_{max}) respectively. The circular tube with palm oil-fibre filler has lowest crushing force efficiency, of the tubes which are filled with filler materials, the corrugated tube with coconut fibre has the maximum crushing force efficiency but the empty corrugated tube has the maximum value of crushing force efficiency as shown in Table 3.

Table 2. Load Parameters.

Specimen	P_i	P_{max}	P_{av}	LR
		kN		kN / kN
ECT	28.90	30.02	27.31	0.962692
CTRH	37.20	48.89	32.5	0.760892
CTW	38.01	38.01	30.94	1
CTCF	27.28	30.56	26.76	0.89267
CTAL	33.39	48.99	30.47	0.681568
CTPOF	27.75	48.69	28.36	0.569932
COE	22.87	48.52	31.94	0.471352
CORH	42.50	48.89	39.9	0.869298
COW	48.77	48.765	42.08	1
COC	46.65	55.64	43.06	0.83848
COAL	42.71	46.56	41.5	0.917354
COPOF	48.44	50.65	43.81	0.956328

Table 3. Crashworthiness Parameters

Specimen	EA(kJ)	SEA(kJ/kg)	EV (kg/m3)	CSR	CFE
ECT	2.411	18.982	3218.496	0.883	0.910
CTRH	2.358	11.729	3147.684	0.725	0.665
CTW	2.522	13.275	3367.586	0.815	0.814
CTCF	2.322	13.046	3100.444	0.868	0.876
CTAL	2.153	7.176	2874.371	0.707	0.622
CTPOF	2.405	11.034	3211.609	0.848	0.582
COE	2.803	19.883	3571.258	0.878	0.658
CORH	3.517	16.284	4480.592	0.882	0.816
COW	3.691	18.184	4702.293	0.877	0.863
COC	3.778	20.646	4813.066	0.877	0.774
COAL	3.663	11.704	4666.503	0.883	0.891
COPOF	3.856	16.270	4912.133	0.880	0.865

3.3. Conclusions

The 36 specimens were fabricated using woven roving fibre glass E-600 with epoxy. Two different geometries have been used in this study circular and corrugated with different types of fillers wood chips, rice husk (RH), coconut fibre (C), aluminium chips (AL) and palm-oil fibre (P) by comparing these fillers with a void tube(E). These specimens where subjected to a quasi-static compression load. The results show that the corrugated tubes are better than the circular tubes for the crashworthiness parameters, initial load, maximum load, average load, energy absorption, crush force efficiency and crush strain relation. All these parameters are better in corrugated tubes when compare with circular tubes as shown in Tables 1, 2. The corrugated tube which is filled with coconut fibre has a higher value of specific energy absorption. The effect of the filler is to increase the area under the curve in load-displacement curves. The corrugated tube which is filled with coconut-fibre filler (COC) has the highest energy absorption of all the fillers.

Acknowledgments
The authors would like to thank Curtin University Sarawak, for the support in this research program.

References

[1] Darshil U. Shah, Peter J. Schubel, Mike J. Clifford, Peter Licence, *Fatigue life evaluation of aligned plant fibre composites through S–N curves and constant-life diagrams*, Composites Science and Technology 74 (2013) 139–149.
[2] Hakim S. Sultan Aljibori, Rahezar Ramle, Haidar Al-qrimli, Chong W. P, *Quasi-Static Axial Crushing of a ±45° Fiber Orientation Angle with a Different Number of Layers: Part 1*, Experimental Study American Journal of Scientific Research ISSN 1450-223X, EuroJournals Publishing, Issue 4, (2009) 72-84.
[3] Giovanni Belingardi, Maria Pia Cavatorta, Davide Salvatore Paolino. *Repeated impact response of hand lay-up and vacuum infusion thick glass reinforced laminates*. Mechanical Engineering Department, Politecnico di Torino, Corso Duca degli Abruzzi, 24, 10129 Torino, Italy, International Journal of Impact Engineering 35, (2008) 609–619.
[4] Liang S, Gning PB, Guillaumat L, *A comparative study of fatigue behaviour of flax/epoxy and glass/epoxy composites*, Compos Sci Technol 72 (2012) (5): 535–43.
[5] Hakim S. Sultan Aljibori, Haidar F. Al-Qrimli, Rahizar Ramli, E. Mahdi, Faris Tarlochan and Chong W. P, *A Comparative Analysis of Experimental and Numerical Investigations of Composite Tubes under Axial and Lateral Loading*, Australian Journal of Basic and Applied Sciences, ISSN 1991-8178, 48 (2010) 3077-3085.
[6] Derek Hull, prof. R. Cahn, A. Davis, professor I. M. Ward, *An introduction to composite materials (Cambridge solid state science series)*, Cambridge. ISBN 0 521 239911111 5, (1990)
[7] Tao Yin, Min Zhi Rong, Jingshen Wu, Haibin Chen and Ming Qiu Zhang, *Healing of impact damage in woven glass fabric reinforced epoxy composites*, J Applied Science and Manufacturing, 39 (2008) 1479-1487.
[8] Denial Gay, Suong V. Hoa, Stephen W. Tsai, *Composite materials design and applications*. CRC Press printed in united states of America. No. 1-58716-084-6, (2003).
[9] Warrior N.A, T.A. Turner, F. Robitaille, C.D. Rudd, *Effect of resin properties and processing parameters on crash energy absorbing composite structures made by RTM*, School of Mechanical, Materials, Manufacturing Engineering and Management, University of Nottingham, Nottingham NG7 2RD, UK, (2003)
[10] Bank, L.C. and Gentry, T.R, *Development of a pultruded composite material highway guardrail*. Composites Part A: Applied Science and Manufacturing, 32 (2001) 1329-1338.
[11] Mallick P.K, *Fiber Reinforced Composites Materials, Manufacturing, and Design*, By Taylor and Francis Group, Third Edition, ISBN 10987654321, (2007)
[12] Menachem Lewin., *Fiber Chemestry*, 3rd edition. New York, Taylor & Frances Group. (2007)
[13] A.S. Abosbaia, E. Mahdi, A.M.S. Hamouda, B.B. Sahari, A.S. Mokhtar, *Energy absorption capability of laterally loaded segmented composite tubes*, Composite Structures, 70 (2005) 356-373.
[14] L.S. Sutherland, C. Guedes Soares, *Effects Of Laminate Thickness And Reinforcement Type On The Impact Behaviour Of E-Glass/Polyester Laminates*, Composites Science Technology, (1999).
[15] Andrew Makeev, *Interlaminar shear fatigue behavior of glass/epoxy and carbon/epoxy composites*, Composites Science and Technology 80 (2013) 93–100.

Experimental Investigation and 3D Finite Element Prediction of Temperature Distribution during Travelling Heat Sourced from Oxyacetylene Flame

Adam Umar Alkali[1], Turnad Lenggo Ginta[2] and Ahmad Majdi Abdul –Rani[3]
[1]Department of Mechanical Engineering, Universiti Teknologi PETRONAS, Bandar Seri Iskandar, Tronoh, Malaysia
[2, 3]The Rehabilitation Rehabilitation and Biomedical – Mechanic Research, Bio – Medical Technology MOR, Universiti Teknologi PETRONAS, Bandar Seri Iskandar, Tronoh, Malaysia

E-mail: adamalkalee@yahoo.com, turnad.ginta@petonas.com.my, majdi@petronas.com.my

Abstract. This paper presents a 3D transient finite element modelling of the workpiece temperature field produced during the travelling heat sourced from oxyacetylene flame. The proposed model was given in terms of preheat–only test applicable during thermally enhanced machining using the oxyacetylene flame as a heat source. The FEA model as well as the experimental test investigated the surface temperature distribution on 316L stainless steel at scanning speed of 100mm/min, 125mm/min 160mm/min, 200mm/min and 250mm/min. The parametric properties of the heat source maintained constant are; lead distance L_d =10mm, focus height F_h=7.5mm, oxygen gas pressure P_{oxy}=15psi and acetylene gas pressure P_{acty}=25psi. An experimental validation of the temperature field induced on type 316L stainless steel reveal that temperature distribution increases when the travelling speed decreases.

1. Introduction

Several research efforts continue to reveal promising technique to improve machinability of difficult-to-cut materials. Thermally enhanced machining (TEM) has demonstrated effective potentials which provide satisfactory results in terms of large material removal rate, prolong tool life, improved surface finish and reduction of cutting forces among others [1-3] Various heat sources are used during the TEM process [4-6]. However, parameters of heat source like temperature distribution, spot diameter of the heat source, and heat affected zone on the substrate (HAZ) have been the major challenges in TEM processes. Current research efforts continue to proposed promising solution to overcome this set back. Singh et al [7] developed a 3D Gaussian laser heat source to predict the temperature distribution in the workpiece material at different scanning speed. Their investigation revealed that the thermal model can be used to resolve the laser parameters that will yield no residual HAZ in the material after TEM process. Dandekar et al [2] developed a 3D finite volume thermal model to predict the material removal temperature to access machinability improvement via LAM. The study predicts the optimum material removal temperature for LAM. Evidently, research trend is dominant on laser heat source as a test bench mechanism. Other heat sources like oxyacetylene flame, liquid petroleum gas flame,

[1] adamalkalee@yahoo.com.

Induction heating etc. had been underprivileged. The potentials of other heat sources were not fully explored. Even so, the oxyacetylene flame has revealed excellent potentials for thermal softening during TEM process. Davami and Zadshakoyan [8] used oxyacetylene gas flame for TEM on AISI 1060 steel using centre lathe. His report observed decrease in tool wear and improved surface finish. Additionally, oxyacetylene flame has revealed significant machinability improvement during turning AISI 1060 nickel chromium steel by Mukherjee and Basu [9]. Despite these potentials, the oxyacetylene flame was not privileged to benefit from numerical control of thermal properties in order to enhance its productivity. Generally, there has been a difficulty with the visual access to the entire thermal field covered by the impinging flame. Numerical Estimation of the oxyacetylene heat flux alongside predicting the temperature distribution will enable better utilization and enhanced its productivity in TEM process. This study intends to unveil the potentials of oxyacetylene flame and presents a numerical procedure for controlling the temperature distribution during a moving heat typically employable while thermally enhanced machining. The goal is to propose the technique which will be used to enhanced machinability of difficult to cut materials.

2. Thermal Modelling

Ansys V.14 was used to develop a 3D transient finite element model. A solid 70 element was used with element edge size of 0.002mm. The thermal model that has been developed determines the temperature distribution from a travelling heat source from oxyacetylene flame impinging on a work piece surface during a preheat-only test. This was meant to predict the thermal field that was developed due to flame preheats at different scanning speed. The model was developed based on certain assumptions as summarized below:

- Heat generation due to machining did not occur.
- Physical and thermal properties of type 316L stainless steel were considered. Thermal conductivity, specific heat capacity and density of the material are obtained elsewhere in [10]. Even so, the thermal conductivity was assumed to be isotropic.
- A full model was considered which mimic the actual workpiece dimension that was considered during the experimental test (200mm (L) x100mm (W) x 12mm (T)) which corresponded to x, y, and z direction respectively.
- Localized preheat sourced from travelling oxyacetylene flame induces non-uniform temperatures on the workpiece. For this reason, an average convection coefficient was used.
- Heat flux impinging on type 316L stainless steel was assumed to be constant around the elements covered at any particular spot diameter.
- Other five faces of the solid exchange heat by convection and radiation with the surrounding environment.
- A 3D transient heat conduction equation for the thermal problem takes the form given as follows:

$$\underbrace{\int_{A_1}\left(q_l^{''}d(A_T)\right)}_{\text{Energy from Oxyacetylene flame}}+\underbrace{\int_{vol}\begin{pmatrix}\frac{\partial}{\partial x}\left(k(T)\frac{\partial T}{\partial x}\right)+\\ \frac{\partial}{\partial y}\left(k(T)\frac{\partial T}{\partial y}\right)\\ +\frac{\partial}{\partial z}\left(k(T)\frac{\partial T}{\partial z}\right)\end{pmatrix}d(vol)}_{\text{Energy as a result of conduction heat transfer}} \tag{1}$$

Where A_I is the area covered by the oxyacetylene flame diameter and A_T are the workpiece areas which are open to free convection and radiation heat transfer. Equation 1 represents the heat flux from the combustion flame of oxyacetylene and the resulting heat transfer in the bulk material. However, the area A_T is associated with free convection and thermal radiation and heat flux from the oxyacetylene flame source. Thus the boundary condition can be defined as given in equation 2.

$$k\frac{\partial T}{\partial f}A_{1t>0} = q_1'' - q_{con}'' - q_{rad}'' \tag{2}$$

Where q_1'' is heat flux from the oxyacetylene heat flame and

q_{con}'' is convective heat given as

$$q_{con}'' = h(T_s - T_\infty) \tag{3}$$

While q_{rad}''

$$q_{rad}'' = \varepsilon\sigma(T_s^4 - T_\infty^4) \tag{4}$$

Figure 1 shows a flame administered region of a three dimensional finite element model. Simulations are performed with 4 different scanning speed for the travelling heat sourced from oxyacetylene flame. Heat flux was used as the surface load and was estimated from actual flame heat of oxyacetylene at the ratio of 1.52 as calculated from the stoichiometric equation 5 and 6 also found in [11, 12].

$$2C_2H_2 + 5O_2 = 4CO_2 + 2H_2O + HEAT \tag{5}$$

$$-Q = \sum[\Delta H_f^0]_p \times n_p] - \sum[\Delta H_f^0]_f \times n_f] \tag{6}$$

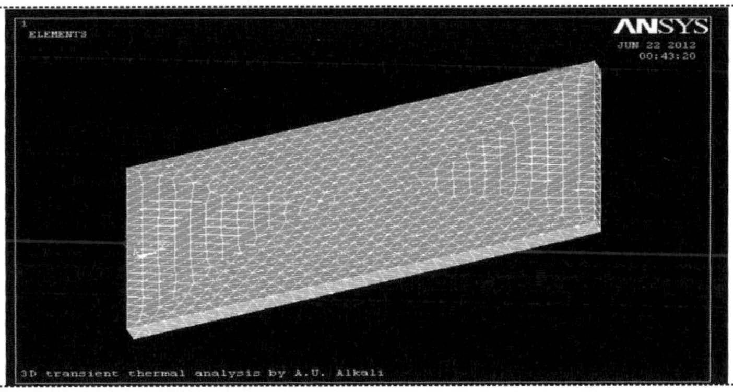

Figure 1. Isotropic Three Dimensional FEA Model

Figure 2 predicted a temperature distribution on preheated sample of 316L stainless steel during a travelling oxyacetylene flame heat source. The FEA model reveals that the maximum temperature occur at the region where the heat flux is imposed. It was discovered that when the oxyacetylene heat flux impinged on the work material, the temperature was instantly elevated within the initial few seconds. Thereafter, the temperature steadily increases while the heat source travels along the span of the work material. The heat approaching region (which will accommodate the flame spot) pre-absorbs the heat energy through conduction heat transfer and as a result, temperatures increases and distributed inwardly whereas the scanned region behind the flame spot gradually cool down due to free convection and thermal radiation. In this case as shown in figure 3, the maximum temperature distributed region occurred at the end of the scanned time. The FEA models show that when the time step size is low, the maximum temperature induced will also be low just as the temperature distribution appears to be less. The maximum temperature region is represented in the model with a symbol MX and appeared reddish in colour print. This region fades outwards the surroundings, indicating a conduction heat to the bulk material.

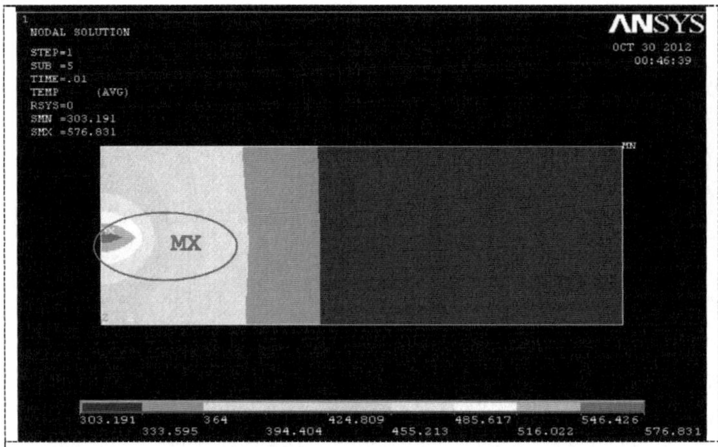

Figure 2. FEA Temperature Distribution at 10 Seconds during a Travelling Heat Source at f=100mm/min

It was notable in the two models shown that maximum temperature distribution induced on the work piece material was influenced by the resident time. This corresponds to the speed of the travelling heat source.

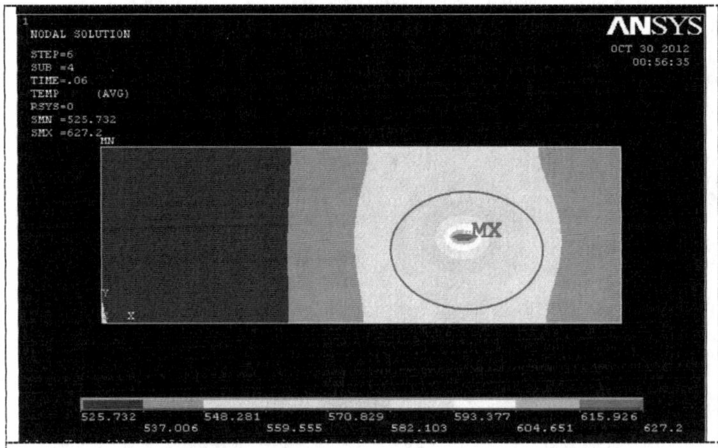

Figure 3. FEA Temperature Distribution at 60 Seconds during a Travelling Heat Source at f=100mm/min

3. Experimental Set up

A combustion flame of oxyacetylene was made to travel along the span of type 316L stainless steel material at varying table feed on vertical milling centre VMC. The chemical composition of the 316L stainless steel used was given in table 1. The nozzle of the impinging oxyacetylene flame was set to strike on the workpiece surface at an angle of 90^0. The parameters used during the experimental test are given in table 2. T640 Flir thermal imaging camera was used to capture the temperature history as the flame travelled along the span of the work piece material. An insulation of parallel strips was used to avoid the clamped workpiece material from having direct contact with the machine vice. This was purposely cushioned to avoid heat losses from the preheated workpiece to an external machine member.

Table 1. Nominal Composition of 316L Stainless Steel

C	Mn	Si	P	S	Cr	Mo	Ni	N	Fe
.03	2.00	.75	.045	.03	18.67	3.16	10.00	.10	Bal

Table 2. Experimental Test Condition

Factors	Travelling Speed (mm/min)	Lead Distance (mm)	Focus Height (mm)	Oxygen Pressure (psi)	Acetylene Pressure (psi)	Incidence Angle(^{0})
Fixed		10	7.5	25	15	90
Varied	100, 125, 160, 250					

4. Experimental Result and Discussion of FEA Model Validation

Figure 4 shows the schematic of moving heat sourced from oxyacetylene gas flame during the preheating-only test using different table feed rate as the scanning speed of the travelling heat source. Experimental record shows that temperature distribution drops from 6600C to 6580C with slight differences of travelling speed of the heat source when f = 100 mm/min and f = 125 mm/min when other parameters of the heat source were maintained constant. The parametric properties of the heat source are lead distance Ld =10mm, focus height Fh=7.5mm, oxygen gas pressure Poxy=15psi and acetylene gas pressure Pacty=25psi.

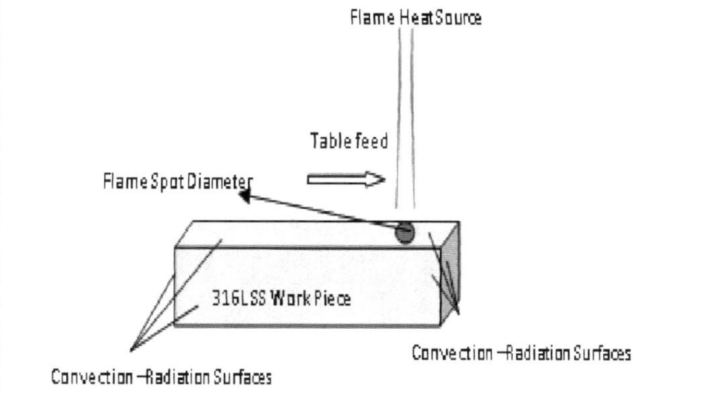

Figure 4. Schematic of The Travelling Heat Source from Oxyacetylene Flame

During the preheating period shown in figure 5, the workpiece was at room temperature (25^{0}C). At the initial few seconds of preheating the temperature rises quickly. Consequently, the temperature grows steadily. The maximum temperature occurs at the end of the preheat time. Indeed, the maximum temperature occurred at the flame covered region. This phenomenon concurs well with the FEA model. When the workpiece advances forward, the oxyacetylene flame heat source gradually moves away from one end along the span of the work material to the other end. Amidst the span of the workpiece, the temperatures slightly become constant.

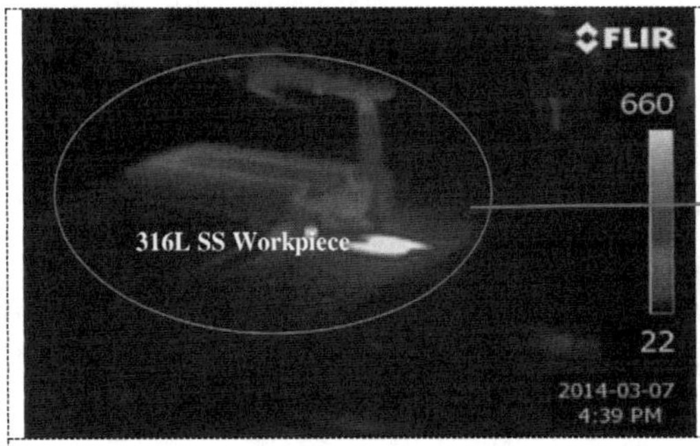

Figure 5a. Experimental Validation of a Preheat-only Travelling Heat Sourced from Oxyacetylene Flame at f = 100 mm/min the caption.

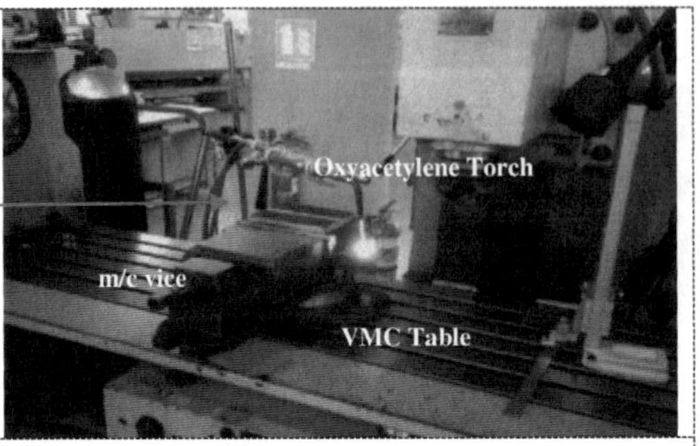

Figure 5b. Experimental Validation of a Preheat-only Travelling Heat Sourced from Oxyacetylene Flame at f = 100 mm/min. Justify the caption.

It was concluded that during this situation, the temperature induced was transitionally conducting to the bulk material. However, when the travelling speed was varied wider to f = 250mm/min, the differences in the temperature distribution increased accordingly. It was evidently discovered that the distributed temperature increased when the travelling speed decreased. This pattern was equally the same case revealed in the FEA model for the travelling heat source from oxyacetylene flame, wherein the faster the heat source travels across the span of the workpiece the lesser temperature will be distributed in the workpiece material. Figure 6 revealed that when the travel speed is low, the resident time taken to preheat the workpiece material becomes longer and results to higher temperatures induced during the preheating period. When the feed rate increases, less heat energy is induced into the workpiece thereby resulting to low temperature distribution. Previous research on the TEM using other heat source [13] have reported that longer resident time spent during the preheat will possibly influences the maximum temperature on the scanned work piece material during thermally enhanced machining.

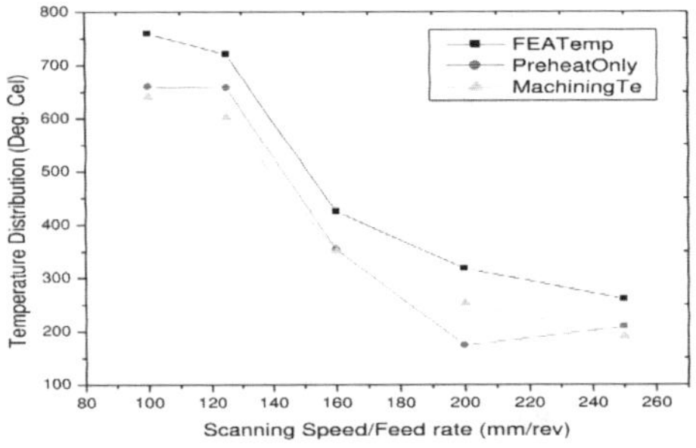

Figure 6. Influence of Travelling Speed on Temperature Distributed during a Preheat –only Test on Type 316L Stainless Steel

Even so, this study has discovered that the FEA simulated temperature history was slightly higher than the measured results with thermal imaging camera. Table 3 shows a contrast between the simulated and the measured temperature peak. A strong correlation was evident in the two investigated mode. It was discovered that the FEA simulated temperature prevails with a slightly higher peak than the experimented temperatures within the scope of the travelling speeds investigated. Although the variation was insignificant, it was concluded that the pattern of the temperature distribution remained similar at both conditions. The discrepancy was believed to be due to the constant terms of thermal losses which were used in the model. The convective and radiation heat transfer might have varied during the actual experimental test. Eventually, the FEA model could justifiably be used to predict and evaluate the behaviour of oxyacetylene flame during a travelling heat process. It should be noted that the value of the feed rate is the same as the flame travelling speed while using the vertical milling centre.

Table 3. Summary of the Simulation and Experimental Temperature Distribution During Travelling Heat Sourced From Oxyacetylene Flame

Travelling Speed (mm/min)	100	125	160	200	250
Predicted Temperature Distribution (^{0}C)	660	658	354	274	209
Experimented Temperature Distribution (^{0}C)	640	602	351	253	191

5. Conclusion
The parametric study on preheating-only experiments reveals that low scanning speed was characterized with high preheat time which results to high temperature distributed on AISI 316L stainless steel workpiece material. The investigation shows that when oxyacetylene flame is maintained with a constant heat flux, varying temperature distribution can parametrically be achieved with either increasing or decreasing the speed of the travelling heat source. In other words, while using the vertical milling centre (VMC), the feed rate influences the travelling speed along the span of the work piece material which consequently control the resident time of the flame preheat.

Acknowledgement
The authors wish to thank the center for graduate studies of Universiti Teknologi PETRONAS as well as bio–medical Technology MOR, Universiti Teknologi PETRONAS for their kind sponsorship during the conduct of this study.

References
[1] Ginta T. L, Amin A, Lajis M. A, Karim A, Daud C, and Radzi M 2009 *Eur. J. Sci. Res.* **27** 384-391
[2] Dandekar C. R, Shin Y. C, and Barnes J 2010 *Int. J. Mach. Tool Manu.* **50** 174-182
[3] Arriola I, Gonzalo O, and Leunda J 2013 *Int. J. Mach. Tool Manu.***74** 19-28
[4] Lajis M. A, Nurul Amin A, Karim A, and Hafiz A 2010 *Adv. Mat. Res.* **83** 56-66
[5] Leshock C. E, Kim J.-N, and Shin Y. C 2001 *Int. J. Mach. Tool Manu.* **41** 877-897
[6] Maity K. P and Swain P. K 2008 *J. Mater. Process Tech.***198** 344-349
[7] Singh R, Alberts M. J and Melkote S. N 2008 *Int. J. Mach. Tool Manu.* **48** 994-1004
[8] Davami M and Zadshakoyan M 2008 *WASET* **22**
[9] Mukherjee P and Basu S 1979 *Int. J. Prod. Res.***11** 21-36
[10] Metals C 2013 *Stainless Steel 316L in Sheet, Plate Bar*
[11] Alkali A. U, Ginta T. L and Abdul-Rani A. M 2014 *Adv. Mater. Res.* **845** 790-794
[12] Ghajar A. J and Bang K 1993 *Heat transfer eng.* **14** 48-59
[13] Shen X and Lei S 2009 *J. manuf. Science eng.* **131** 051007

Bad Actors Criticality Assessment for Pipeline system

Meseret Nasir [1], Kit wee Chong[2] Sabtuni Osman [3] Wee Siaw Khur[4]
[1,2] Department of Mechanical Engineering
Curtin University Sarawak Miri, Malaysia
[3]Department of Reliability and Maintenance
Petronas Carigali Sarawak Operations Miri, Malaysia
[4]Department of Chemical Engineering
Universiti Malaysia Sabah, Malaysia

E-mail: meseretnasir@curtin.edu.my

Abstract. Failure of a pipeline system could bring huge economic loss. In order to mitigate such catastrophic loss, it is required to evaluate and rank the impact of each bad actor of the pipeline system. In this study, bad actors are known as the root causes or any potential factor leading to the system downtime. Fault Tree Analysis (FTA) is used to analyze the probability of occurrence for each bad actor. Birnbaum's Importance and criticality measure (BICM) is also employed to rank the impact of each bad actor on the pipeline system failure. The results demonstrate that internal corrosion; external corrosion and construction damage are critical and highly contribute to the pipeline system failure with 48.0%, 12.4% and 6.0% respectively. Thus, a minor improvement in internal corrosion; external corrosion and construction damage would bring significant changes in the pipeline system performance and reliability. These results could also be useful to develop efficient maintenance strategy by identifying the critical bad actors.

1. Introduction

Over the past few decades, the pipeline has been widely used in the oil and gas industry to transport unrefined fossil fuels from upstream to downstream for further refinement process prior to the delivery of purified fuels to consumer [1]. It is coherent that the majority of the existing pipelines in the world are aged which subjected to several external and internal factors. One of the major causes of pipeline failures is due to internal and external corrosion. These can also be caused by the extreme weather conditions, mechanical failures, manufacturing defects, external forces as well as operational errors. According Mustaffa [2], corrosion contributes approximately 49% to the cause of pipeline failures in USA while failures due to other hazards, maritime activities and natural forces are about 25%, 14% and 12% respectively. The failure of a pipeline may cause damage to the society, environment as well as economic impacts due to financial losses [3]. For instance, the explosion of pipeline in North Sea oil production platform resulted in a total loss of approximately £1.7 billion. Further, the explosion of gas pipeline located somewhere in between Lawas and Long Sukang in the northernmost district of Sarawak resulted in an estimated loss of RM4 billion due to temporary shutdown. This incident also causes for the evacuation of nearby villagers with some houses and vehicles reportedly destroyed [4].

[1] Corresponding author- Meseret Nasir, meseretnasir@curtin.edu.my.

IOP Conf. Series: Materials Science and Engineering **78** (2015) 012026 doi:10.1088/1757-899X/78/1/012026

The failure and downtime of pipeline system will greatly reduce the financial benefit of the plant in terms of downtime cost, maintenance cost and production loss. Therefore, in order to reduce pipeline failure, it is also crucial to understand which bad actors contribute more to the overall system performance and reliability. In reliability terms, Bad Actors generally represent for plant equipment tags or process systems which experience huge loss, such as repair cost and production loss of repetitive failures [5]. In other words, bad actors are also known as the root cause or any potential factor leading to the system downtime. Several researches have been conducted to evaluate the failure of pipeline systems. Sun, et al, [6] predicts the reliability of pipeline with different preventive maintenance strategies by using a split system approach. However, this paper discusses only for a general reliability prediction model without considering possible bad actors. Ossai [7] used Monte Carlo Simulation and degradation models to predict corrosion rates and reliability of oil and gas pipelines. Despite achieving high accuracy in predicting the corrosion rate of pipelines, this combined methodology does not analyze for other factors of pipeline failure. This is also observed in other study proposed by Lee and Kim [8] where only failure prediction of pipeline due to corrosion is addressed through Finite Element Analysis (FEA). To efficiently mitigate the failure of pipeline, it is essential to address all possible bad actors instead of just focusing on corrosion. One of the decent way to understand which failure modes contribute more to the overall system performance and reliability is by incorporating Fault Tree Analysis (FTA). FTA was extensively used as a failure analysis tool by reliability experts in system safety, reliability engineering and in all major fields of engineering such as Nuclear Power industry, Chemical industry, Robotics industry and Software industry [9]. However, it has limited application in the oil and gas industry.

In this paper, FTA is used to evaluate the pipeline system failure. FTA is able to generate quantified evidence from historical database with the aids of mathematical computation. Further, the obtained results are able to rank the failure cause according to the likelihood of occurrence which will lead to application of the decent maintenance procedure, thus reduced downtime and smoother commissioning. Thus, in this study, FTA and BICM were used to evaluate the probability of occurrence and rank the impact for each bad actor in the pipeline system.

2. Methodology

2.1. Fault Tree Analysis (FTA)

This study used FTA to analyze the probability of each bad actor. FTA is a graphical approach which describes the interaction of failures and other events in a system. Figure 1 shows the structure of FTA. Basic events are usually depicted at the bottom of the FTA and are connected to the top events through logic symbols, known as gates. Common events and gates are shown in Table 1. The top event denotes the detected hazards or system bad actors.

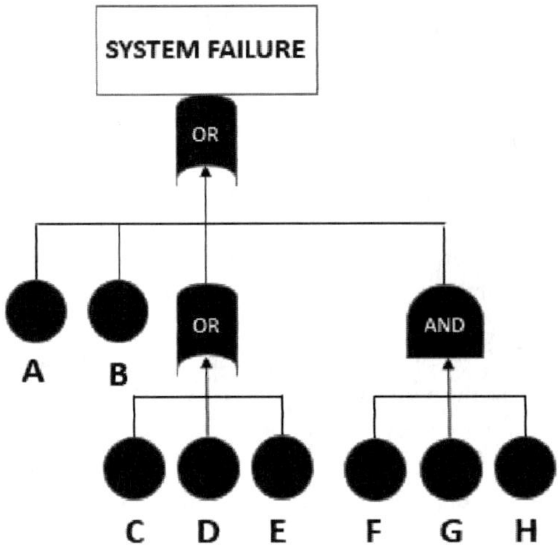

Figure 1: Fault Tree Approach

Table 1: Common symbols and logic gates used in FTA (Adapted from [10])

Symbol Name	Symbol	Descriptions
Basic Event		A basic initiating fault (A, B, C, D.....)
Undeveloped Event		A basic event without further developed
Top Event		An event that is depending on the logic of input events
OR Gate		The output occurs if at least one of the event occurs
AND Gate		The output occurs only if all the input events occur

According to Nouri Gharahasanlou, et al. [5], it is essential to estimate the probability of occurrence for the logic gate output fault events in order to estimate the probability of failure for the top event. The corresponding equations to estimate the probability of occurrence of "OR" and "AND" logic gates output fault events are presented in Equation (1) and (2) respectively.

$$Q_o(t) = 1 - \prod_{i=1}^{n} \left(1 - q_i(t)\right) \tag{1}$$

$$Q_o(t) = \prod_{i=1}^{n} \left(q_i(t)\right) \tag{2}$$

where $Q_o(t)$ is the probability of occurrence of output fault or top event, $q_i(t)$ is the probability of occurrence for each bad actor with a function of time, for i = 1, 2, 3, …,n.

2.2. Birnbaum's Importance and Criticality Measure (BICM)

Importance measure is used to evaluate and rank the impact of individual bad actors within a system. These importance measures provide a numerical rank to determine which bad actors are more important or critical to system failure. BIM is a partial derivative of system reliability with respect to individual failure mode probability [11] . The index gives an indication of how the pipeline failure probability will change with changes in each failure mode probability. Analytically, this is defined by Equation (3).

$$I_K^B(t) = \frac{\partial Q_o(t)}{\partial q_i(t)} \tag{3}$$

$I_K^B(t)$ is reliability importance of the kth failure mode,

Whereas the Birnbaum importance provided the probability that a given failure mode would be responsible for the failure at time t, another measure, "criticality importance" is used to determine the probability that the given failure was responsible for system failure before time t. This measure is given by Equation (4).

$$I_K^C(t) = I_K^B(t)\frac{\partial q_k(t)}{\partial Q_o(t)} \tag{4}$$

$I_K^C(t)$ is criticality of the kth failure mode

3. Results and discussion

3.1. Case study

In this study, the Liquids Pipeline Delivery Network is considered. Table 2 presents the pipeline failure data in Alberta from year of 1990 to 2012 [12]. There are a total of 13 causes of pipeline failure being analyzed such as the internal corrosion, external corrosion, failures due to weld, construction damage, overpressure, pipe, joint, earth movement, valve and fitting, operator error, miscellaneous, damage by others and some undetectable causes. By using failure incidents data for 23 years, Mean Time Between Failure (*MTBF*) can be calculated using Equation (5).

$$MTBF = \frac{Total\ Operating\ time}{Number\ of failure\ Incident} \tag{5}$$

Based on *MTBF*, the two parameters Weibull distribution was used to estimate the shape (β) and scale (η) parameters. Weibull distribution is the most efficient distribution in life data analysis and in reliability calculations which involve time related failures is employed to model the wear out or fatigue failure of pipeline system. The β and η can be estimated by using Maximum likelihood estimation method [13]. The likelihood function is given by:

$$L = \prod_{i=1}^{n} \frac{\beta}{\eta^{\beta}} t^{\beta-1} \exp\left[-(\frac{t}{\eta})^{\beta}\right] \tag{6}$$

where β, beta is the shape parameter & η, eta is the scale parameter or characteristic life

Differentiating Equation (6) with respect to β and η, and equating to zero can yield Equation (7) and Equation (8).

$$\frac{\partial L}{\partial \beta} = \frac{n}{\beta} + \sum_{i=1}^{n} ln^{t_i} - \frac{1}{\eta}\sum_{i=1}^{n} t_i^{\beta} lnt_i = 0 \tag{7}$$

$$\frac{\partial L}{\partial \beta} = -\frac{n}{\eta} + \frac{1}{\eta^2}\sum_{i=1}^{n} t_i^{\beta} = 0 \tag{8}$$

where t_i is the time between failures & n is the sample size

The estimated value of β and η were presented in Table 2. As can be seen from Table 2, the value of β is less than one which means most of the failures occur in the early stage or the first few kilometers of the pipeline.

Table 2: Parameter estimation for each Bad actors

Causes of Failure	Total Time, T (years)	Number of Incident, N	*MTBF*	β	η
Internal Corrosion	23	9024	2.548×10-3	0.857682	0.002695
External corrosion	23	2098	0.01096	0.869435	0.011639
Third Party Damage	23	745	0.03087	0.877282	0.032817
Weld	23	489	0.04703	0.874761	0.050022
Construction Damage	23	888	0.02590	0.874461	0.02759
Overpressure	23	281	0.08185	0.875936	0.086991
Pipe	23	509	0.04519	0.874257	0.048019
Joint	23	614	0.03746	0.87746	0.039833
Earth Movement	23	281	0.08185	0.875936	0.086991
Valve/ Fitting	23	558	0.04122	0.873781	0.043826
Operator Error	23	271	0.08487	0.875316	0.09025
Miscellaneous	23	353	0.06516	0.875066	0.069225
Unknown	23	378	0.06085	0.875711	0.06467

3.2. Development of FTA

Figure 2 shows a fault tree diagram constructed based on the failure causes stated in Table 2. The failures of the pipeline are classified into five categories, mainly due to the external interference, mechanical failure, corrosion, incorrect operation and also other type of failure. These components are further divided into circular blocks known as basic events. External Interference is broken up into earth movement, construction damage and third party damage. Examples of earth movement are landslides, earthquakes, or any natural disaster that could lead to pipeline failure, whereas third party damages are generally caused by other parties such as third party excavation and interference. Construction damage are losses suffered during the construction process due to improper techniques applied, faulty alignment or inappropriate backfilling. Mechanical Failures are separated into pipe, joint, weld, valve and fitting. Pipe failure could be hydrogen induced cracking (HIC), fatigue, whereas joint failure are such as gasket failure, O-ring failure, internal joint coupling failure and mechanical coupling failure. Examples of weld defects are girth weld failure or seam rupture whereas valve and fitting failure are such as seal blowouts, pig trap failures and packing leaks. Corrosion could be of internal corrosion or external corrosion that occurs on the internal or external surface of the pipe, respectively, while failure due to incorrect operation are divided into operator error and overpressure failure. Lastly, in the category of others, pipeline failure can be caused by miscellaneous reasons such as failure of pump or compressors and some unknown causes are undetectable. For simplicity of analysis, only OR gate is applied in the development of fault tree model for this study, therefore, the pipeline will fail if any of these basic events occurred.

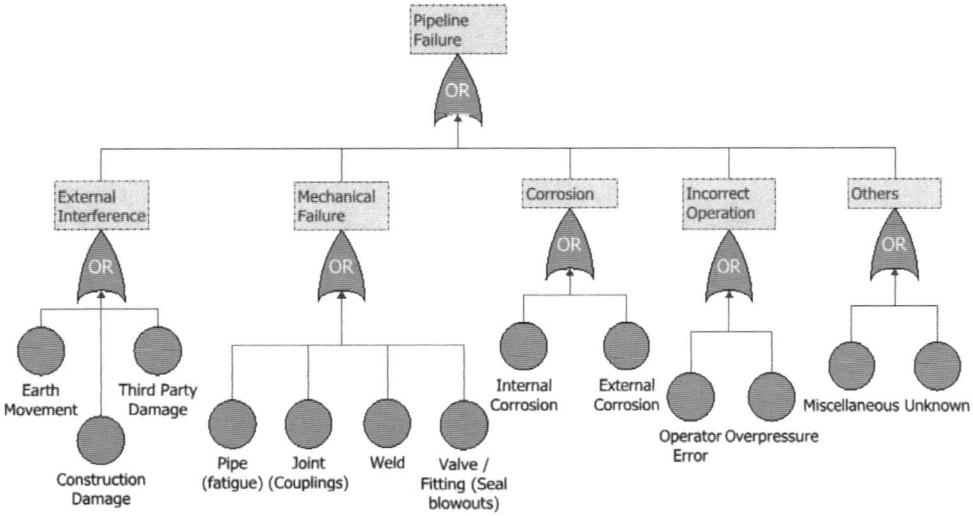

Figure 2: Fault Tree Model for Pipeline Failure

Once the fault tree is developed, various analytical plots can be produced for analysis. Figure 3 shows the system unreliability through time. As can be observed from Figure 3, the failure probability of the overall system increases as the time increases. At the beginning of operation, the unreliability of the system remains at zero and gradually increases. This indicates that the pipeline may experience higher probability of failure when the operation time is getting longer. In other word, as operation time increases, the pipeline system may expose to high failure risks due to internal corrosion, external corrosion, external interference, manufacturing defects, mechanical failure, incorrect operation and also some other causes.

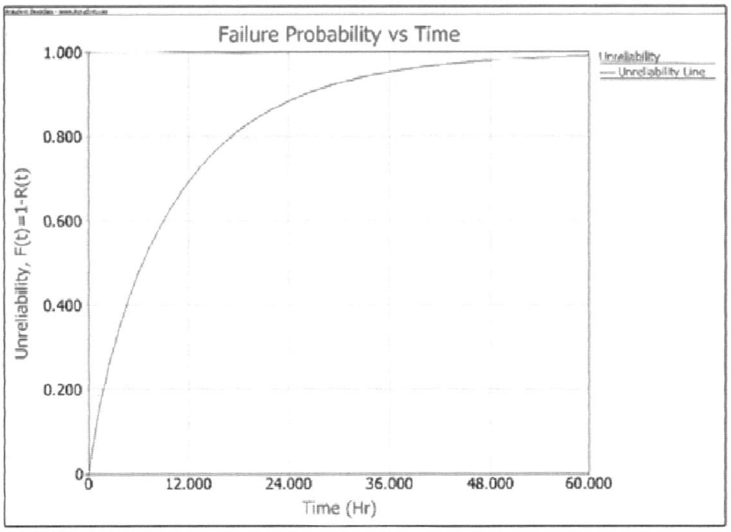

Figure 3: Probability of Failure for pipeline system

The pipeline can be failed due to several factors as presented in Figure 2 and Table 2. However, it is very important to identify which causes contribute more to the system reliability. Thus, in this study, the reliability importance analysis was conducted. In general, reliability importance can be defined as a measure of the relative importance of each component in a system with respect to the overall reliability of the system. Figure 4 shows the static reliability importance of twelve bad actors in the pipeline system at operation time of 50 hours. The result indicates that internal corrosion has a higher reliability importance of the blocks in the system at the constant time. It is then followed by external corrosion and construction damage being ranked as secondly and thirdly important. The remaining bad actors or components constitute almost the similar ranking of reliability importance as presented in the graph. Similarly, internal corrosion which is shown in red represents lowest reliability, blocks with yellow and green denote medium reliability and highest reliability. This explains that a small improvement in the internal and external corrosion could bring high system reliability compared to other factors. Therefore, it is very important to adopt an effective maintenance strategy to mitigate the internal and external corrosion in order to have high performance and reliability of pipeline systems.

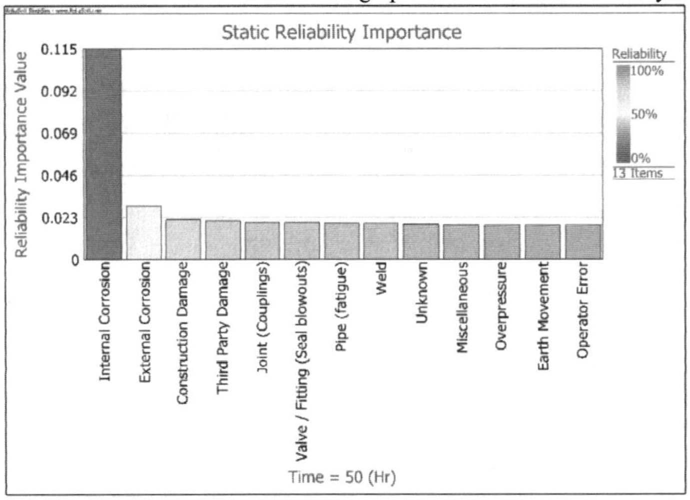

Figure 4: Static Reliability Importance for bad actors

Figure 5 shows the failure probability through time for each failure cause. At the beginning of the system operation, the failure probability of the system due to each bad actor remains at zero and gradually increases as time increases. As observed from Figure 5, internal corrosion, external corrosion and construction damage are three failure cases with the highest unreliability over time compared to other factors. It is then followed by third party damage, joint failure, valve, fitting, pipe, weld, unknown failures, miscellaneous while earth movement and operator error are the first two failure causes which achieve unreliability of one. This means that internal corrosion and external corrosion are the two most critical failure causes which have the lowest reliability over time among all the factors being analyzed and therefore should be prioritized in risk mitigation processes.

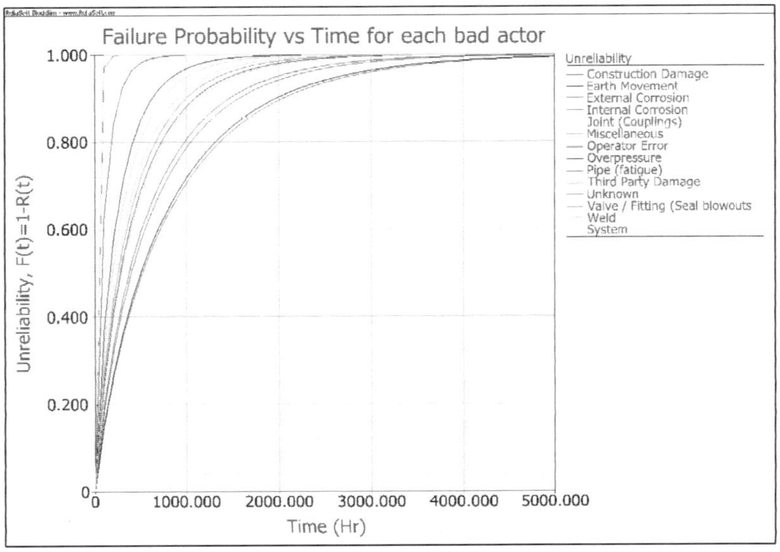

Figure 5: Failure probability for each Bad Actor

Table 3 shows the failure criticality index for each bad actor. The results indicate that internal corrosion and external corrosion are the most frequent causes of failure with percentage of 48.4% and 12.7% respectively. It is then followed by construction damage by 6.5%, whereas the remaining failure causes contribute only for negligible effects to the system overall failure. Therefore, failure causes which fall under higher failure occurrence ranking must be treated with extra caution and immediate corrective actions must be taken in order to reduce the risks to as low as reasonably practicable. Despite a failure ranking for other causes, any possible control measure such as periodic maintenance and inspection of the pipeline system are encouraged in order to mitigate risk or maintain the risk at acceptable levels.

Table 3: Failure criticality index for each Bad Actors

Failure Mode		Failure Criticality Index %	Expected Probability of Failure	Failure Occurrence Ranking
External Interference	Earth Movement	2.90	0.029	10
	Construction Damage	6.50	0.065	3
	Third Party Damage	4.20	0.042	6
Mechanical Failure	Pipe	4.30	0.043	5
	Joint	3.30	0.033	8
	Weld	4.70	0.047	4
	Valve Fitting	3.50	0.035	7
Corrosion	Internal Corrosion	48.40	0.484	1
	External Corrosion	12.70	0.127	2
Incorrect Operation	Operator Error	1.90	0.019	12
	Over Pressure	1.90	0.019	12
Others	Miscellaneous	2.70	0.027	11
	Unknown	3.00	0.030	9

4. Conclusion

In this paper, the impact and probability of occurrence for bad actors were elevated using FTA and BICM. The findings show that internal and external corrosions are highly contributed to unavailability of the pipeline system. The result of the study could also suggest which bad actor to be treated immediately to mitigate the overall risk using the Failure Criticality Index as indicated in Table 3. These results could also help to develop efficient maintenance strategy to increase the reliability and performance of the pipeline system.

References

[1] P. Hopkins, "Oil and Gas Pipelines: Yesterday and Today," Pipeline Systems Division (PSD), International Petroleum Technology Institute, American Society of Mechanical Engineers (ASME). Available online at http://www. engr. mun. ca/~ spkenny/Courses/Undergraduate/ENGI8673/Reading_List/2007_Hopkins. pdf, 2007.

[2] Z. Mustaffa, "System reliability assessment of offshore pipelines," PhD, Faculty of Civil Engineering and Geosciences, Delft Uni. of Technology, The Netherlands, 2011.

[3] A. Shahriar, R. Sadiq, and S. Tesfamariam, "Risk analysis for oil & gas pipelines: A sustainability assessment approach using fuzzy based bow-tie analysis," Journal of Loss Prevention in the Process Industries, vol. 25, pp. 505-523, 2012.

[4] StephenThen, "Blast rips Sabah-Sarawak gas pipeline," in The star vol. 2014 ed. Kuching, Malaysia Star Publication 2014.

[5] A. Nouri Gharahasanlou, A. Mokhtarei, A. Khodayarei, and M. Ataei, "Fault tree analysis of failure cause of crushing plant and mixing bed hall at Khoy cement factory in Iran," Case Studies in Engineering Failure Analysis, vol. 2, pp. 33-38, 2014.

[6] Y. Sun, L. Ma, and J. Morris, "A practical approach for reliability prediction of pipeline systems," European Journal of Operational Research, vol. 198, pp. 210-214, 2009.

[7] C. I. Ossai, "Pipeline Corrosion Prediction And Reliability Analysis: A Systematic Approach With Monte Carlo Simulation And Degradation Models," International Journal of Scientific & Technology Research, vol. 2, 2013.

[8] O.-S. Lee and D.-H. Kim, "Reliability of Buried Pipelines with Corrosion Defects under Varying Boundary Conditions," in The 15th European Conference of Fracture - advanced fracture mechanics for life and safety assessments, , Stockholm, Sweden, 2004.

[9] C. A. Ericson and C. Ll, "Fault tree analysis," in System Safety Conference, Orlando, Florida, 1999, pp. 1-9.

[10] D. W. Vesley, D. J. Dugan, J. Fragole, J. Minarik II, and J. Railsback, "Fault tree handbook with aerospace applications," NASA Office of Safety and Mission Assurance, NASA Headquarters, Washington DC, vol. 20546, 2002.

[11] M. Rausand and A. Høyland, System reliability theory: models, statistical methods, and applications vol. 396: John Wiley & Sons, 2004.

[12] "Report 2013-B: Pipeline Performance in Alberta, 1990–2012," Alberta Energy Regulator, Calgary, Alberta August 2013 2013.

[13] Guure, Chris Bambey, and Noor Akma Ibrahim. "Methods for Estimating the 2-Parameter Weibull Distribution with Type-I Censored Data." (2013).

Probabilistic Life Cycle Cost Model for Repairable System

Meseret Nasir [1], **H.Y. Chong**[2]
[1,2]Department of Mechanical Engineering
Curtin University Sarawak Miri, Malaysia

E-mail: meseretnasir@curtin.edu.my

Sabtuni Osman [3]
Department of Reliability and Maintenance
Petronas Carigali Sarawak Operations
Miri, Malaysia

Abstract. Traditionally, Life cycle cost (LCC) has been predicted in a deterministic approach, however; this method is not capable to consider the uncertainties in the input variables. In this paper, a probabilistic approach using Adaptive network-based fuzzy inference system (ANFIS) is proposed to estimate the LCC of repairable systems. The developed model could handle the uncertainties of input variables in the estimation of LCC. The numerical analysis shows that the acquisition and downtime cost could have a high effect towards the LCC compared to repair cost. The developed model could also provide more precise quantitative information for decision making process.

1. Introduction

LCC has been used widely as a decision making tool in various industries as the industries have experienced high maintenance cost, limitation of budgets, high cost of products and higher cost of ownership [1]. Generally, LCC can be estimated using two approaches, namely deterministic and probabilistic approaches [2]. The deterministic LCC analysis uses fixed and distinct input variables. Korpi and Alarisku [3] found that manufacturers prefer deterministic method compared to probabilistic which 83% of the manufacturer uses the deterministic approach while 13% of them uses the probabilistic method of LCC. Although this method has been practiced by various industries, it is unable to consider the uncertainties of the input variables. In other words, estimation of LCC using deterministic method may not reflect the actual phenomena of the system [4]. In fact, probabilistic LCC analysis is capable to handle the uncertainties of input variables such as maintenance cost, downtime and failure rate [5], however; the application of probabilistic method is very limited. Thus, it is required to develop the probabilistic LCC model to improve the estimation of LCC.

Probabilistic method of LCC can be estimated using soft computing technology. This technology is able to compromise the imprecise, uncertainty and vague data. There are three common soft computing techniques such as genetic algorithm (GA), artificial neural network (ANN) and fuzzy logic system (FL). In order to handle uncertainties, hybrid approaches are recommended [4]. In this study, adaptive network-based fuzzy inference system (ANFIS) is proposed to evaluate LCC which

[1] Corrsponding author- Meseret Nasir, meseretnasir@curtin.edu.my.

Probabilistic method of LCC can be estimated using soft computing technology. This technology is able to compromise the imprecise, uncertainty and vague data. There are three common soft computing techniques such as genetic algorithm (GA), artificial neural network (ANN) and fuzzy logic system (FL). In order to handle uncertainties, hybrid approaches are recommended [4]. In this study, adaptive network-based fuzzy inference system (ANFIS) is proposed to evaluate LCC which is the combination of ANN and FL system. This method is capable to handle a set of complicated data by using hybrid learning rules and it has the strength of the ANN which is the learning capability [6].

The developed ANFIS LCC model can be integrated with reliability, availability and maintainability (RAM) to manage the uncertainty which can be caused by maintenance. Integrating RAM analysis in LCC model, it can mitigate the uncertainties due to maintenance [7]. As a result, the accuracy of LCC estimation could be increased. Furthermore, when reliability and availability of the system are increased, failure rate and downtime are minimized so that LCC can be reduced [8]. Hence, by integrating RAM analysis with LCC model, the accuracy of the model could be improved. Therefore, in this paper, the probabilistic LCC model is proposed to evaluate the overall cost of the repairable system.

2. Methodology

2.1. LCC Model

LCC model has played an important role in the decision making process and to evaluate the effectiveness of the repairable system. The governing equation for LCC model can be defined as (1):

$$LCC = C_{aq} + C_m + C_{op} + C_d \tag{1}$$

where, C_{aq} = acquisition cost; C_m = maintenance cost; C_{op} = operation cost; and C_d = downtime cost.

2.2. Acquisition Cost

Acquisition cost is consists of fixed (C_{FA}) and replacement cost. Fixed cost is the initial cost of equipment while replacement cost (C_{RA}) is the purchased cost of sub components when the equipment is failed. The acquisition cost can be defined using Equation (2):

$$C_{aq} = C_{FA} + C_{RA} \tag{2}$$

2.3. Maintenance Cost

Maintenance can run anytime when there is a failure occurs. The present value of maintenance cost can be estimated as [9]:

$$C_m = \int_0^t C_r \frac{\beta}{\alpha^\beta} t^{\beta-1} \exp\left[-\left(\frac{t}{\alpha}\right)^\beta - \theta t\right] dt \tag{3}$$

where, C_r = cost of repair per failure; t = number of years; θ = continuous discount rate; α = scale parameter; and β = shape parameter.

2.4. Operation Cost

Operation cost is the related cost to the operation and it can be referred to the cost to operate a certain equipment or device. Operation cost can be estimated using (4).

$$C_{op} = C_{fo} + C_v \tag{4}$$

where, c_{fo} = fixed operating cost; and c_v = variable operating cost.

The present value of operating cost can be expressed using (5):

$$C_{op} = [C_{fo} + C_v] \cdot \left[\frac{(1+i)^n - 1}{i(1+i)^n}\right] \tag{5}$$

2.5. Downtime Cost

When the system is down, the production is either reduced or totally zero, this causes huge financial consequence. Since the system failure is stochastic in nature, the downtime cost is a random variable and it can be expressed:

$$C_D = D_t \cdot Q \cdot C \tag{6}$$

where, D_t = cumulative downtime due to failure; Q = production per hour; and C = production lost per unit. The cumulative downtime can be expressed as:

$$D_t = [1 - A(t)] \cdot T \tag{7}$$

where, $A(t)$ = availability of the system and T = total operating time for the system.
The availability of the system can be determined by using the mean time between failure (MTBF) and mean time to repair (MTTR) [10].

$$A(t) = \left[\frac{MTBF}{MTTR + MTBF}\right] \tag{8}$$

Thus, the present value of downtime cost for the system can be expressed as:

$$C_D = \left[1 - \frac{MTBF}{MTTR + MTBF}\right] \cdot T \cdot Q \cdot C \; e^{-\theta t} \tag{9}$$

Hence, the probabilistic LCC model for repairable system is expressed as:

$$LCC = \left[C_F + C_{RA} \cdot \frac{1}{(1+i)^n}\right] + \int_0^t C_r \frac{\beta}{\alpha^\beta} t^{\beta-1} \exp\left[-\left(\frac{t}{\alpha}\right)^\beta - rt\right] dt + [C_{fo} + C_v] \cdot \left[\frac{(1+i)^n - 1}{i(1+i)^n}\right] + \left[1 - \frac{MTBF}{MTTR + MTBF}\right] \cdot T \cdot Q \cdot C \cdot e^{-\theta t} \tag{10}$$

2.6. ANFIS LCC Model

ANFIS is used to simulate the LCC of the repairable system for 20 years. It is able to solve data sets by using hybrid learning rules and it is capable to learn the training data [6]. Four numerical data sets are used to train the ANFIS as shown in Table 1. Each data set is used to calculate the cumulative LCC for the next 20 years with a discount rate of 10%.

Table 1: Train and test data set for ANFIS

No.	Fixed C_{aq}, $ (10^6)	Annual C_{aq}, $ (10^6)	Annual C_m, $ (10^6)	Annual C_{op}, $ ((10^6)	Annual C_d, $ (10^6)
1	21.0	2.0	0.6	3.1	4.5
2	13.0	1.5	0.4	2.0	2.8
3	31	3	1.2	4.9	6.0
4	20	1.8	0.5	2.9	4.2

Based on 20 years life span, 21 data sets are produced for each data set in Table 2. Data set 1 to 3 are used as training data and data set 4 is for testing data. Testing data is important in generating

IOP Conf. Series: Materials Science and Engineering **78** (2015) 012027 doi:10.1088/1757-899X/78/1/012027

ANFIS as it can prevent overfitting when it is training the data [11]. Thus, ANFIS LCC model has five inputs and layers which contributes to estimate LCC as indicated Fig 1.

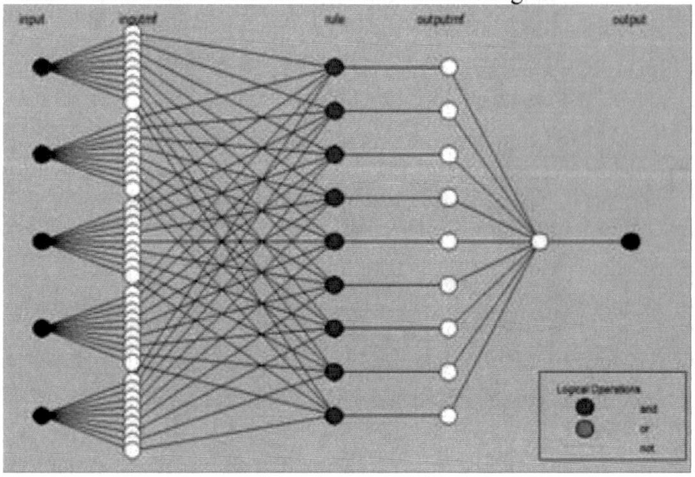

Figure 1: ANFIS LCC Model Structure

- Layer 0 is the input parameters for ANFIS which are acquisition cost, maintenance cost, operation cost, downtime cost and year.
- Layer 1 consists of the input membership function of the five input parameters. Each input parameter is subjected to five Gaussian membership functions. The Sub-clustering method is used to generate the membership function for all the input parameters.
- Layer 2 consists of nine rules. In this layer, the input parameters multiplied by their own member-ship functions and the output of this multiplication could firing strength of the rules.
- Layer 3, the nodes are used to calculate the ratio of rule's firing strength to the total sum of rule's firing strength. The output of this layer is known as normalized firing strength of rules.
- Layers 4, the nodes in this layer are known as adaptive node and it is the output membership function. In this stage, the input parameters with its own membership function and normalized firing strength of rules will be further integrated with the IF-THEN fuzzy rules.
- In Layer 5, there is only one node which is the final output (LCC) of the ANFIS.

Once the ANFIS-LCC model developed, the input data train is conducted to minimize the error of the model. In this study, the model has been trained for 100 epochs with a targeted error tolerance of 0 values. As a result, the training error has decreased linearly when the epoch is increasing. After 100 epochs, the training error has reached 0.24 as shown in Fig 2. This is a very small error which it shows that the accuracy of the model has been greatly increased.

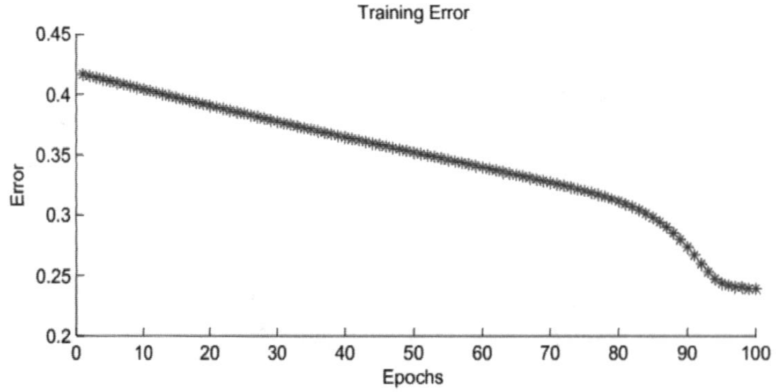

Figure 2: Training Error of ANFIS Model

3. Result and discussion

To illustrate the model, the numerical data used in this paper are indicated in Table 2. The numerical data for each cost value are calculated by using the Equation (2), (3), (5) and (9). Using Table 3, the estimated LCC of a repairable system for 20 years is shown in Fig 3. Based on Fig 3, LCC of the system is increasing linearly for the first 10 year. This is due to the high acquisition cost effect at the beginning of operation. After year 10, LCC of the system increases in a lower gradient compared to a previous 10 years. In year 20, the cumulative LCC of the repairable system is estimated to be $95.8 million.

Table 2. Numerical Data for LCC Prediction

Cost elements	Value
Fixed C_{aq}, \$(million)	21.0
Annual C_{aq}, \$(million)	2.5
Annual C_m, \$(million)	0.6
Annual C_{op}, \$(million)	2.45
Annual C_d, \$(million)	3
Discount Rate, %	10

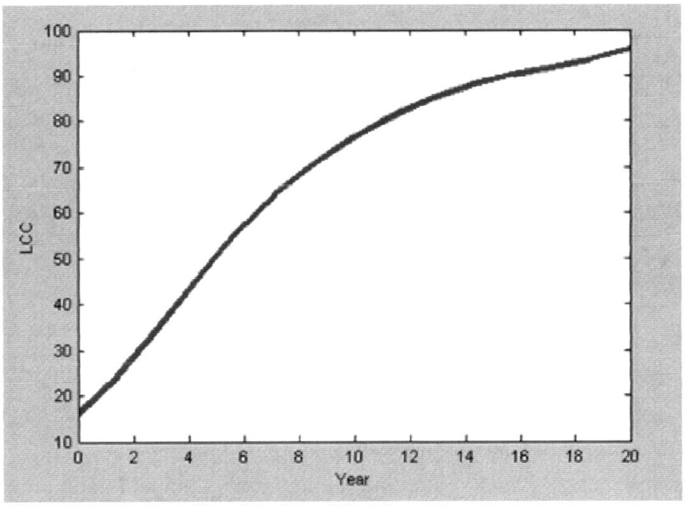

Figure 3: Training Estimation of LCC

The relationships between the input parameters with the LCC are indicated in Fig. 4 and Fig. 5 for 20 year life span of time. Fig. 4 shows the relationship between acquisition cost and maintenance cost with LCC prediction. Fig. 4 shows increasing in acquisition cost can produce a significant increasing of LCC while the maintenance cost only produces a minor effect.

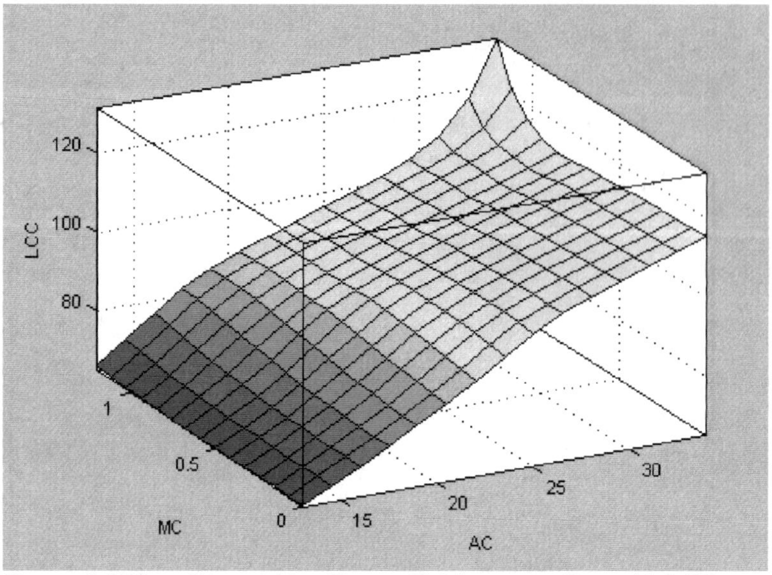

Figure 4: Effect of Acquisition Cost and Maintenance Cost with LCC

Fig. 5 shows the relationship between the operation cost and downtime cost with LCC of the repairable system. The result indicates that downtime cost will have a significant effect on LCC. As the downtime cost is increased, LCC is increased drastically. The LCC is decreased drastically when the operating cost is increased from \$ 3 million to \$ 5 million. Decreases in LCC will increase the financial benefit of the system as they use lesser cost to sustain the system.

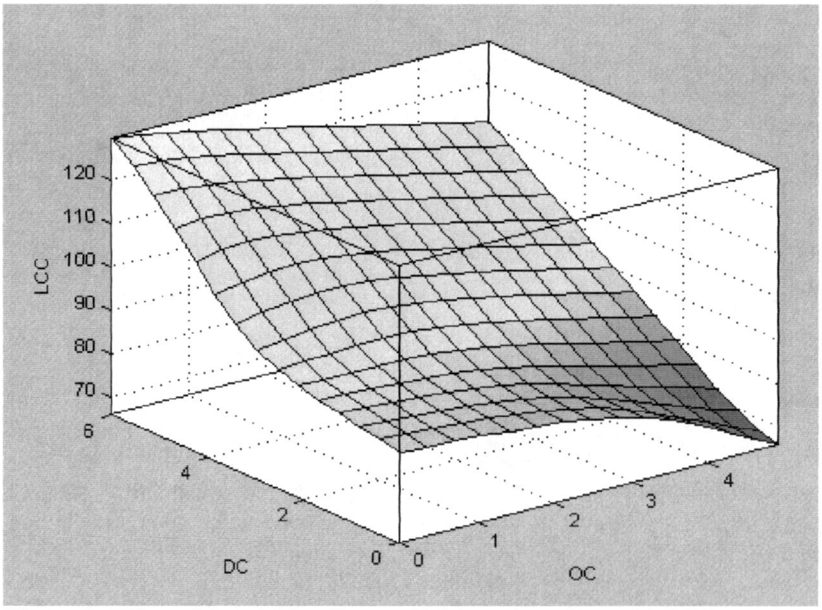

Figure 5: Effect of Acquisition Cost and Maintenance Cost with LCC

3.1. Model Validation

In order to validate the hypothesis, a Monte Carlo (MC) method proposed by Barringer [8] was employed to estimate the LCC. This estimated LCC value compared with developed ANFIS-LCC Model. LCC has been estimated for 20 years by using the same numerical value and the result is compared with the ANFIS prediction. Figure 6 shows the percentage of difference between the prediction of MC and ANFIS. Year 0, 1 and 2 showed high variation which is more than 20% and when the time goes on, the percentage of difference is decreasing. After year 4, the percentage of difference is less than 10% which is acceptable. The comparison indicates that ANFIS model could be used in LCC prediction for the repairable system as an alternative.

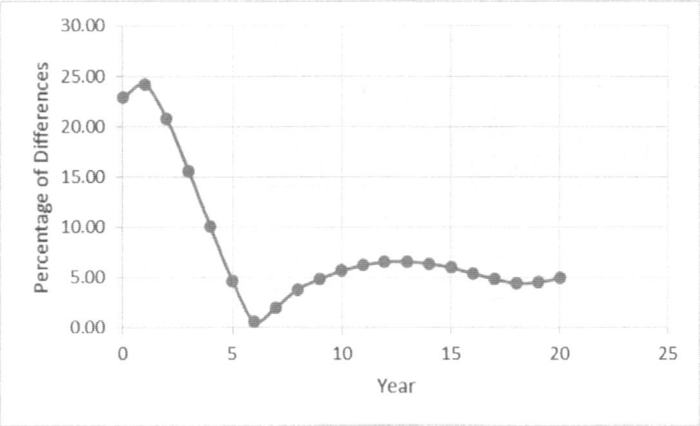

Figure 6: Differences between Prediction of Excel and ANFIS

4. Conclusion

This study developed the probabilistic LCC model using ANFIS to estimate the overall LCC of the repairable system. The developed model also enables easily to incorporate the uncertainties of the input variables. The findings show that repair cost has a minor effect towards the LCC while acquisition cost and downtime cost have a greater effect on LCC. The ANFIS model is validated by comparing Monte carlo approach. The validation shows that the percentage of difference between ANFIS and MC is less than 10%. Hence, ANFIS could be used to predict the LCC as alternatives.

References

[1] B. S. Dhillon, Design reliability: fundamentals and applications: CRC press, 1999.

[2] T. M. Zayed, L.-M. Chang, and J. D. Fricker, "Life-cycle cost analysis using deterministic and stochastic methods: Conflicting results," Journal of Performance of Constructed Facilities, vol. 16, pp. 63-74, 2002.

[3] E. Korpi and T. Ala-Risku, "Life cycle costing: a review of published case studies," Managerial Auditing Journal, vol. 23, pp. 240-261, 2008.

[4] G. W. Flintsch and C. Chen, "Soft computing applications in infrastructure management," Journal of Infrastructure Systems, vol. 10, pp. 157-166, 2004.

[5] S. Rahman and D. Vanier, "Life cycle cost analysis as a decision support tool for managing municipal infrastructure," 2004.

[6] L. H. Martinez, "A neuro-fuzzy decision support system for risk analysis of revenue-dependent infrastructure projects," 2001.

[7] F. Mulubrhan, A. A. Mokhtar, and M. Muhammad, "Integrating Reliability Analysis in Life Cycle Cost Estimation of Heat Exchanger and Pump," Advanced Materials Research, vol. 903, pp. 408-413, 2014.
[8] H. P. Barringer and D. P. Weber, "Life cycle cost & reliability for process equipment," in 8th Annual Energy Week Conference and Exhibition. Houston, Texas, 1997, pp. 1-22.
[9] M. Muhammad, M. Nair, A. A. Mokhtar, and H. Hussin, "Continuous Life Cycle Cost Model for Repairable System," presented at the 8th World Congress on Engineering Asset Management Hong Kong 2013.
[10] T. Helbig, J. Hoos, and E. Westkämper, "A Method for Estimating and Evaluating Life Cycle Costs of Decentralized Component-based Automation Solutions," Procedia CIRP, vol. 17, pp. 332-337, 2014.
[11] J.-S. R. Jang, "Input selection for ANFIS learning," in Proceedings of the fifth IEEE international conference on fuzzy systems, 1996, pp. 1493-1499.

Wind cannot be Directed but Sails can be Adjusted for Malaysian Renewable Energy Progress

C. Palanichamy [1], **Meseret Nasir** [2] **S. Veeramani** [3]

[1,3] Department of Electrical and computer Engineering
Curtin University Sarawak Miri, Malaysia

[2] Department of Mechanical Engineering
Curtin University Sarawak Miri, Malaysia

E-mail: drcpc@curtin.edu.my

Abstract. Wind energy has been the promising energy technology since 1980s in terms of percentage of yearly growth of installed capacity. However the progress of wind energy has not been evenly distributed around the world. Particularly, in South East Asian countries like Malaysia and Singapore, though the Governments are keen on promoting wind energy technology, it is not well practiced due to the low wind speeds. Owing to the recent advancements in wind turbine designs, even Malaysia is well suited for wind energy by proper choice of wind turbines. As evidence, this paper presents successful wind turbines with simulated study outcomes to encourage wind power developments in Malaysia.

1. Introduction

Wind energy is used extensively to generate electricity in various part of the world. In fact, using wind as an energy source, it has a positive impact on the environment and on the cost of electricity [1]. This source is cheap and inexhaustible energy. Depending on availability, wind as a source of energy is one of the cleanest and reliable source of energy. As reported in Global wind statistics 2012, the energy generated using wind was about 282,586 MW [2]. The contribution of each region of the world is indicated in Fig. 1. Europe, Asia and North America accounted the large portion of wind power capacity. Even though Asia and the pacific region contributed high percentage of wind power, the large portion of the wind power capacity was being contributed by China and India which are 77% and 19% respectively. These results show that the usage of wind as an energy source in other Asian countries is very limited. One of the reasons of limited usage of wind as an energy source is the wind speed [3]. The average annual wind speed should be at least 4.0 to 4.5 m/s in order to run the wind turbine effectively. However, some of the Asia countries like Malaysia and Singapore have very low wind speed which is not adequate to generate power effectively. The annual wind speed Malaysia is less than 3m/s at 10 m height from ground level [4] which is too low. Thus, the Malaysia's ability of generating power from wind is insignificant. In this paper, Malaysia wind energy opportunity has been discussed in detail and possible recommendations on suitable technologies and associated cost related to the technologies will be drawn.

1 Corresponding author- Prof. C. Palanichamy, drcpc@curtin.edu.my.

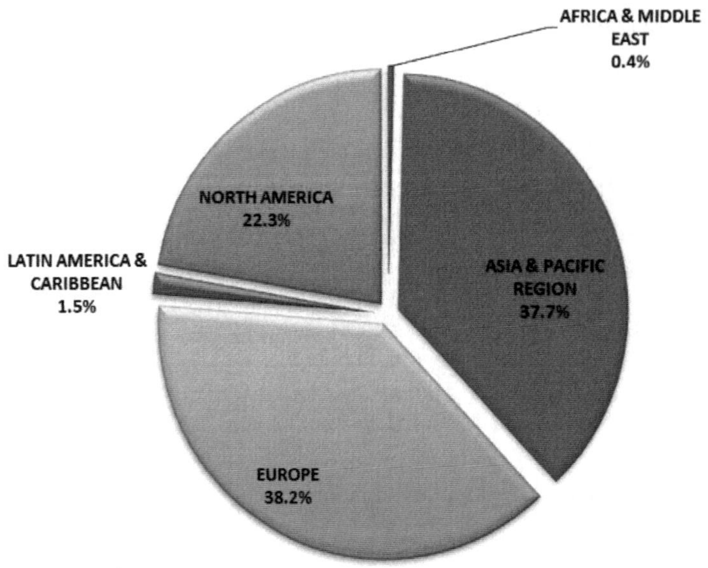

Figure 1: Global installed wind power capacity at the end of 2012 (Extracted from [1])

2. Malaysian energy scenario

2.1. Country Profile

Malaysia, a Southeast Asian nation, has two distinctive land extents; the western area contains the Peninsular Malaysia and the eastern part includes the states of Sarawak and Sabah, which shares the island of Borneo with Indonesia and Brunei [5]. It is situated in the equatorial zone in between 3°10' in the North of latitude and 101°42' in the East of longitude. Malaysia has a land extent of 329,847 km^2 and a population of 30,142,072 in July 2014 [6]. Malaysia's economy strengthened to a sturdy growth of 6.2 per cent, while on a quarter-on-quarter seasonally adjusted, the economy continued a positive momentum by registering 0.8 per cent. In the first quarter of 2014, the GDP in current terms amounted to RM 256.9 billion and the GNI stood at RM 250.5 billion [7].

2.2. Malaysia's Energy Sector

Energy has contributed suggestively towards the rapid growth of the Malaysian economy. As Malaysia headways towards flattering an urbanized nation, energy consumption congruently increases. While meeting the increasing energy demand, strenuous efforts needed to be undertaken to guarantee grander security and sustainability of energy supply. The primary functional components of the energy sector are portrayed in Fig. 2. To facilitate the energy security and sustainability, the National Energy Policy by the Ministry of Energy, Green Technology and Water plays an important role as a guide to the development of the energy sector in the future. The objectives of the National Energy Policy are energy supply, energy utilization and environmental conservation as highlighted in Fig. 3 [8].

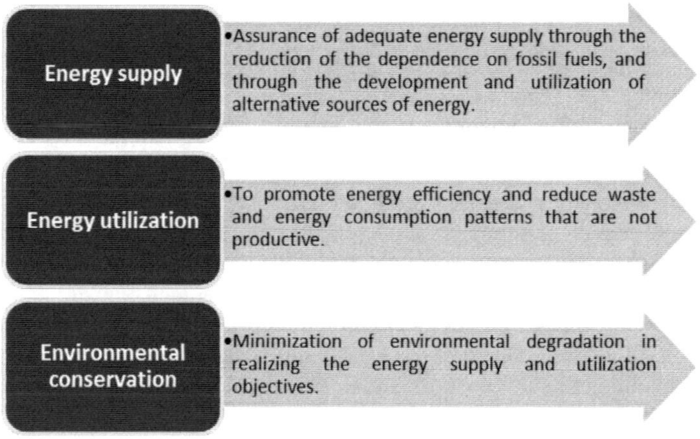

Figure 3: Malaysia's Energy Sector

2.3. Supply and Demand

Energy has been commonly acknowledged as one of the most important inputs for economic growth and human development. There is a resilient two-way relationship between economic development and energy consumption. On one hand, growth of an economy, with its global attractiveness, hinges on the obtainability of economic and environmental friendly energy sources, and on the other hand, the level of economic development has been observed to be contingent on the energy demand [9]. Malaysia is well-endowed with both exhaustible and renewable energy resources. Malaysia's primary energy sources comprise natural gas, coal, oil, biomass even though renewable energy sources such as solar, wind, geothermal, tidal and hydro are currently being exploited. According to Malaysia Economic Planning Unit (EPU) [10, 11], there is an estimated reserves of 5954 billion barrels of crude oil and condensates, 92.122 trillion standard cubic feet of natural gas and 1938.40 million tonnes of coal and coke. As per Economic Planning Unit, Green Technology and Water, Energy Commission, Tenaga Nasional Berhad, Sabah Electricity Sdn Bhd & Sarawak Energy Berhad [12], the primary commercial energy final demand in 2013 was found to be: Petroleum products - 1,063.3 Peta Joules (53.9%), Natural gas - 84.3 Peta Joules (19.5%) and Coal & Coke - 82.8 Peta Joules. Further, it is estimated that Petroleum products will grow at 3.02% p.a., Natural gas will grow at 3.48% p.a. and Coal and coke will grow at 6.04% p.a. As far as electricity is concerned, the installed capacity as of 2013 was found to be around 26400 MW and the peak demand of 19589 MW with a reserve margin of 33%. The reserve margin has been reduced by 17.88% compared to 2012 [13, 14]. Coal has become the largest fuel input in the power sector with a share of 46.6 per cent, followed by natural gas at 39.3 per cent, hydro at 6.6 per cent and oil at 7.5 per cent.

2.4. Challenges and Solutions

Malaysia's energy sector is fronting with momentous challenges that sternly restrict its sustainability.

- Dependence on fossil-fuels: Major portion of Malaysia's electricity generation capacity is natural gas-fired, though gas dearth and mounting electricity demand in recent years have stirred the use of other fuels such as coal, diesel, and hydroelectricity. However, it is reported that the oil reserve in Malaysia will be depleted by this decade and gas by 2048 [15].
- Need for energy imports: Despite government efforts to preserve declining energy reserves, Malaysia will become a net energy importer in the next 20 years [16]. Net import dependency will reach 32 per cent in 2030.
- Optimum Energy Mix Policy: To aim at sustainable energy management, it is imperative that energy supply remains as reliable and as inexpensive as possible. Hence the energy sector should go for a balanced, best value-for-money mix of renewables and fossil fuels from

domestic and foreign sources. The reliance on a single fuel need to be avoided. Nuclear energy is also a necessary part of the mix.

- Energy efficiency: Improved energy efficiency/energy conservation is an additional task of energy sector around the world. The benefits are - reduced investments in energy infrastructure, lower fossil fuel dependency, increased competitiveness and improved consumer welfare. Enhanced efficiency and energy conservation also deliver environmental benefits by reducing greenhouse gas emissions and local air pollution. By improving energy efficiency and energy conservation, the final energy consumption in the industrial, commercial and residential sectors need to be brought down by 10% by 2030.
- Environmental concern: Total CO2 emissions from fuel combustion are projected to reach 264 million tonnes CO2 in 2035, which is 46% higher than in 2010 and 360% higher than in 1990. By 2035, the biggest source of CO2 emissions is the electricity generation sector (33%), followed by the domestic transport sector (24%) and the industry sector (21%) [17]. As national and global concern, the emission due to production and utilization of energy need to be reduced to the safer level considering the frequent hazes in the various parts of the country. Reducing the usage of fossil fuels could be a timely solution.
- Rural and secluded area electrification: A meagre portion of the rural population and secluded communities were not adequately electrified. The limiting factors are
  - High cost of electrification
  - Need far-reaching and complicated installation works
  - Geographical obstacles
  - Long distance from National Grid sources
  - No contribution towards renewable energy

A Solar/wind hybrid power system suitable for future grid integration is a feasible solution for providing electricity to such rural and secluded societies.

3. Renewable energy opportunities

Malaysia's potential for renewable energy generation is significant. According to government targets, renewable energy should contribute at least 6 per cent to the country's generation mix by 2015. The National Renewable Energy Policy and Action Plan aim to draw more of the nation's electricity supply (11 per cent by 2020) from renewable energy. The Ministry of Energy, Green Technology and Water's Green Technology Financing Scheme offers incentives worth of RM1.5 billion to green technologies. Any renewable energy plant (including biomass, biogas, municipal waste, solar, mini-hydropower and wind) may apply to sell energy to the grid on a "willing seller and willing buyer" basis [18].

The Sustainable Energy Development Authority of Malaysia (SEDA Malaysia) is a statutory body formed under the Sustainable Energy Development Authority Act 2011 [Act 726]. The key role of SEDA is to administer and manage the implementation of the feed-in tariff mechanism which is mandated under the Renewable Energy Act 2011 [Act 725] [19]. The Feed-in Tariff (FiT) is simply a guaranteed price established for anyone who wants to sell renewable electricity to the grid and acts as a guarantee that they will have access to the grid to do so. The tariff is set so that a modest profit is ensured and unleashes the collective capital resources of the entire province, state or country to be part of the transition to renewable energy [20]. With a gorgeous FiT rate and profuse natural resources with achievable national goals (Fig. 4), Malaysia is appropriate for foreign investment in renewable energy projects.

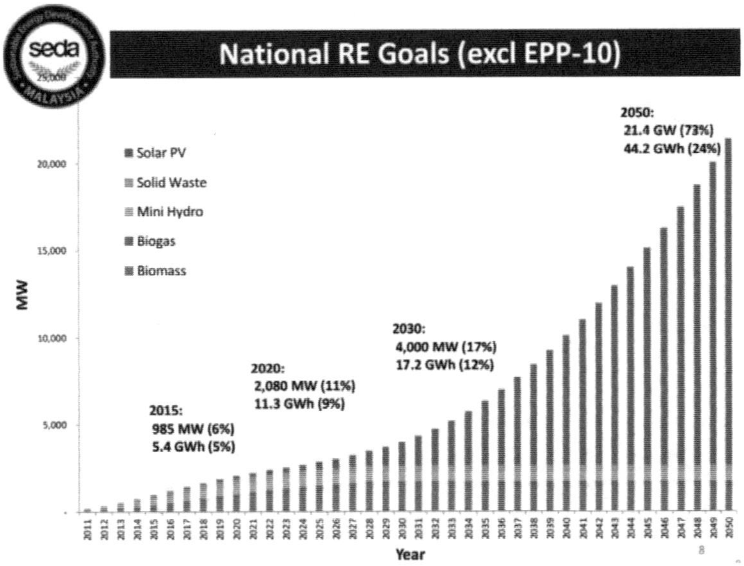

Figure 4: - National Renewable Energy Goals

4. Promising wind energy

Several countries globally recognize that the present-day energy developments are not sustainable and that a better balance must be found between energy security, economic development and protection of the environment. Malaysia has many renewable energy sources such as solar, wind, biomass, hydro, geothermal and tidal wave. However, it is not yet widely developed. Table 1 depicts the installed capacities of the various renewables as of 2014 and the renewable energy on-going projects expected to be functional in 2016 [19].

In particular to wind energy, only meager data is available with universities and research institutions. The Meteorological Department has been recorded the wind data at various stations particularly around airports at a height of approximately 10 meters. Some of the previous studies claim that wind power plant is feasible and offers a high potential based on theoretical simulations [12, 21, 22, 23]. Though not all areas in Malaysia are suitable for wind energy, the general assumption is that some locations have a good potential for wind power generation. However to know the realistic situation, a comprehensive wind map for Malaysia is being prepared to determine the potential for wind energy generation, which would be ready by 2015 [19].

Table 1: Renewable Energy Statistics [18]

Status	Biogas	Biogas (Landfill/ Agri. Waste)	Biomass	Biomass (Solid Waste)	Small Hydro	Solar PV	Total
Operational Plants (2014)	5.38 MW	6.36 MW	43.40 MW	8.90 MW	11.70 MW	118.28 MW	194.02 MW
Plants in progress (functional in 2016)	3.50 MW	56.16 MW	114.75 MW	9.20 MW	215.59 MW	101.13 MW	500.33 MW

IOP Conf. Series: Materials Science and Engineering **78** (2015) 012028 doi:10.1088/1757-899X/78/1/012028

Table 2: Wind Turbine Performances at Low Wind Speed [26-28]

Turbine	Cut in wind speed(m/s)	Total investment (RM)	Net Income/Year (RM)	NPV (%)	(%) Rate of Return (Year)	Cost of generation (RM/kWh)	Capacity Factor (%)
Nordex N27/150	3	433875	48534	5	9.30	0.75	4
Bonus 33.4/300	3	867750	86495	5	7.71	0.83	4
NEG Micon 43/500	3	1235000	133760	5	8.84	0.77	3
Nordex S70/1500	3	5362500	389797	5	3.86	1.09	4
Nordex N29/250	4	723125	52566	5	3.86	1.09	3
Vestas 44/600	4	1482000	83832	5	1.21	1.33	2
NEG Micon 64/1500	4	5362500	284102	5	0.56	1.40	3

Table 3: Wind Turbine Economics

Turbine	Cut in wind speed (m/s)	Total investment (RM)	Net Income/Year (RM)	NPV (%)	(%) Rate of Return (Year)	Cost of generation (RM/kWh)
Nordex N27/150	3	433875	48534	5	9.30	0.75
Bonus 33.4/300	3	867750	86495	5	7.71	0.83
NEG Micon 43/500	3	1235000	133760	5	8.84	0.77
Nordex S70/1500	3	5362500	389797	5	3.86	1.09
Nordex N29/250	4	723125	52566	5	3.86	1.09
Vestas 44/600	4	1482000	83832	5	1.21	1.33
NEG Micon 64/1500	4	5362500	284102	5	0.56	1.40

5. Conclusion

A simulation based study has been performed with several wind turbine manufactures at different cut in wind speeds. The wind turbine performances and the financial aspects are clearly depicted for reference. From the results, low cut in wind speed turbines such as 3 m/s yield a good rate of return at a minimum generation cost of 0.75 RM/kWh which returns a profit of 0.25 RM/kWh. Due to the advancements in wind turbine technology, lower cut in wind speed turbines (1 m/s, SheerWind Involex Technology) are entering into marketing, which in turn would result in potential benefits than the 3 m/s wind turbines. Since wind cannot be directed, our sails with suitable turbines can be adjusted to promote wind energy even in low wind profile regions.

References

[1] R. Saidur, M. Islam, N. Rahim, and K. Solangi, "A review on global wind energy policy," Renewable and Sustainable Energy Reviews, vol. 14, pp. 1744-1762, 2010.
[2] L. Fried, "Global wind statistics 2012," Report, Global Wind Energy Council (GWEC), Brussels, Belgium, 2013.
[3] R. Ali, I. Daut, and S. Taib, "A review on existing and future energy sources for electrical power generation in Malaysia," Renewable and Sustainable Energy Reviews, vol. 16, pp. 4047-4055, 2012.
[4] A. Albani, M. Z. B. Ibrahim, and M. Hamzah, "Assessment of Wind Energy Potential based on METAR data in Malaysia," International Journal of Renewable Energy Research (IJRER), vol. 3, pp. 959-968, 2013.
[5] US EIA 3rd September 2013.
[6] Department of Statistics, Malaysia http://www.statistics.gov.my/portal/index.php?lang=en, retrieved on 12th July 2014
[7] Department of Statistics, Malaysia, Gross Domestic Product (GDP) first quarter of 2014, May 16, 2014
[8] National Energy Policy - Ministry of Energy, Green Technology and Water. 31 January 2008.
[9] C. Palanichamy, "Renewable energy investment opportunities in Mauritius – an investor's perspective," Renewable Energy – An international journal, Vol.29, Issue 5, April 2004, p.p.703-716
[10] C. Palanichamy, "A Visual Package for Educating Preparatory Transmission Line Series Compensation," IEEE Transactions on Education, Vol.48, No.1, Feb. 2005, p.p.16-22
[11] C. Palanichamy and K. Srikrishna, "Simple Algorithm for Economic Power Dispatch," Electrical Power Systems Research, Vol. 121, 1991, p.p. 147 – 153
[12] APEC Energy Demand and Supply Outlook, 5th Edition
[13] Malaysia Energy Resources. http://www.indexmundi.com/malaysia/#Energy
[14] S. M. Lawan, W. A. W. Z. Abidin, W. Y. Chai, A. Baharun and T. Masri, "An Overview of Energy Sector and Wind Power Potential in Malaysia," British Journal of Applied Science & Technology 4(7): 1042-1059, 2014
[15] Malaysia Energy Information Hub - Energy Balance, http//www. meih.st.gov.my
[16] Zaharin Zulkifli, 'Malaysia Country Report' in Kimura, S. (ed.), Analysis on Energy Saving Potential in East Asia, ERIA Research Project Report 2012-19, pp.207-224.ERIA [online]. http:/www.eria.org/RPR_FY2012_No.19_Chapter_11.pdf
[17] Haris, A.H. and Omar A., "A Pilot Project to Study the Performance of Grid-Connected Solar Photovoltaic System in Malaysia", TNB Research Sdn Bhd, Malaysia, Contract No. TNB 973/97 & PTM 3/99, March 2003. Available: http://www.st.gov.my/
[18] APEC Energy Demand and Supply Outlook 2006
[19] APEC Energy Demand and Supply Outlook 2013
[20] Tildy Bayar, Malaysia Explores Its Renewables Options, Renewable Energy World September 16, 2011

[21] Sustainable Energy Development Authority of Malaysia, www.seda.gov.my
[22] http://azhamvosovic.blogspot.com, visited Novomber 2014
[23] E.P. Chiang, Z.A. Zainal, P.A. Aswatha Narayana and K.N. Seetharamu, "Potential of renewable wave and offshore wind energy sources in Malaysia", Marine Technology Seminar, 2003
[24] M. R. Islam, R, Saidur, N. A. Rahim, and K. H. Solangi, "Renewable energy research in Malaysia", Engineering e-Transaction (ISSN 1823-6379), Vol. 4, No. 2, Dec. 2009, p.p. 69-72
[25] A. Albani, M.Z. Ibrahim, M.H.M Hamzah, "Assessment of Wind Energy Potential based on METAR data in Malaysia", International journal of renewable energy research, Vol.3, No.4, 2013
[26] Nordex, "The Spica Retrofit Controller was the best alternative to an obsolete controller,"N27 150Kw, Decowicon A/S, Denmark, available at:http://spicatech.dk/references/retrofit-cases/nordex-n27-150kw/
[27] Bonus Wind Energy, http://www.wind-energy-market.com/en/wind-turbines/big-plants/details/details/bp/an-bonus-30033/
[28] NEG MICON, http://www.wind-energy-market.com/en/wind-turbines/big-plants/details/details/bp/neg-micon-nm-43500-micon-m-1500-500125/

IOP Conf. Series: Materials Science and Engineering **78** (2015) 012029 doi:10.1088/1757-899X/78/1/012029

Robust H_∞ stabilization of a hard disk drive system with a single-stage actuator

Hendra G Harno[1] **and Raymond Song Kiin Woon**

Department of Mechanical Engineering, Faculty of Engineering and Science, Curtin University, CDT 250, 98009 Miri, Sarawak, Malaysia

[1] E-mail: h.g.harno@curtin.edu.my(@gmail.com)

Abstract. This paper considers a robust H_∞ control problem for a hard disk drive system with a single stage actuator. The hard disk drive system is modeled as a linear time-invariant uncertain system where its uncertain parameters and high-order dynamics are considered as uncertainties satisfying integral quadratic constraints. The robust H_∞ control problem is transformed into a nonlinear optimization problem with a pair of parameterized algebraic Riccati equations as nonconvex constraints. The nonlinear optimization problem is then solved using a differential evolution algorithm to find stabilizing solutions to the Riccati equations. These solutions are used for synthesizing an output feedback robust H_∞ controller to stabilize the hard disk drive system with a specified disturbance attenuation level.

1. Introduction

Since hard disk drives (HDD) have become portable data storage devices, disturbance rejection has been one of important features of the HDD in order to prevent shock and vibration from causing data loss and corruption; e.g., see [1]. This relates to HDD system's ability to detect the occurrence of disturbances and then to suspend reading/writing operation. One way is to apply a feedforward controller to compensate the disturbances measured by an accelerometer where phase shift of accelerometer associated with phase delay correction is used to measure disturbances; e.g., see [2]. This method has been effective in improving tracking accuracy of the HDD. Also, dual piezoelectric accelerometers has been used to measure angular acceleration when compensating disturbance effects in the HDD; see [3]. Despite the fact that the dual accelerometers may have varying sensitivity, the latter method has performed satisfactorily in rejecting rotational disturbances.

In addition to disturbance perturbation, track misregistration may also become an important concern when operating the HDD systems. This issue can be addressed by applying a dynamic output feedback controller for track-following control of a magnetic read/write head in the HDD systems. Two optimal robust control techniques: mixed H_2/H_∞ and H_2/μ-synthesis have been proposed in [4, 5] to provide different robustness. The mixed H_2/H_∞ method is a controller design method for reconciling performance and robustness against exogenous disturbances; whereas mixed H_2/μ-synthesis is to deal with structured uncertainties; see [4, 5].

[1] To whom any correspondence should be addressed.

However, these techniques result in the dynamic output feedback controllers which only deal with norm-bounded uncertainties.

In this paper, a single-stage actuator HDD is considered and represented as a linear time-invariant uncertain system. Uncertainties of the HDD system considered consist of uncertain parameters and high-order dynamics. The uncertainties are required to satisfy integral quadratic constraints (IQC), which are suitable to represent nonlinear time-varying dynamic uncertainties including the norm-bounded uncertainties; see [6]. This representation leads to the application of an IQC-based robust H_∞ control method to achieve a stable closed-loop HDD control system with a specified disturbance attenuation level; e.g., see [7]. In this method, a set of scaling constants is applied to convert a robust H_∞ control problem with IQC into a standard H_∞ control problem; see [7]. A solution the latter problem involves stabilizing solutions to a pair of algebraic Riccati equations parameterized by the scaling constants. Thus, the resulting stabilizing solutions are then used to synthesize an output feedback robust H_∞ controller having the same order as that of the HDD model.

To obtain the desired output feedback controller, the given robust H_∞ control problem is transformed into a nonlinear optimization problem with nonconvex constraints and is then solved using an evolutionary optimization approach; see [8]. This approach is realized using a differential evolution (DE) algorithm as presented in [9, 10]. It has been shown in [8] that this approach is effective in handling such a nonlinear optimization problem with nonconvex constraints arising from the robust H_∞ control problem with IQC. Note that this control problem is usually difficult to solve using a conventional nonlinear optimization method.

The DE algorithm is chosen because it is more robust against parameter variation and is simpler in implementation than other types of evolutionary algorithm such as a genetic algorithm; see [9]. Moreover, the DE algorithm tends to be more reliable and results in a simpler formulation than an linear-matrix-inequality approach with rank constraints as presented in [11] although the former algorithm is a heuristic algorithm. These considerations thus lead to the application of the DE algorithm to solve the robust H_∞ control problem for the HDD system with the evolutionary optimization approach as has also been considered in, for instance, [12].

The rest of this paper proceeds as follows. Section 2 presents the formulation of the robust H_∞ control problem for the HDD system. Section 3 describes the DE approach used to solve the robust control problem and discusses about the performance of the resulting controller. Finally, Section 5 presents concluding remarks.

2. Problem Formulation

The single-stage actuator of the HDD system is referred to as a voice coil motor (VCM) actuator which exhibits dynamical characteristics of double integrators with high-frequency resonance modes. According to [13], the dynamics of the VCM actuator can then be represented by a transfer function as follows:

$$G_v(s) = \frac{k_v k_y}{s^2} \prod_{i=1}^{4} G_{v_i}(s) \tag{1}$$

where k_y is a position measurement gain and $k_v = k_t/m$ with k_t is a current-force conversion coefficient and m is the mass of the VCM actuator. Each $G_{v_i}(s)$ is a second-order transfer function having eigenvalues with negative real parts and thus, it is stable. The product of all $G_{v_i}(s)$ then results in an eighth-order transfer function which only significantly represent the VCM actuator's dynamics in a high-frequency region, which may be beyond the operational frequency range of the VCM actuator. Although the transfer function (1) does not represent nonlinear dynamics of the VCM actuator due to mechanical friction, it is sufficient for linear controller design.

2.1. Nominal plant and uncertain system model

For the purpose of designing the robust H_∞ controller as mentioned in Section 1, the VCM actuator can nominally be modeled as a linear system with double integrators as follows:

$$G_v(s) = \frac{k_v k_y}{s^2}. \tag{2}$$

In terms of state equations, the transfer function (2) can then be written as follows (see [13]):

$$\begin{bmatrix} \dot{y}(t) \\ \dot{v}(t) \end{bmatrix} = \begin{bmatrix} 0 & k_y \\ 0 & 0 \end{bmatrix} \begin{bmatrix} y(t) \\ v(t) \end{bmatrix} + \begin{bmatrix} 0 \\ k_v \end{bmatrix} u(t) \tag{3}$$

where y and v represent the position and the velocity of read/write (R/W) head, respectively; and u is the actuator (control) input. Thus, in the robust H_∞ control framework, the high-frequency resonance modes and the nonlinear dynamics of the VCM actuator can be considered as uncertainties. Note that, k_y and k_v might also be uncertain to some extent.

Based on (3), a linear time-invariant uncertain system to represent the VCM actuator can be expressed as follows:

$$\begin{aligned} \dot{x}(t) &= Ax(t) + B_1 w(t) + B_2 u(t) + B_3 \xi(t); \\ z(t) &= C_1 x(t) + D_{12} u(t); \\ \zeta(t) &= Gx(t); \\ y(t) &= C_2 x(t) + D_{21} w(t) + D_3 \xi(t) \end{aligned} \tag{4}$$

where

$$\begin{aligned} & x = \begin{bmatrix} y \\ v \end{bmatrix}; \quad w = \begin{bmatrix} w_1 \\ w_2 \end{bmatrix}; \quad \xi = \begin{bmatrix} \xi_1 \\ \xi_2 \end{bmatrix}; \quad \zeta = \begin{bmatrix} \zeta_1 \\ \zeta_2 \end{bmatrix}; \\ & A = \begin{bmatrix} 0 & 1 \\ 0 & 0 \end{bmatrix}; \quad B_1 = \begin{bmatrix} 0 & 0 \\ 1 & 0 \end{bmatrix}; \quad B_2 = \begin{bmatrix} 0 \\ 6.4013 \times 10^7 \end{bmatrix}; \\ & B_3 = \begin{bmatrix} 0.5 & 0 \\ 0 & 0 \end{bmatrix}; \quad C_1 = \begin{bmatrix} 1 & 0 \\ 0 & 0 \end{bmatrix}; \quad D_{12} = \begin{bmatrix} 0 \\ 1 \end{bmatrix}; \quad C_2 = \begin{bmatrix} 1 & 0 \end{bmatrix}; \\ & D_{21} = \begin{bmatrix} 0 & 1 \end{bmatrix}; \quad D_3 = \begin{bmatrix} 0 & 0.5 \end{bmatrix}; \quad G = \begin{bmatrix} 0 & 1 \\ 1 & 0 \end{bmatrix}. \end{aligned} \tag{5}$$

In the state equations (4), $x \in \mathbb{R}^2$ is the state vector; $w \in \mathbb{R}^2$ is the disturbance input; $u \in \mathbb{R}$ is the control input; $\xi \in \mathbb{R}^2$ is the uncertainty input; $z \in \mathbb{R}^2$ is the controlled output; $\zeta \in \mathbb{R}^2$ is the uncertainty output; and $y \in \mathbb{R}$ is the measurement output. Moreover, in order to be admissible, the uncertainty input $\xi(t)$ and the uncertainty output $\zeta(t)$ in (4) are also required to satisfy integral quadratic constraints as follows (see [7, 6]):

$$\int_0^\infty \|\xi_i(t)\|^2 dt \le \int_0^\infty \|\zeta_i(t)\|^2 dt + d_i, \quad \text{for } i = 1, 2 \tag{6}$$

where $d_i \ge 0$.

2.2. The closed-loop HDD control system

For the uncertain system (4), (6), the robust H_∞ output feedback controller is of the form

$$\begin{aligned} \dot{x}_c(t) &= A_c x_c(t) + B_c y(t); \\ u(t) &= C_c x_c(t). \end{aligned} \tag{7}$$

and has the same order as that of the nominal plant (3). The closed-loop HDD control system is obtained by interconnecting the controller (7) and the uncertain system (4), (6) in order to achieve absolute stabilization with a specified disturbance attenuation level $\gamma > 0$.

The uncertain system (4), (6) is thus said to be absolutely stabilizable with a specified disturbance attenuation level $\gamma > 0$ if there exists an output feedback controller (7) and constants $c_1 > 0$ and $c_2 > 0$ such that the following conditions hold (see [7]):

(i) For any initial conditions $[x(0), x_c(0)]$, any admissible uncertainty input $\xi(\cdot)$ and any disturbance input $w(\cdot) \in \mathbb{L}_2[0,\infty)$, then

$$[x(\cdot), x_c(\cdot), u(\cdot), \xi_1(\cdot), \xi_2(\cdot)] \in \mathbb{L}_2[0,\infty)$$

and

$$\begin{aligned} &\|x(\cdot)\|_2^2 + \|x_c(\cdot)\|_2^2 + \|u(\cdot)\|_2^2 + \|\xi_1(\cdot)\|_2^2 + \|\xi_2(\cdot)\|_2^2 \\ &\leq c_1 \left[\|x(0)\|^2 + \|x_c(0)\|^2 + \|w(\cdot)\|_2^2 + d_1 + d_2\right]. \end{aligned} \tag{8}$$

(ii) The H_∞ norm-bound condition is satisfied: If $x(0) = 0$ and $x_c(0) = 0$, then

$$\mathcal{J} := \sup_{w(\cdot)\in\mathbb{L}_2[0,\infty)} \sup_{\xi_1(\cdot),\xi_2(\cdot)\in\Xi} \frac{\|z(\cdot)\|_2^2 - c_2[d_1 + d_2]}{\|w(\cdot)\|_2^2} < \gamma^2. \tag{9}$$

Note that $\|f(\cdot)\|_2$ denotes the $\mathbb{L}_2$-norm of the function $f(\cdot)$, that is $\|f(\cdot)\|_2^2 := \int_0^\infty \|f(t)\|^2 dt$; and Ξ is the set of all admissible uncertainty inputs $\xi_1(\cdot)$ and $\xi_2(\cdot)$.

If an absolutely stabilizing controller of the form (7) for the HDD system exists, then the resulting closed-loop system (as illustrated in Figure 1) can be written as follows:

$$\begin{aligned} \begin{bmatrix} \dot{x}(t) \\ \dot{x}_c(t) \end{bmatrix} &= \begin{bmatrix} A + B_3\Delta G & B_2C_c \\ B_cC_2 + B_cD_3\Delta G & A_c \end{bmatrix} \begin{bmatrix} x(t) \\ x_c(t) \end{bmatrix} + \begin{bmatrix} B_1 \\ B_cD_{21} \end{bmatrix} w(t); \\ z(t) &= \begin{bmatrix} C_1 & D_{12}C_c \end{bmatrix} \begin{bmatrix} x(t) \\ x_c(t) \end{bmatrix} \end{aligned} \tag{10}$$

where

$$\Delta = \begin{bmatrix} \Delta_x & 0 \\ 0 & \Delta_y \end{bmatrix}; \quad \Delta_x, \Delta_y \in \mathbb{R}. \tag{11}$$

Note that $\xi(t) = \Delta\zeta(t)$; and $\|\Delta\|^2 \leq 1$, where $\|\cdot\|$ is an induced matrix norm.

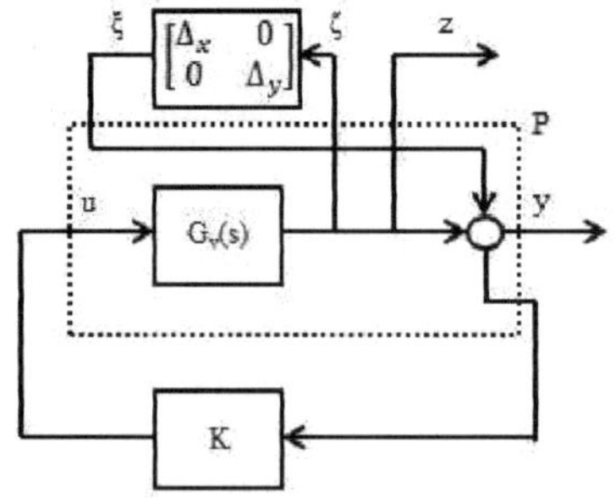

Figure 1. A closed-loop control system of the HDD with a single-stage VCM actuator.

3. Synthesis of the Robust H_∞ Controller

Synthesizing the robust H_∞ controller (7) for the uncertain system (4), (6) involves stabilizing solutions to a pair of parameterized algebraic Riccati equations presented as follows (see [7, 14]):

$$\begin{aligned}&(A - B_2E_1^{-1}\hat{D}_{12}'\hat{C}_1)'X + X(A - B_2E_1^{-1}\hat{D}_{12}'\hat{C}_1)\\&\quad + X\left(\hat{B}_1\hat{B}_1' - B_2E_1^{-1}B_2'\right)X + \hat{C}_1'(I - \hat{D}_{12}E_1^{-1}\hat{D}_{12}')\hat{C}_1 = 0; \qquad (12)\\&(A - \hat{B}_1\hat{D}_{21}'E_2^{-1}C_2)Y + Y(A - \hat{B}_1\hat{D}_{21}'E_2^{-1}C_2)'\\&\quad + Y\left(\hat{C}_1'\hat{C}_1 - C_2'E_2^{-1}C_2\right)Y + \hat{B}_1(I - \hat{D}_{21}'E_2^{-1}\hat{D}_{21})\hat{B}_1' = 0 \qquad (13)\end{aligned}$$

where

$$\begin{aligned}&\hat{B}_1 = \left[\gamma^{-1}B_1 \quad \sqrt{\tau_1}^{-1}B_{31} \quad \sqrt{\tau_2}^{-1}B_{32}\right]; \quad \hat{D}_{21} = \left[\gamma^{-1}D_{21} \quad \sqrt{\tau_1}^{-1}D_{31} \quad \sqrt{\tau_2}^{-1}D_{32}\right];\\&\hat{C}_1 = \begin{bmatrix} C_1 \\ \sqrt{\tau_1}G_1 \\ \sqrt{\tau_2}G_2 \end{bmatrix}; \quad \hat{D}_{12} = \begin{bmatrix} D_{12} \\ 0 \\ 0 \end{bmatrix}; \quad E_1 = \hat{D}_{12}'\hat{D}_{12}; \quad E_2 = \hat{D}_{21}\hat{D}_{21}'. \qquad (14)\end{aligned}$$

Note that $\tau_1 > 0$, and $\tau_2 > 0$. In this regard, the solutions to the Riccati equations (12) and (13) are required to be positive definite, that is, $X > 0$ and $Y > 0$.

Solving the Riccati equations (12) and (13) tends to be challenging as their coefficient matrices are parameterized, and thus, lead to a nonconvex robust H_∞ control problem. Such a problem is considered to be difficult to solve using a conventional nonlinear optimization or an LMI-based approaches commonly used in robust H_∞ control applications. This concern thus becomes an underlying motivation to apply an evolutionary optimization method based on the DE algorithm (see [9, 10]) in order to find the solutions to the Riccati equations (12) and (13); see [8]. In this case, the given robust H_∞ control problem has to be transformed into a nonlinear constrained optimization problem describe as follows:

$$\min_{\gamma,\tau_1,\tau_2\in\mathbb{R}_+} f(\gamma) \qquad (15)$$

subject to the Riccati equations (12) and (13), and the stability of the resulting closed-loop system (10). Here, $f(\gamma)$ is a polynomial function of γ (for instance, $f(\gamma) = \gamma^2$) and is to be minimized with respect to γ, τ_1, and τ_2. Note also that $\mathbb{R}_+$ denotes the set of positive real numbers.

Since the DE algorithm is an evolutionary algorithm, it is heuristic and involves cross-over, mutation, and selection routines which operate on a population of candidate solutions; see [9]. Each candidate solution is evaluated through a fitness test procedure consisting of all constraints involve in the optimization problem (15), (12), (13), (10). The derivation of the fitness test procedure then follows the approach presented in [8]. Thus, solving the optimization problem (15),(12), (13), the controller matrices in (7) can be determined as follows:

$$\begin{aligned}A_c &= A + B_2C_c - B_cC_2 + (\hat{B}_1 - B_c\hat{D}_{21})\hat{B}_1'X;\\B_c &= (I - YX)^{-1}(YC_2' + \hat{B}_1\hat{D}_{21}')E_2^{-1};\\C_c &= -E_1^{-1}(B_2'X + \hat{D}_{12}'\hat{C}_1). \qquad (16)\end{aligned}$$

4. Numerical Results and Discussion

Solving the optimization problem (15), (12), (13), (10) through the DE algorithm implemented on MATLAB yields the following solution:

$$\gamma = 2.7378; \quad \tau_1 = 0.0012; \quad \tau_2 = 1.3717. \qquad (17)$$

Thus, the resulting robust H_∞ controller of the form (7) is given as follows:

$$A_c = \begin{bmatrix} -4.1592 \times 10^4 & 1.0000 \\ 1.1383 \times 10^9 & -2.2409 \times 10^7 \end{bmatrix}; \quad B_c = \begin{bmatrix} 4.1604 \times 10^4 \\ 1.2264 \times 10^3 \end{bmatrix};$$
$$C_c = \begin{bmatrix} -17.7827 & -0.3501 \end{bmatrix}. \tag{18}$$

Interconnecting the robust H_∞ controller (7), (18) to the uncertain system (4), (6) will then provide a stable closed-loop system. As an example, for

$$\Delta = \begin{bmatrix} 1 & 0 \\ 0 & 0 \end{bmatrix}, \tag{19}$$

the resulting closed loop system (10) is asymptotically stable as its poles λ_{cl} are all negative real numbers which are shown as follows:

$$\lambda_{cl} = \begin{bmatrix} -2.24086 \times 10^7 \\ -0.00415 \times 10^7 \\ -0.00001 \times 10^7 \\ -0.0221 \end{bmatrix}. \tag{20}$$

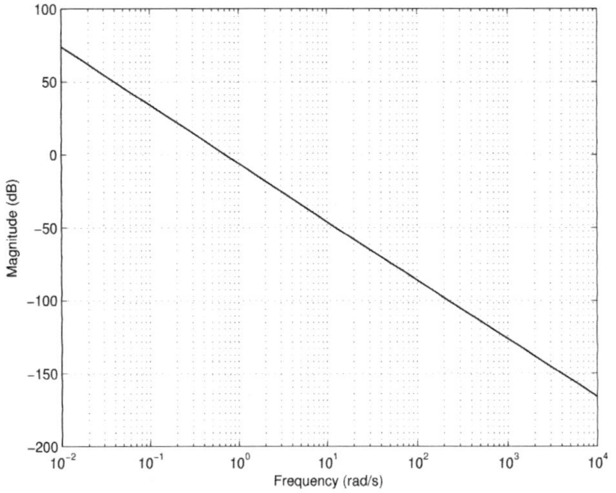

Figure 2. An open-loop frequency response of the HDD system.

The open-loop and closed-loop characteristics of the HDD control system considered can now be compared by examining the corresponding frequency responses. From Figure 2, it is apparent that the magnitude of the open-loop HDD frequency response is relatively large within the low-frequency range ($\omega < 0.7\ rad/s$) and has constant roll-off within the high-frequency range. This implies that the HDD system significantly amplifies disturbance input signals, which is undesirable when the R/W head performs track-seeking and track-following operations. This concern has been addressed through the application of the robust H_∞ controller (7), (18) to the uncertain system (4), (6), which leads to the closed-loop HDD frequency response as depicted in Figure 3. It is shown that the magnitude of the closed-loop HDD frequency response is relatively

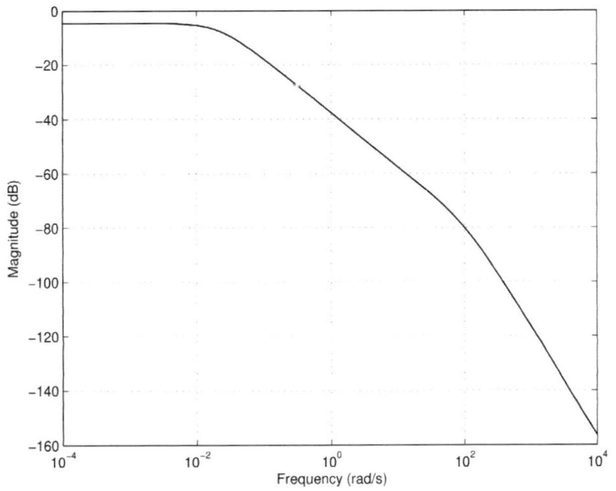

Figure 3. A closed-loop frequency response of the HDD system.

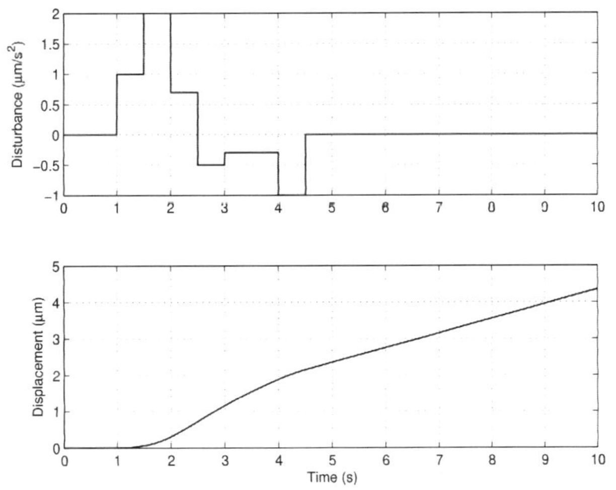

Figure 4. An open loop time response of the HDD system.

small within the low-frequency range and has steeper roll-off as the frequency getting higher. This implies that the robust H_∞ controller (7), (18) is effective in attenuating the disturbance input signals in the whole range of frequency, and thus the the closed-loop HDD system is absolutely stabilized.

Simulated open-loop and closed-loop time responses shown in Figure 4 and Figure 5 respectively confirm the open-loop and closed-loop characteristics of the HDD control system. In Figure 4, it is evident that the R/W head displacement of the open-loop HDD system continuously grows although the accelerating disturbance input has ceased. This is due to the fact that the nominal HDD system (2) has double integrators. Meanwhile, as shown in

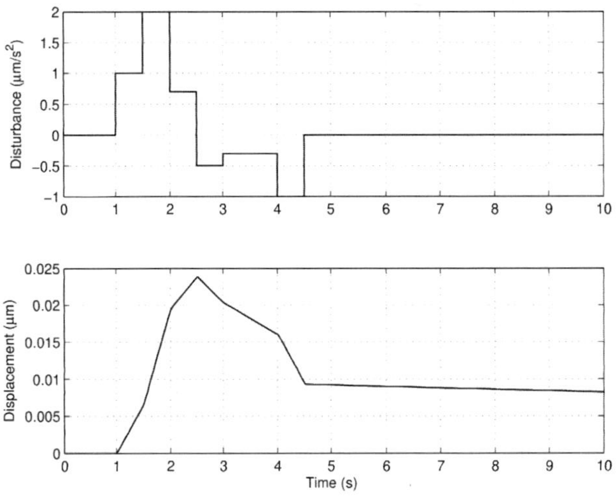

Figure 5. A closed loop time response of the HDD system.

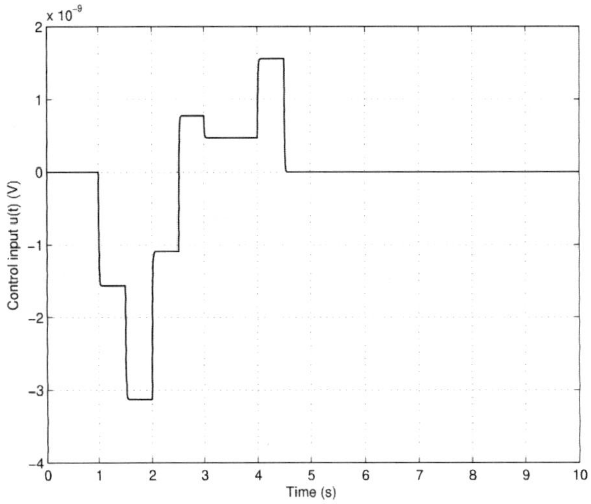

Figure 6. A control input $u(t)$ of the HDD system.

Figure 5, when the accelerating disturbance input has stopped perturbing the HDD system, the R/W head displacement of the closed-loop HDD system continuously decreases toward a much smaller magnitude than that of the open-loop HDD system. Moreover, Figure 6 also shows that the robust controller (7), (18) is efficient and has spent sufficiently small effort in order to absolutely stabilize the closed-loop HDD system perturbed by the accelerating disturbance input.

5. Conclusion
In this paper, the IQC-based robust H_∞ control problem for the HDD system has been presented. From the frequency and time responses, it has been shown that the resulting robust controller has been effective in absolutely stabilizing the closed-loop HDD system while attenuating any exogenous disturbance inputs perturbing the HDD system. Such a controller can be synthesized by using the stabilizing solutions to the pair of parameterized algebraic Riccati equations. Since the control problem considered is numerically nonconvex, the DE algorithm has been employed to solve the Riccati equations. Moreover, as a future research direction, a tracking control problem for the HDD system can be considered by extending the current results.

References
[1] Du C, Ge S S and Lewis F L 2008 *International Journal of Adaptive Control and Signal Processing* **22** 374—387
[2] Abramovitch D Y 1997 *Control Engineering Practice* **5** 1517–1524
[3] Jinzenji A, Sasamoto T, Aikawa K, Yoshida S and Aruga K 2001 *IEEE Transactions on Magnetics* **37** 888–893
[4] Du C, Xie L, Teoh J N and Guo G 2005 *IEEE Transactions on Control Systems Technology* **13** 832–839
[5] Nagamune R, Huang X and Horowitz R 2010 *Journal of Dynamic Systems, Measurement, and Control* **132** 021005
[6] Petersen I R, Ugrinovskii V A and Savkin A V 2000 *Robust Control Design using H^∞ Methods* (Springer-Verlag London)
[7] Savkin A V and Petersen I R 1996 *Journal of Mathematical Systems, Estimation and Control* **6** 339–342
[8] Harno H G and Petersen I R 2011 *International Journal of Control* **84** 665–678
[9] Price K V, Storn R M and Lampinen J A 2005 *Differential Evolution - A Practical Approach to Global Optimization* (Berlin, Germany: Springer)
[10] Price K V 2008 *Advances in Differential Evolution* (*Studies in Computational Intelligence* vol 143) ed Chakraborty U K (Berlin Heidelberg: Springer-Verlag) pp 33–88
[11] Orsi R, Helmke U and Moore J B 2006 *Automatica* **42** 1875–1882
[12] Tan K C, Sathikannan R, Tan W W and Loh A P 2007 *Soft Computing* **11** 131–139
[13] Chen B M, Lee T H, Peng K and Venkataramanan V 2006 *Hark Disk Drive Servo Systems* 2nd ed (London, the UK: Springer-Verlag)
[14] Petersen I R, Anderson B D O and Jonckheere E A 1991 *International Journal of Robust and Nonlinear Control* **1** 171–185

Effect of Machining Velocity in Nanoscale Machining Operations

Sumaiya Islam*[1], Raafat Ibrahim[2], Noman Khondoker[1]

[1]Department of Mechanical Engineering, Faculty of Engineering & Science, Curtin University Sarawak, CDT 250 Miri, Sarawak 98009, Malaysia.

[2] Department of Mechanical and Aerospace Engineering, Monash University, Clayton, Victoria 3800, Australia.

*E-mail: sumaiya.islam@curtin.edu.my

Abstract. The aim of this study is to investigate the generated forces and deformations of single crystal Cu with (100), (110) and (111) crystallographic orientations at nanoscale machining operation. A nanoindenter equipped with nanoscratching attachment was used for machining operations and in-situ observation of a nano scale groove. As a machining parameter, the machining velocity was varied to measure the normal and cutting forces. At a fixed machining velocity, different levels of normal and cutting forces were generated due to different crystallographic orientations of the specimens. Moreover, after machining operation percentage of elastic recovery was measured and it was found that both the elastic and plastic deformations were responsible for producing a nano scale groove within the range of machining velocities from 250-1000 nm/s.

Keywords-Nano machining; Single crystal; Elastic recovery; Pile up

1. Introduction

From the view point of learning from nature, the controlling of crystal orientation is accounted to be a major subject for material processing. In nano machining, the machining length and depth of cut are usually less than the grain size of a polycrystalline aggregate [1] and machining is performed along the direction of the grain. The polycrystalline work piece material may be considered to be an isotropic and homogenous continuum in the conventional analysis and must be treated as a series of single crystal with random orientations and anisotropic properties. Therefore, the crystallographic orientation of the material being cut exerts a great influence on the cutting behaviour and machined surface. Most studies of the cutting mechanism are performed under the assumption that the material is isotropic and is a homogenous continuum [2] and that the effect of material anisotropy is often not included in the theoretical analysis.

Atomic Force Microscope (AFM) is a well recognised tool employed for nano machining operations for measuring mechanical characteristics [3-4].The convention of the AFM based lithographic technique is that the machined surface scale of the structure is solely determined by the geometry of the AFM probe [5-6]. The nanoindenter produced by Hysitron utilises the same mechanism to create a scratch. However, a diamond probe is placed with the tool holder instead of the cantilever attachment used in AFM. The advantage of using a nano indenter is to avoid the effect of the cantilever's geometry, stiffness and their respective calibration reliability on the measured nano machining parameters.

Using the indentation method, the deformation behaviour of various materials have been extensively reported by varying the tip geometry and tip orientation [7], and in most cases the material behaviour was assumed to be rigid plastic in nature [8]. However, some research have been conducted to analyse the deformation behaviour with the variation of the depth of cut [9-12] considering different crystallographic orientation, however, variation of the machining velocity was not considered. Therefore, in this paper, the effect of the machining velocity on the generated forces as well as the deformation behaviour of Cu samples with different crystallographic orientations were analysed with nano indenter.

2. Experimental details

Nano machining was conducted on the (100), (110) and (111) orientated single crystal Copper surfaces. These three single crystal Cu surfaces with different orientations were prepared for the nano machining tests with a dimension of 10 mm diameter and 1 mm thickness. These Cu samples used herein were produced by the MaTeck Company in Germany. The orientation accuracy of the Cu single crystal is <.01°.

In these experiments, nano machining was performed on a Hysitron Tribolab (2008) nano mechanical testing instrument with the scratching method, using the displacement control mode. As a machining parameter, the machining velocities were varied from 250nm/s to 1000 nm/s, to investigate the generated forces (normal and cutting forces) and elastoplastic deformation behavior during nano machining.The diamond Conical tool having a tip radius of 100 nm and 60° cone angle was used for the machining purpose and in situ observation was performed. All experiments were performed under room temperature and normal atmospheric conditions with the temperature range of 20-24°C and relative humidity range of 45-50%. The topographic scanning method was employed to analyse the depth of the machined surface. The area of the topographic measurement was 10x10 μm^2. During scanning, 20 μm/s tip velocity and 1 Hz scan rate were used.

3. Results and discussion

To evaluate the effect of the machining velocity on the generated forces, different crystallographic orientations have been chosen. Figure 1 shows the variation in the normal force with a change in the machining velocity at a 150 nm depth of cut. It was shown that there was not enough variation in the generation of the normal forces due to the variation of the machining velocities, however; the largest normal forces were observed at 333nm/s machining velocity.

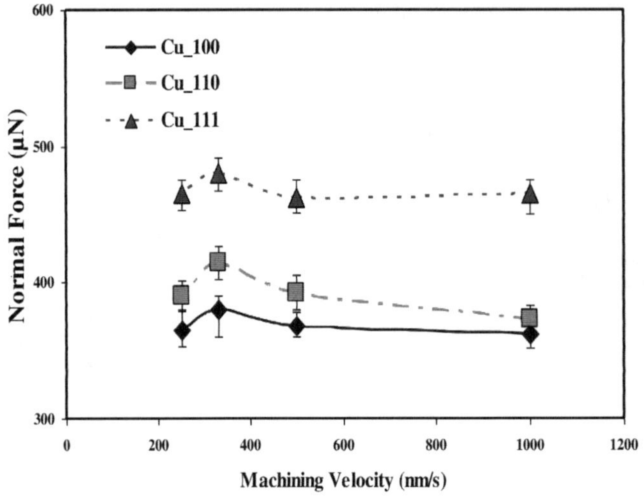

Figure 1: Variation in the normal force with change in machining velocity at 150 nm depth of cut on single crystal copper of (100), (110) and (111) planes.

Beyond this velocity, the generated normal forces decreased and remained constant with a further increase in the machining velocity. These results also showed that the generated normal forces were higher when machining was performed on the Cu (111) plane than on the Cu (100) and Cu (110) planes. This result is also consistent with the results [12] when the depth of cut was the variable parameter to measure the generated forces.

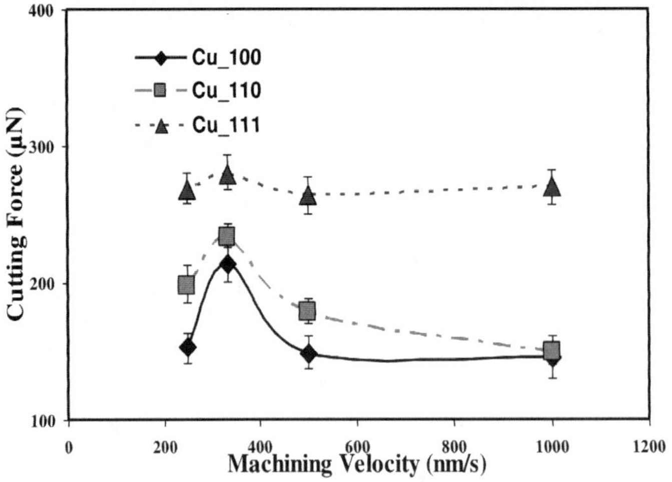

Figure 2: Variation in the cutting force with change in machining velocity at 150 nm depth of cut on single crystal copper of (100), (110) and (111) planes.

Figure 2 shows the variation in the generated cutting forces with the change in the machining velocity on different crystallographic orientations at a 150 nm depth of cut. It was observed that the trend of the generated cutting forces were similar to the obtained normal forces. Furthermore, the generated cutting forces were larger when machining was performed on the Cu (111) plane than that of other crystallographic orientations. The normal and cutting forces data showed that both normal and cutting forces were higher in the Cu (111) plane than that in other planes; therefore it reveals that the Cu (111) plane is harder than the Cu (110) and Cu (111) planes.

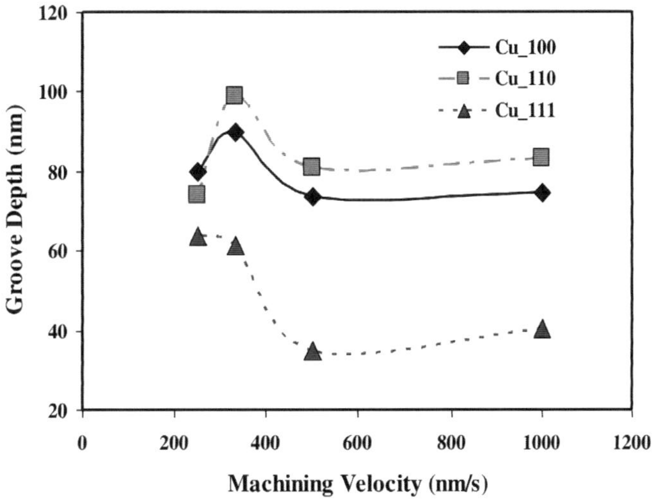

Figure 3: Variation in the obtained groove depth with the machining velocity at 150 nm depth of cut on single crystal copper of (100), (110) and (111) planes.

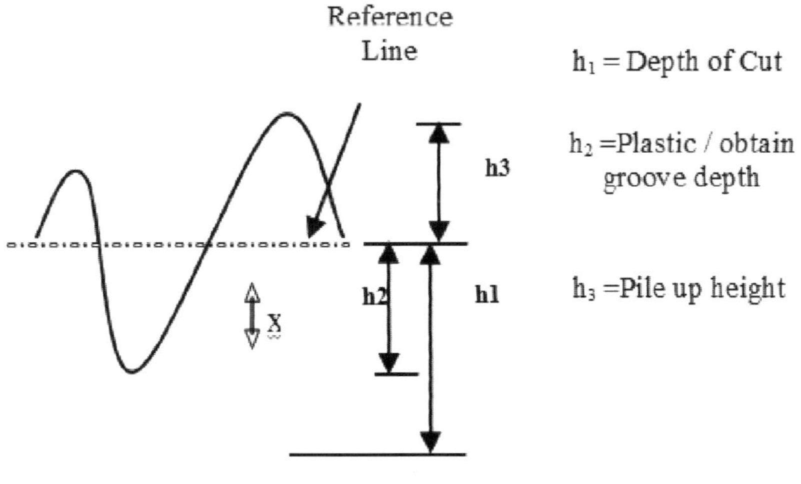

Figure 4: Schematic diagram of the machined profile.

To determine the elastic recovery at different machining velocities, the obtained groove depths were measured. Figure 3 shows the variation in the obtained groove depth with a change in the machining velocity at 150 nm depth of cut. The obtained groove depth increased from 250 to 333 nm/s machining velocity at 150 nm depth of cut. Above a machining velocity of 333nm/s, the groove depths decreased and remained almost constant with a further increase of the machining velocities. The obtained groove depths were larger when machining was performed on the Cu (110) plane than that on the Cu (100) and Cu (111) planes at different machining velocities. The generated forces were higher and obtained depth value were lower on Cu (111) plane than the other planes and this result also reveal that the hardness of the Cu (111) plane is higher than that of other planes due to the higher atomic density of the Cu (111) plane.

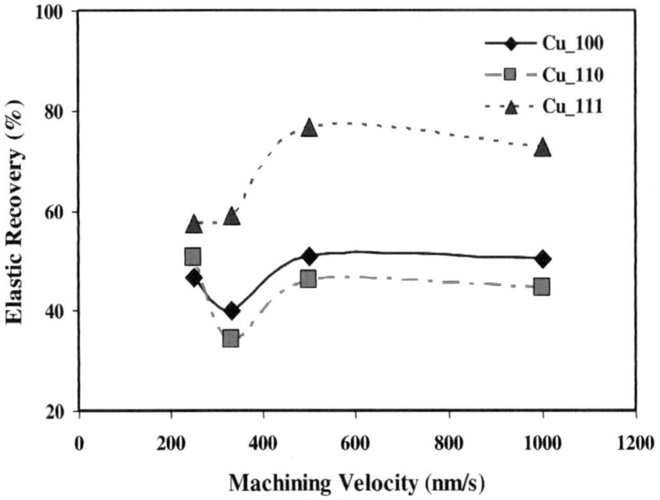

Figure 5: Variation in the elastic recovery with change in machining velocity at 150 nm depth of cut on single crystal copper of (100),(110) and (111) planes.

These obtained depth values were utilized to calculate the percentage of elastic recovery using the following expression:

$$\text{Elastic Recovery}\,(\%) = \frac{h_1 - h_2}{h_1} \times 100$$

The details of h_1 and h_2 were described in Figure 4.

Figure 5 shows the variation in the elastic recovery (%) with a change in the machining velocity at 150 nm depth of cut. The elastic recovery decreased from 250 to 333 nm/s machining velocity. From a 333 nm/s machining velocity, the value of elastic recovery increased and remained constant with the further increase of the machining velocities. As before, the percentage value of elastic recovery was inversed to the obtained groove depth. Therefore, the higher value of elastic recovery was observed on the machining of the Cu (111) plane by comparison to the other crystallographic orientations. These percentage values reveal that after a single scratching operation the primary deformation behaviour for these three crystal orientations were relatively highly elastic in nature. Moreover, the pile up were observed along the scratched profile after the machining operations and shown in Figure 6. These Pile up occurred due to the plastic deformation of the material. Therefore, this non-traditional nano scale machining condition responses to the percentage of elastic recovery and pile up for these materials indicate the mode of elastoplastic deformation mechanism as opposed to fully plastic deformation characteristics for the traditional machining operation.

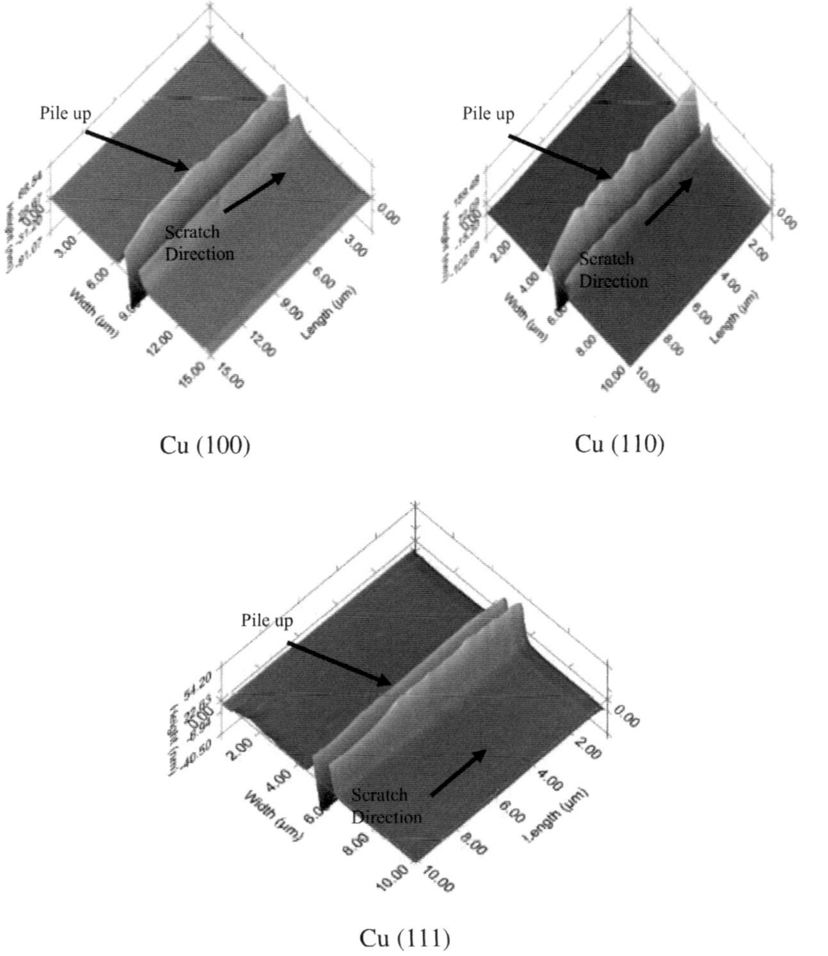

Figure 6: AFM-3D view of the machined surface morphology of the Cu single crystal after nano machining at the 150 nm depth of cut and 333nm/s machining velocity.

4. Conclusion

In this research, nano machining was performed on single crystal copper materials in different orientations. It was found that the largest amounts of normal and cutting forces were generated at 333 nm/s machining velocity. Beyond this velocity, both normal and cutting forces were decreased with a further increase in the machining velocity for Cu (100), (110), and (111) planes. The obtained groove depth was larger when machining was performed on the Cu (110) plane than the Cu (111) and Cu (100) planes at 333nm/s machining velocity. Therefore elastic recovery (%) of the crystal orientation was lower than the other two crystal orientations. Therefore, 333nm/s machining velocity can be the recommended velocity to do such type of nano machining operations.

References

[1] Shigeo A., Ken-Suke S.& Tahshi M. (2003), *Science and Technology of Advanced Materials,* vol.4, pp.455-460

[2] Liu Y., Varghese S., Maa J., Yoshino M., Lua H.& Komanduri R. (2008), *International Journal of Plasticity,* vol.24, pp.1990-2015

[3] Bhushan B., Israelachvili J. N. and Landman U. (1995), *Nature*, vol.374, pp. 607-616

[4] Fang T. H., Chang W. J. and Tsai S. L. (2005), *Microelectronic Journal*, vol. 36, pp.5-59

[5] Fang T.-H., Chang W. J. and Weng C. I. (2006), *Material Science and Engineering .A.* vol.430, pp. 332-340

[6] Fang T. H., Chang W. J. (2003), *Journal of Physics and Chemistry of Solids,* vol.64, pp. 913-918

[7] Youn S. W.and Kang C. G. *Material Science and Engineering A,* vol.390, pp.233-239, 2005

[8] Kato K. (2002), *Journal of Engineering Tribology,* vol.216, pp.349-355

[9] Islam S., Ibrahim R. (2011), *Tribology Letters*, vol. 42, pp.275-284

[10] Islam S., Ibrahim R., Das R. (2011), *Key Engineering Materials*, vol. 462-463, pp. 931-936

[11] Islam S., Ibrahim R. (2011), *International Journal of Surface Science and Engineering (IJSurf Se)*, vol.5 (1), pp. 38-50

[12] Islam S, Ibrahim R., Khandoker N. (2011), *Procedia Engineering*, vol.10, pp. 2369- 2374

IOP Conf. Series: Materials Science and Engineering **78** (2015) 012031 doi:10.1088/1757-899X/78/1/012031

Some advantages of implementing an adaptive moving mesh for the solution to the Burgers equation

Sudi Mungkasi
Department of Mathematics, Faculty of Science and Technology,
Sanata Dharma University, Mrican, Tromol Pos 29, Yogyakarta 55002, Indonesia

E-mail: sudi@usd.ac.id

Abstract. The Burgers equation is considered. The equation is solved using finite difference methods. The standard finite difference method may lead to inaccurate solutions, unless a very fine mesh is used, which results in expensive computations. Therefore, we implement an adaptive finite difference moving mesh method as an alternative numerical method to solve the equation. The advantages of implementing the adaptive method are investigated.

1. Introduction

The Burgers equation has been widely used to model some problems in gas dynamics and traffic flows. Therefore, accurately solving the equation is useful. It is well-known that the behaviour of the solutions is influenced by the viscosity factor involved in the equation [1].

Based on the viscosity factor the Burgers equation is of two types, namely, viscous and inviscid [2]. When the viscosity factor is nonzero, the equation is viscous and so have a source term. Otherwise it is inviscid and does not have any source terms. This paper deals with the viscous type of the equation.

We solve the Burgers equation using finite difference methods and refer to the work of Huang and Russell [1] (see [3]-[5] for kinds of adaptive strategy). The standard finite difference method may lead to inaccurate solutions for coarse mesh. Therefore, an alternative finite difference method is considered. We implement an adaptive moving mesh finite difference method as the alternative. In this paper, this numerical method is called the adaptive moving mesh method or simply the adaptive method. Our contribution is identifying the advantages of this adaptive method.

This paper is structured as follows. We first write the problem and numerical methods that shall be used to solve the problem. Numerical results are then presented and discussed. Some concluding remarks are given at the final part.

2. The considered problem and solvers

In this section, we present the Burgers equation and the numerical methods to solve the equation.

2.1. The considered problem

The Burgers equation is well known to be a simplification of the Navier-Stokes equation. The equation is named after Johannes Martinus Burgers, a Dutch scientist, for his fundamental work on the theory of turbulence [6]. The Burgers equation is

$$u_t = \varepsilon u_{xx} - \left(\frac{u^2}{2}\right)_x \quad (1)$$

where the space domain is all $x \in (0,1)$ and the time domain is all $t > 0$. The notation u_t means the first partial derivative of the quantity $u(x,t)$ with respect to t. The notations u_{xx} and $(u^2/2)_x$ are understood similarly. The parameter ε is positive, so that the Burgers equation (1) is viscous.

Following Huang and Russell [1], we take the boundary condition

$$u(0,t) = u(1,t) = 0 \quad (2)$$

and the initial condition

$$u(x,0) = \sin(2\pi x) + \frac{1}{2}\sin(\pi x)\,. \quad (3)$$

Equations (1)-(3) form an initial-boundary value problem.

2.2. The standard finite difference method

To solve the Burgers equation (1), we discretize the space domain pointwisely into N equidistant points, as

$$x_j = (j-1)\,h \quad (4)$$

for $j = 1,\ldots,N$. When we use central finite difference approximations for all spatial derivatives, we obtain the semi-discrete finite difference method

$$\frac{du_j}{dt} = \frac{\varepsilon}{h^2}(u_{j+1} - 2u_j + u_{j-1}) - \frac{1}{4h}(u_{j+1}^2 - u_{j-1}^2) \quad (5)$$

for $j = 2,\ldots,N-1$. This is a standard finite difference method. The numerical boundary conditions are $u_1(t) = 0$ and $u_N(t) = 0$ for $t > 0$. The numerical initial conditions are

$$u_j(0) = \sin(2\pi x_j) + \frac{1}{2}\sin(\pi x_j) \quad (6)$$

for $j = 1,\ldots,N$.

With the initial conditions (6), we can solve the semi-discrete equation (5) using an ODE solver. One can use ODE solvers available in some computer programming language, such as MATLAB, Python, or any others. In this paper, for simplicity, we use the Euler method to solve equation (5). We present the numerical results later in Section 3.

2.3. An adaptive moving mesh finite difference method

Alternatively, we use an adaptive moving mesh method (see mainly [1] and further [3-5]). We discretize the space domain pointwisely into N points as follows. We consider the transformation of the quantity

$$\hat{u}(\xi,t) = u(x(\xi,t),t)\,, \quad (7)$$

with the transformation of the space

$$x_j(t) = x(\xi_j,t) \quad (8)$$

for $j = 1,\ldots,N$. Here

$$\xi_j = \frac{j-1}{N-1} \quad (9)$$

for $j = 1,\ldots,N$. We then have

$$\hat{u}_\xi = u_x x_\xi \quad \text{and} \quad \hat{u}_t = u_t + u_x x_t\,. \quad (10)$$

Note that $x_t = \partial x/\partial t$, where $x = x(\xi,t)$. Substitution of the transformed variables into the Burgers equation leads to

$$\hat{u}_t - \frac{\hat{u}_\xi}{x_\xi} x_t = \frac{\varepsilon}{x_\xi}\left(\frac{\hat{u}_\xi}{x_\xi}\right)_\xi - \frac{1}{x_\xi}\left(\frac{\hat{u}^2}{2}\right)_\xi . \tag{11}$$

When we use central difference approximations for all spatial derivatives, we obtain the adaptive finite difference moving mesh method

$$\frac{du_j}{dt} - \frac{(u_{j+1}-u_{j-1})}{x_{j+1}-x_{j-1}}\frac{dx_j}{dt} = \frac{2\varepsilon}{(x_{j+1}-x_{j-1})}\left[\frac{(u_{j+1}-u_j)}{(x_{j+1}-x_j)} - \frac{(u_j-u_{j-1})}{(x_j-x_{j-1})}\right] - \frac{1}{2}\frac{(u_{j+1}^2-u_{j-1}^2)}{(x_{j+1}-x_{j-1})} \tag{12}$$

for $j = 2,...N-1$, where $u_j(t) \approx \hat{u}(\xi_j,t) := u(x_j(t),t)$. The time derivative in the computational domain is defined by

$$x_t = \frac{1}{\rho\tau}(\rho x_\xi)_\xi \tag{13}$$

with $x(0,t)=0$ and $x(1,t)=1$. In addition, we take

$$\rho = \left(1 + \frac{1}{\alpha}|u_{xx}|^2\right)^{\frac{1}{3}}. \tag{14}$$

Here

$$\alpha = \max\left\{1,\ \left[\int_0^1 |u_{xx}|^{2/3}\,dx\right]^3\right\}. \tag{15}$$

Therefore, we obtain

$$\frac{dx_j}{dt} = \frac{1}{\rho_j\,\tau\,\Delta\xi^2}\left[\frac{\rho_{j+1}+\rho_j}{2}(x_{j+1}-x_j) - \frac{\rho_j+\rho_{j-1}}{2}(x_j-x_{j-1})\right] \tag{16}$$

for $j = 2,...,N-1$, and also, $\frac{dx_1}{dt} = 0$ as well as $\frac{dx_N}{dt} = 0$. Note that $\Delta\xi = 1/(N-1)$. In the computations, we use the discretization of ρ as

$$\rho_j = \left(1 + \frac{1}{\alpha_h}|u_{xx,j}|^2\right)^{\frac{1}{3}} \tag{17}$$

for $j = 1,...,N$, where

$$\alpha_h = \max\left\{1,\ \left[\sum_{j=2}^{N}\frac{(x_j-x_{j-1})}{2}\left(|u_{xx,j}|^{2/3} + |u_{xx,j-1}|^{2/3}\right)\right]^3\right\}. \tag{18}$$

The spatial derivatives are taken as

$$u_{xx,j} = \frac{2}{(x_{j+1}-x_{j-1})}\left[\frac{(u_{j+1}-u_j)}{(x_{j+1}-x_j)} - \frac{(u_j-u_{j-1})}{(x_j-x_{j-1})}\right] \tag{19}$$

for $j = 2,...,N-1$, and also,

$$u_{xx,1} = \frac{2\left[(x_2-x_1)(u_3-u_1)-(x_3-x_1)(u_2-u_1)\right]}{(x_3-x_1)(x_2-x_1)(x_3-x_2)} \tag{20}$$

as well as

$$u_{xx,N} = \frac{2\left[(x_{N-1}-x_N)(u_{N-2}-u_N)-(x_{N-2}-x_N)(u_{N-1}-u_N)\right]}{(x_{N-2}-x_N)(x_{N-1}-x_N)(x_{N-2}-x_{N-1})}. \tag{21}$$

Furthermore,

$$\rho_j := \frac{1}{4}\rho_{j-1} + \frac{1}{2}\rho_j + \frac{1}{4}\rho_{j+1} \tag{22}$$

for $j = 2,...,N-1$, and also,

$$\rho_1 := \frac{1}{2}\rho_1 + \frac{1}{2}\rho_2 \tag{23}$$

as well as

$$\rho_N := \frac{1}{2}\rho_{N-1} + \frac{1}{2}\rho_N. \tag{24}$$

We solve equation (12) using the ode15i MATLAB function, which is an implicit method [1]. The numerical results are presented in the next section.

3. Numerical results

We shall compare the results of the adaptive moving mesh method with the standard finite difference method in this section.

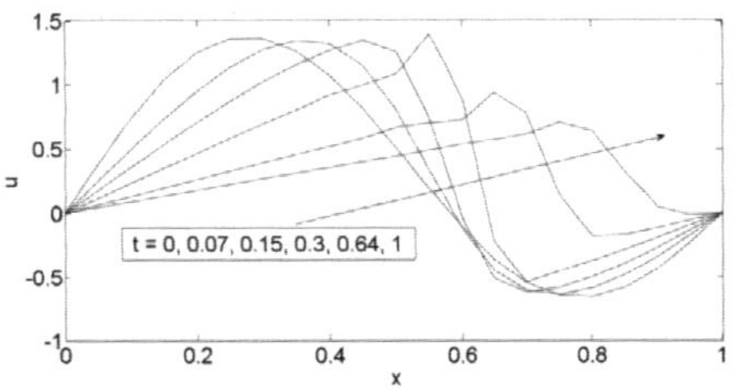

Figure 1. Results of the standard method with 21 points and viscosity 10^{-2} for time instants $t = 0, 0.07, 0.15, 0.3, 0.64, 1$.

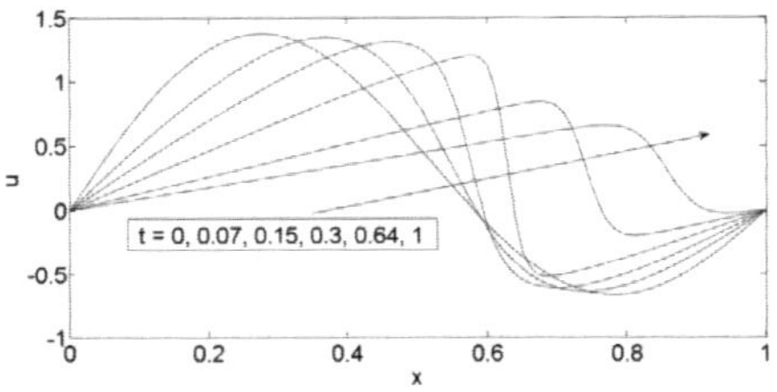

Figure 2. Results of the standard method with 101 points and viscosity 10^{-2} for time instants $t = 0, 0.07, 0.15, 0.3, 0.64, 1$.

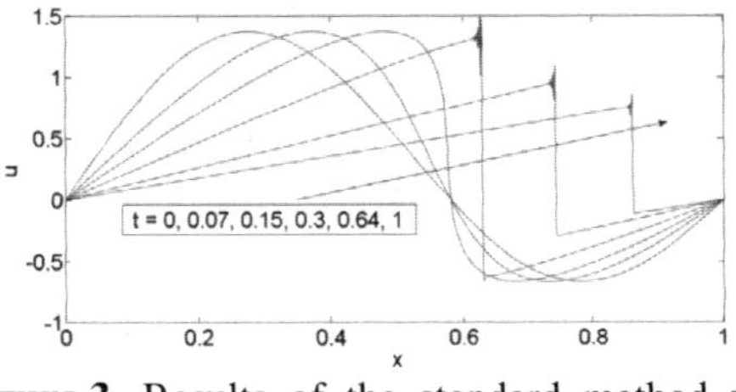

Figure 3. Results of the standard method with 1001 points and viscosity 10^{-4} for time instants $t = 0, 0.07, 0.15, 0.3, 0.64, 1$.

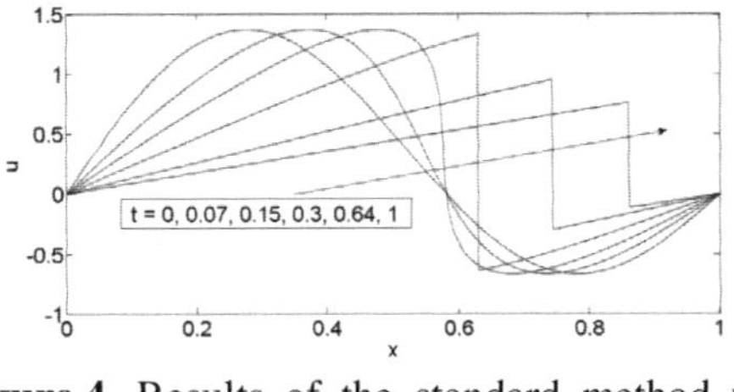

Figure 4. Results of the standard method with 5001 points and viscosity 10^{-4} for time instants $t = 0, 0.07, 0.15, 0.3, 0.64, 1$.

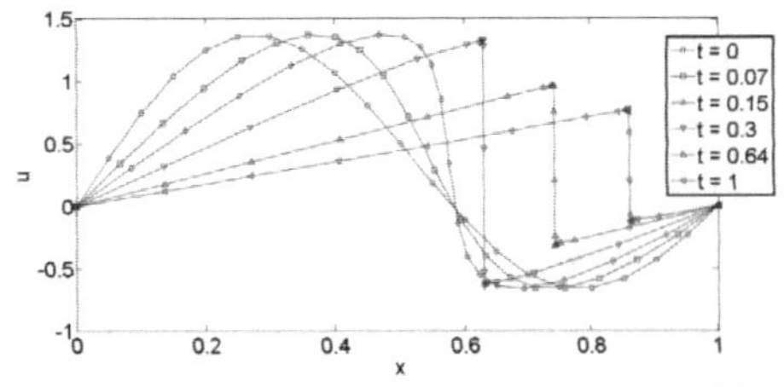

Figure 5. Results of the adaptive method with 21 points and viscosity 10^{-4} for time instants $t = 0, 0.07, 0.15, 0.3, 0.64, 1$.

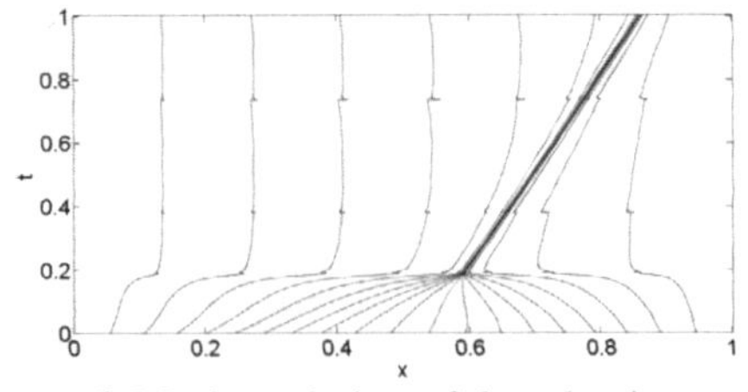

Figure 6. Mesh evolution of the adaptive method with 21 points and viscosity 10^{-4} for time instants $t = 0, 0.07, 0.15, 0.3, 0.64, 1$.

Table 1. Results for the largest magnitude of the solutions produced by the adaptive method with 21 points and the standard method with 5001 points. Here the results of the standard method is considered as the reference solution, so the absolute and relative errors can be computed.

The considered time (second)	The maximum magnitude of the solution produced		Absolute error	Relative error
	by the adaptive method with 21 points	by the standard method with 5001 points		
0.00	1.3555650	1.3679074	0.0123424	0.0090228
0.07	1.3648793	1.3678883	0.0030090	0.0021997
0.15	1.3657246	1.3678663	0.0021417	0.0015657
0.30	1.3228471	1.3350550	0.0122078	0.0091440
0.64	0.9630831	0.9497996	0.0132834	0.0139855
1.00	0.7684366	0.7553905	0.0130460	0.0172706
			The average absolute error is 0.0093384	The average relative error is 0.0088647

We consider two cases. The first is the Burgers equation with the viscosity factor $\varepsilon = 10^{-2}$. The second is with the viscosity factor $\varepsilon = 10^{-4}$. A smaller viscosity factor results in a solution containing a sharper "discontinuity-like" as time evolves. We call a discontinuity-like, because it resembles a shock discontinuity, but in fact the exact solution is continuous. If the viscosity factor is zero, then the exact solution contains a shock discontinuity, which is not a discontinuity-like solution.

When the first case with $\varepsilon = 10^{-2}$ is solved using the standard finite difference method, there is no obvious a discontinuity-like in the solution. The results for the first case is shown in Figure 1 and Figure 2. Figure 1 is produced using 21 uniform spatial points. We see in Figure 1 that the solution is oscillatory, because the number of spatial points does not give enough resolution for this case. However if we refine the space discretization by taking more number of points we have the solution with the correct behaviour. That is, no oscillation appears in the solution, as shown in Figure 2. Here Figure 2 is produced using 101 uniform spatial points.

The problem becomes more difficult to solve when the viscosity factor is smaller, such as where $\varepsilon = 10^{-4}$ which we consider as the second case. Even when we discretize the spatial domain into 1001 points, the resolution is not enough. This is illustrated in Figure 3, where oscillation appears around the position of the discontinuity-like. This artificial oscillation does not occur in the solution if we discretize the space into much finer mesh, such as 5001 uniform points. With this excessive number of points, the solutions are very accurate. This is shown in Figure 4. However, this fine mesh makes the computation is very expensive. The average computational time with 5001 points using the standard method is 10 seconds.

In contrast, fine mesh is not necessary in the adaptive moving mesh method. For example we solve the second case, which has the viscosity factor $\varepsilon = 10^{-4}$. We use only 21 computational points to obtain relatively the same accurate solutions. The numerical solutions are shown in Figure 5. The corresponding moving meshes are shown in Figure 6. The average of relative L^{∞} errors between the reference solutions in Figure 4 and the numerical solutions in Figure 5 is 0.89 % (see Table 1 for detailed results). Therefore, the adaptive moving mesh method results in very accurate solutions. Furthermore we note that the average computational time with 21 points using the adaptive method is about 3 seconds. This means that the adaptive method is not only accurate, but also efficient in terms of computer memory as well as the computational time. These give great advantages for us, as accurate results and fast computation are always desired in practice.

4. Conclusion
The adaptive moving mesh method is very accurate and efficient, and in addition, it requires small computer memory. We have demonstrated these advantages by solving the Burgers equation using the adaptive method. That is, the adaptive moving mesh method has better performance than the standard finite difference method. Extending the adaptive moving mesh method to solve higher dimensional problems could be a future direction of this research.

Acknowledgment
The author thanks Professor Weizhang Huang at The University of Kansas and Professor Robert Russell at Simon Fraser University for sharing some of their computer codes. An early version of this paper was titled "On the advantage of implementing an adaptive moving mesh to solve the Burgers' equation" presented in the 9th Curtin University of Technology Science and Engineering International Conference (CUTSE 2014) at Curtin University Sarawak Malaysia on 3-4 December 2014.

References
[1] Huang W and Russell R D 2011 *Adaptive Moving Mesh Methods* (New York: Springer)
[2] LeVeque R J 2002 *Finite Volume Methods for Hyperbolic Problems* (Cambridge: Cambridge University Press)
[3] Budd C J, Huang W and Russell R D 2009 Adaptivity with moving grids *Acta Numer.* **18** 1–131
[4] Remski J, Zhang J and Du Q 2014 On balanced moving mesh methods *J. Comput. Appl. Math.* **265** 255–63
[5] Tang H and Tang T 2003 Adaptive mesh methods for one- and two-dimensional hyperbolic conservation laws *SIAM J. Numer. Anal.* **41** 487–515
[6] Burgers J M 1948 A mathematical model illustrating the theory of turbulence *Advances in Applied Mechanics* **1** 171–99

Phenol Biodegradation by Free and Immobilized *Candida tropicalis RETL-Cr1* on Coconut Husk and Loofah Packed in Biofilter Column

D Shazryenna, R Ruzanna, M S Jessica and M T Piakong
Faculty of Science and Natural Resources, Universiti Malaysia Sabah, 88400 Kota Kinabalu, Sabah, Malaysia.

E-mail: shazryennadalang@gmail.com

Abstract. Phenols and its derivatives are environmental pollutant commonly found in many industrial effluents. It is toxic in nature and causes various health hazards. However, they are poorly removed in conventional biological processes due to their toxicity. Immobilization of microbial cells has received increasing interest in the field of waste treatment and creates opportunities in a wide range of sectors including environmental pollution control. Live cells of phenol-degrading yeast, *Candida tropicalis RETL-Cr1*, were immobilized on coconut husk and loofah by adsorption. The immobolized particle was packed into biofilter column which used for continuous treatment of a phenol with initial phenol concentration of 3mM. Both loofah and coconut husk have similar phenol biodegradation rate of 0.0188 $gL^{-1}h^{-1}$ within 15 hours to achieve a phenol removal efficiency of 100 %. However loofah have lower biomass concentration of 4.22 gL^{-1} compared to biomass concentration on coconut husk, 4.39 gL^{-1}. Coconut husk contain higher biomass concentration which makes it better support material than loofah. Fibrous matrices such as loofah and coconut husk provide adequate supporting surfaces for cell adsorption, due to their high specific surface area. Therefore, coconut husk and loofah being an agricultural waste product have the potential to be used as low-cost adsorbent and supportmatrix for microbial culture immobilization for the removal of organic pollutant from wastewater.

1. Introduction

Phenols are among the most common water pollutants. They impact taste and odour to water and are highly toxic to aquatic life, animal and human beings [1]. Consequently, many treatment processes have been applied for the removal of phenols from waste waters. Some of these processes include: adsorption [2], photo-fenten degradation [3], photocatalytic degradation [4], and biodegradation [5].Combined methods like biochemical, electrochemical, physicochemical or simultaneous adsorption and biodegradation is gaining importance. Simultaneous adsorption and biodegradation increases the life of adsorbent, as biofilm on adsorbent degrade the adsorbate [6].

This adsorption-biodegradation process using suitable adsorbent has shown high efficiency for phenol removal. The adsorption-biodegradation process involves immobilization of microbial culture on solid porous support matrix [7]. Adsorbents used as solid support matrix for immobilization should be stable both physically and chemically and have a high mechanical strength or resistance [8]. Activated carbon is most widely used as adsorbent and solid support matrix for microbial cell in the removal of heavy metals, phenols and other hazardous chemicals [9] which may be found in waste waters, but its high cost and difficulty in regeneration limits its commercial application in large-scale waste water treatment [10].

In recent years, extensive research has been undertaken to develop alternative and economic adsorbents from microbial biomass [11] and agricultural by- products which is a ubiquitous green waste in the environment and may cause some serious environmental pollution when filling at a fixed site. Some of the agricultural waste products that has been developed as adsorbents include, orange and banana peels [12] spent tea leaves [9], tamarind fruit shell [10], soya bean hull [13], cotton seed hull and corn cobs [14], rubber fruit pericarp [15] and pineaple peel [16]. More and more interests are focused on developing these agricultural wastes as adsorbent for wastewater treatment due to their relative high sorption affinity, ubiquitous presence in the environment, and the ease of being modified to materials with higher efficiency [17].

However, the use of raw agricultural waste adsorbent for the immobilization of cells has rarely been reported for the biodegradation of phenol. The objective of this study is to investigate the potential of loofah and coconut husk, an abundantly available agricultural by-products as a non-conventional adsorbent used as support matrix for *Candida tropicalis* RETL-Cr1 immobilization to degrade phenol.

2. Materials and methods

2.1 Miccroorganism

A pure strain of *Candida tropicalis* RETL-Cr1 was obtained from the Environmental Microbiology Laboratory of university Malaysia Sabah. The culture was maintained on nutrient agar (NA) slants at 4ºC.

2.2 Medium and culture conditions

Yeast was grown on nutrient agar. For adaptation experiments of the cell to phenol, which was also used as a sole carbon source, a Ramsay media was used as described by Ramsay (1983) containing (g/L): 2.0g NH4NO3 , 0.5g KH2 PO4 , 1.0g K2 HPO4 , 0.5g MgSO4 .7H2 O, 0.01g CaCl .2H20, 0.1g KCl and 0.06g yeast extract. Loop full activated culture on NA was inoculated into an Erlenmeyer flask containing 100 ml of Ramsay broth. The inoculum was incubated at 30 ºC in rotary shaker (200 rpm) for 9 to 10 hour prior to analysis

2.3 Adsorbent (support material)

There are two adsorbents had been prepared prior to this phenol biodegradation study; coconut husk and loofah. Both materials were cut into desired sizes of 1.0 x 1.0 x 0.5 cm and 1.5 x 1.5 x 0.5 cm for coconut husk and loofah respectively. All materials were soaked in distilled water and sterilized by autoclaved at 121 ºC for 15 minutes.

2.4 Phenol biodegradation

The study of phenol biodegradation by *C.tropicalis* RET-Cr1 was done by using the column packed with loofah and coconut husk. The schematic drawing of biofilter set-up is shown in figure 1.

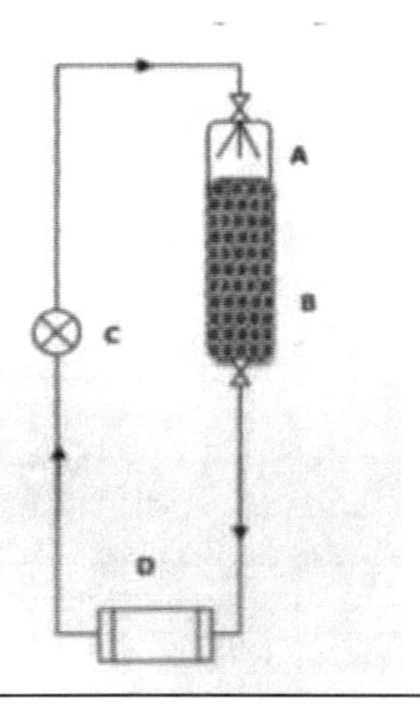

Figure 1. Schematic drawing of biofilter set-up for the degradation of phenol by C.tropicalis RETL-Cr1. (A) biofilter column (length = 1m, working volume = 1 L); (B) coconut husk (size ≈ 1.0 cm x 1.0 cm x 0.5cm, weight ≈ 334.07g); (C) tap; (D) peristaltic pump (250mL min^{-1})

2.5 Analytical method

2.5.1 Determination of phenol concentration

Phenol will be determined by isocratic elution high performance liquid chromatography (HPLC) (W600 2487) using a Waters Hypersil C18 5μm (4.6mmx 250 mm) column with UV detector at 280 nm

2.5.2 Determination of biomass concentration

Biomass concentrations will be determined by monitoring the optical density (OD). The OD will be monitored spectrophotometrically by measuring the absorbance at wavelength 600nm using the UV–Visible spectrophotometer [Model E 1011, 1000 series].

2.5.3 Determination of phenol biodegradation rate

According to [18], the phenol biodegradation rate can be calculated by dividing the total amount of phenol consumed with the time required for the consumption.

Phenol degradation rate = Amount of phenol consumed / Time required for consumption

2.5.4 Determination of specific growth rate

The exponential increase in biomass after inoculation is measured as a function of time and analyzed to obtained specific growth rate for that substrate concentration [19]. The specific growth rate was measured from the slope of the biomass curved by delineating points between the log growth phase represented by the equation (1).

$$\mu = (\ln X_t - \ln X_0) / t \quad (1)$$

where, X_t and X_0 is biomass concentration at time t and 0, and t is operational time.

3. Result and discussion

For the purpose of result accuracy, the control treatment for the phenol degradation (suspension condition) was run twice and simultaneously with the phenol degradation by loofah and coconut husk. As the purpose of simplification, the control experiments in the suspension condition, with absence of loofah would be referred as LC, whereas that in the absence of coconut husk is referred as CC. Similar for experiments run with the presence of adsorbents, it would be referred as L1 for experiment with the presence of loofah, whereas that in the presence of coconut husk is C1.

3.1 Growth pattern of C.tropicalis RETL-Cr1

The growth pattern of the yeast *C.Tropicalis* RETL-Cr1 was conducted inside a biofilter column with the absence and presence of both support material. The presence of support material affects the growth pattern of the yeast. Generally, at the first 3 hours, all four experiments LC, L1, CC and C1 experienced slow cell growth as the cells are adapting to the new environment. Apparent increase in cell growth can be seen at the following hours until 12th to 14th hour, where the cells are consuming the available nutrients. The cell growth declines after the 15th hour because of the nutrient depletion. Figure 2 shows the absence of loofah causes higher cell density at the 12th hour of 0.477 and eventually declined. It is higher than L1, where the presence of loofah reached highest cell density of 0.422 on the 13th hour. The different amount of cell density achieved is because of the physical properties of loofah. Loofah has many network lines which provide large surface area to the cells to bud more. According to [20], the biomass grows thicker on the surface on which it attached and consequently becomes difficult for the innermost cells to receive enough nutrients. Eventually, this leads to the loss of biomass contained on the surface area

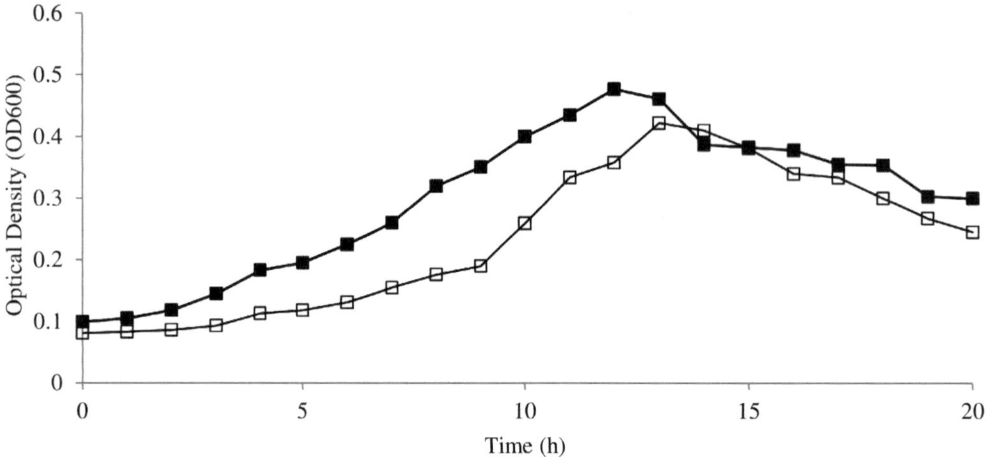

Figure 2. Growth pattern of yeast *C.tropicalis* RETL-Cr1 on Ramsay media broth with phenol (3mM) in biofilter column with the absence of loofah, LC (■) and the presence of loofah, L1 (□) as support material.

Figure 3 shows the growth curve of the yeast with the presence of coconut husk is slightly lower than with the absence of coconut husk. The highest cell density for CC was at 13th hour by 5.03, where as the highest cell density achieved by C1 was 4.39 at 15th hour. Coconut husk are known to have many voids which provide large surface area for cell budding and sustain growth and for yeast activity to occur [21]. Biomass grows on the exterior surface of the coconut husk which prevents nutrient and oxygen from going into the interior of the coconut husk [22] which makes it more difficult for the innermost cell to receive sufficient nutrient as the competition is increased. Cells are attached to the support material resulting on low cell density. Whereas, in suspension LC and CC, cell density is higher.

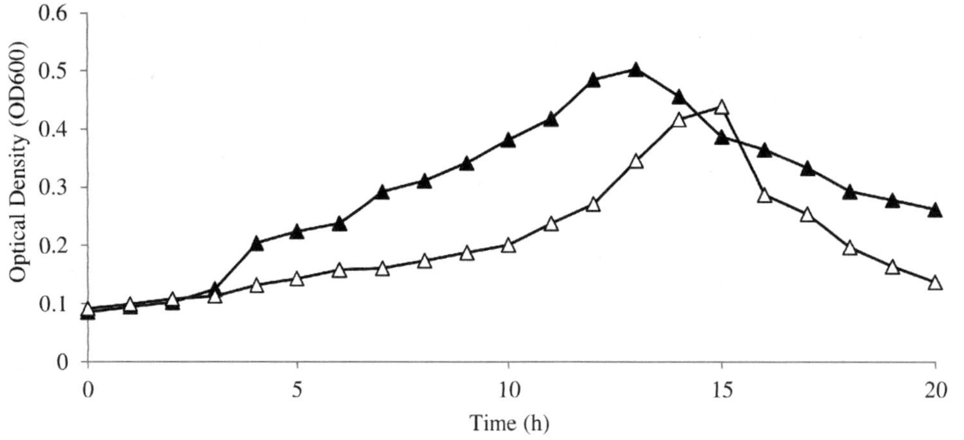

Figure 3. Growth pattern of yeast *C.tropicalis* RETL-Cr1 on Ramsay media broth with phenol (3mM) in biofilter column with the absence of coconut husk, CC (▲) and the presence of coconut husk, C1 (Δ) as support material

3.2 Phenol biodegradation

Phenol biodegradation was studied in batch experiments with initial concentration of 3mM by immobilized *C. tropicalis* RETL-Cr1. The phenol concentration in the medium decreased clearly when the adsorbed microorganism started to grow (Figure 4 and 5). Phenol biodegradation were 100% completed for all four treatments, LC, L1, CC and C1.

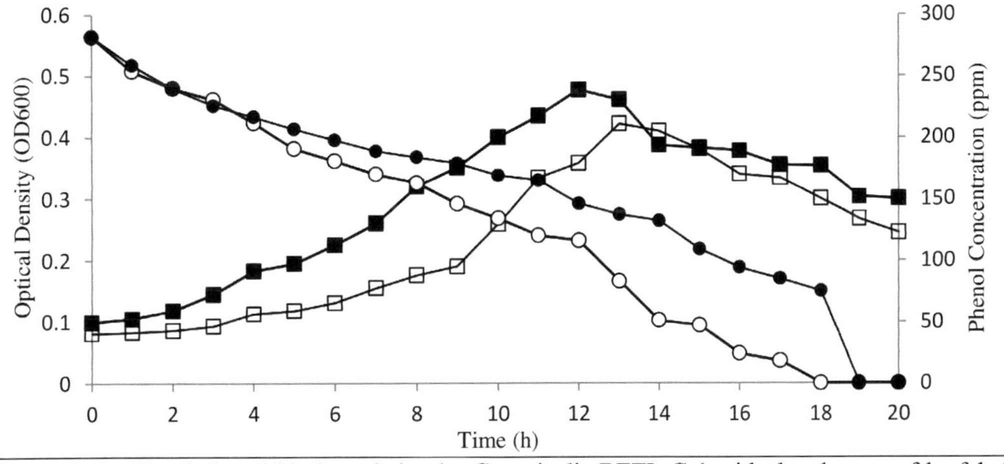

Figure 4. Profile of phenol biodegradation by C.tropicalis RETL-Cr1 with the absence of loofah (●) and with the presence of loofah (○) against time and growth pattern with the absence of loofah (■) and with the presence of loofah (□).

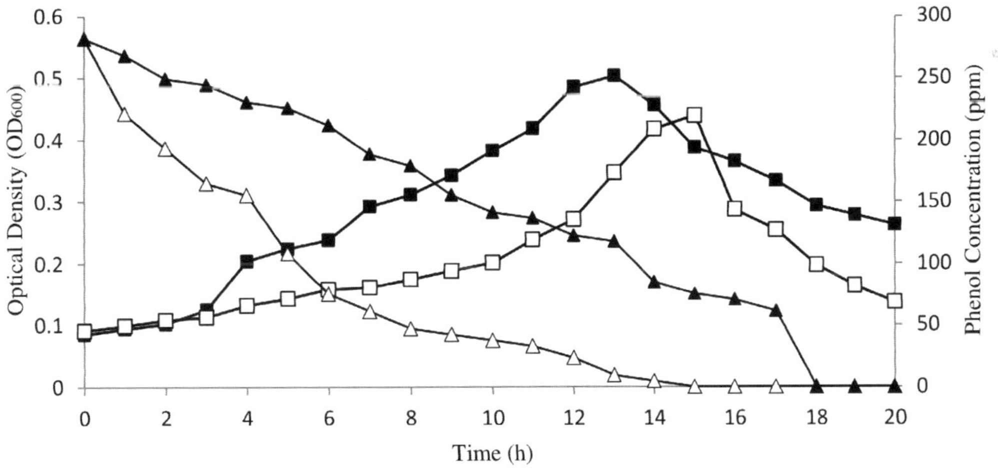

Figure 5. Profile of phenol biodegradation by C.tropicalis RETL-Cr1 with the absence of coconut husk (▲) and with the presence of coconut husk (Δ) against time and growth pattern with the absence of loofah (■) and with the presence of loofah (□).

Based on Table 1, L1 and C1 degraded phenol earlier by 1 and 2 hours respectively. The presence of support materials increase the contact between phenol and the yeast *C.tropicalis* RETL-Cr1, and hence degrade phenol faster than with the absence of support material. Phenol biodegradation rate was higher in the present of support material. L1 degrades phenol faster by the rate of 0.0188 $gL^{-1}h^{-1}$ compared to LC with the rate of 0.0176$gL^{-1}h^{-1}$. Whereas C1 degrades phenol at a rate of 0.0188 $gL^{-1}h^{-1}$ which also faster than CC as it rate is 0.0166 $gL^{-1}h^{-1}$.

Table 1. Performance of phenol biodegradation rate by *C.tropicalis* RETL-Cr1 in biofilter column with the absence and presence of support materials.

Performance	LC (Absence of loofah)	L1 (Presence of loofah)	CC (Absence of coconut husk)	C1 (Presence of coconut husk)
X_{max} (gL^{-1})	4.77	4.22	5.03	4.39
Specific growth rate, μ (h^{-1})	0.3928	0.3000	0.4906	0.3405
Operational time (h)	19	18	18	15
Lag time (h)	3	3	1	0
Biodegradation time (h)	16	15	17	15
Phenol biodegradation rate ($gL^{-1}h^{-1}$)	0.0176	0.0188	0.0166	0.0188
Phenol removal efficiency (%)	100	100	100	100

4. Conclusion

The biodegradation of phenol by free and immobilized *Candida tropicalis RET-Cr1* on loofah and coconut husk packed in biofilter column has been investigated. Biodegradation for all experiments of LC, L1, CC and C1 were 100% completed, with shorter biodegradation time for L1 and C1. Both loofah and coconut husk have similar phenol biodegradation rate of 0.0188 $gL^{-1}h^{-1}$ within 15 hours. However, It can be conclude that coconut husk is a better support material than the loofah as it enhanced the yeast *C.tropicalis* RETL-Cr1 growth better than loofah. Loofah have lower biomass concentration of 4.22 gL^{-1} compared to biomass concentration on coconut husk, 4.39 gL^{-1}. Overall, coconut husk and loofah is agricultural by-product which shows a potential as an alternative and economic adsorbent.

5. References

[1] Annadurai A, Ling L Y and Lee J F 2007 Biodegradation of phenol by Pseudomonas pictorum on immobilized with chitin *Afr. J. Biotechnol.* **6(3)** 296-303.

[2] Zhao G H, Li Y F, Liu X X and Liu X L 2010 Preparation of capsules containing 1- nonanol for rapidly removing high concentration phenol from aqueous solution *J. Hazard Mater.* **175** 715 – 725.

[3] Babuponnusami A and Muthukumar K 2011 Degradation of phenol in aqueous solution by fenton, sono-fenton and sono-photo fenton methods *Clean-Soil, Air, Water,* **39** 142- 147.

[4] Devipriya S P and Yesodharan S 2010 Photo catalytic degradation of phenol in water using TiO2 and ZnO *J. Environ. Biol.* **31** 247-249.

[5] Agarry S E, Audu T O K and Solomon BO 2009 Substrate inhibition kinetics of phenol degradation by Pseudomonas fluorescence from steady state and washout data *Int. J. Environ. Sci. Tech.* **6(3)** 443 – 450

[6] Dabhade M A, Saidutta M B and Murthy D V R 2009 Adsorption of phenol on granular activated carbon from nutrient medium: equilibrium and kinetic study *Int. J. Environ. Res.* **3(4)** 557–568.

[7] Baoshan X, William B M and Martin J D 1994 Sorption of phenol by selected biopolymers isothanus, energetics and polarity *Environ. Sci. Technol.* **28** 466-473

[8] Annadurai A, Babu S R, Mahesh K P O and Murugesan T 2000 Adsorption and biodegradation of phenol by chitosan-immobilized Pseudomonas putida (NICM 2174) *Bioprocess Engineering.* **2** 493-501

[9] Hameed B H 2009 Spent tea leaves: A new non-conventional and low-cost adsorbent for removal of basic dye from aqueous solution *J. Hazard. Mater.* **161** 253-259

[10] Popuri S R, Jammala A, Reddy K V N and Abburi K 2007 Biosorption of hexavalent chromium using tamarind (Tamarinddus indica) fruit shell - a comparative study *Electronic J. Biotechnol.* **3** 358-367

[11] Texier A-C, Andres Y and Lecloirec P 1999 Selective biosorption of lanthanide (La, Eu, Yb) ions by Pseudomonas aeruginosa *Environ. Sci. Technol.* **33** 489-495

[12] Annadurai G R S, Juang and Lee D J 2002 Biodegradation and adsorption of phenol using activated carbon immobilized with pseudomonas putida *J. Environ. Sci. Health Part A, Toxic/Hazardous Substances Environ. Eng.* **37** 1133-1146

[13] Marshall W E, Wartelle L H, Boler D M and Toles C A 2000 Metal ion adsorption by soybean hulls modified with citric: a comparative study *Environmental Technology.* **21(6)** 601-607

[14] Reddad Z, Gerente C, Andres Y and Lecloirec P 2002 Adsorption of several metal ions onto a low cost biosorbent: Kinetic and equilibrium studies *Environ. Sci. Technol.* **36** 2067-2073

[15] Agarry S E and Owabor C N 2005 Utilization of rubber fruit pericarp – activated carbon for abattoir waste water treatment and absorption of iron (III) ions form aqueous solution *J. Sci. Technol. Res.* **4(2)** 49 -54

[16] Agarry S E and Aremu M O 2012 Batch Equilibrium and Kinetic Studies of Simultaneous Adsorption and Biodegradation of Phenol by Pineapple Peels Immobilized *Pseudomonas aeruginosa* NCIB 950 *British Biotechnology Journal.* **2(1):** 26-48

[17] Chen B, Yuana M and Liu H 2011 Removal of polycyclic aromatic hydrocarbons from aqueous solution using plant residue materials as a biosorbent *J. Hazard Mater.* **188** 436 – 442

[18] Kar S, Swaminathan T and Baadarajan A 1996 Studies on biodegradation of a mixture of toxic and non-toxic pollutant using arthrobacter species *Bioproc. Eng.* **15 (4):** 195 – 199

[19] Yoong E T, Staib C I and Lant P A 2004 Kinetic coefficient of high strengthphenolic wastewater biodegradaion. Environmental biotechnology: Advancement in water and wastewater application in the topics *Wat. Env. Manag. Ser.* 35 – 42.

[20] Levin M A and Gealt MA 1993 *Biotreatment of Industrial and Hazardous Waste* (eds). McGraw-Hill New York.

[21] Ortiz I, Revah S, and Auria R 2003. Effects of packing material on the biofiltration of benzene, toluene and xylene vapours. *Environmental Technology.* **24(3):** 265 – 275.

[22] Levstek M and Plazi I 2009 Influence of carrier type on nitrification in the moving-bed biofilm process *Water science and technology.* **59 (5)** 875 – 882

Isolation and Characterization of Surface and Subsurface Bacteria in Seawater of Mantanani Island, Kota Belud, Sabah by Direct and Enrichment Techniques

L D Benard[1], **P M Tuah, E G Suadin and N Jamian**

[1]Faculty of Science and Natural Resources, Universiti Malaysia Sabah

E-mail: laurenciadb@gmail.com

Abstract. The distribution of hydrocarbon-utilizing bacterial may vary between surface and subsurface of the seawater. One of the identified contributors is the Total Petroleum Hydrocarbon. The isolation and characterization of bacteria using Direct and Enrichment techniques helps in identifying dominant bacterial populations in seawater of Mantanani Island, Kota Belud, Sabah, potential of further investigation as hydrocarbon degrader. Crude oil (5% v/v) was added as the carbon source for bacteria in Enrichment technique. For surface seawater, the highest population of bacteria identified for both Direct and Enrichment technique were 2.60×10^7 CFU/mL and 3.84×10^6 CFU/mL respectively. Meanwhile, for subsurface seawater, the highest population of bacteria identified for both Direct and Enrichment technique were 5.21×10^6 CFU/mL and 8.99×10^7 CFU/mL respectively. Dominant species in surface seawater were characterized as *Marinobacter hydrocarbonoclasticus*–RMSF–C1 and RMSF–C2 and *Alcanivorax borkumensis*–RMSF–C3, RMSF–C4 and RMSF–C5. As for subsurface seawater, dominant species were characterized as *Pseudomonas luteola*–SSBR–W1, *Burkholderia cepacia*–SSBR–C1, *Rhizobium radiobacter*–SSBR–C3 and *Leuconostoc–cremois* –SSBR–C4.

1. Introduction

Microorganisms' population is numerous in the open ocean. There are many different types of environmental-relevant microorganisms (ERM) mostly aerobes that are known to be able to degrade oil. The most important (based on frequency of isolation) genera of hydrocarbon utilisers in aquatic environment are of *Pseudomonas*, *Achromobacter*, *Micrococcus*, *Nocardia*, *Vibrio*, *Acinetobacter*, *Brevibacterium*, *Corynebacterium*, *Candida*, *Rhodutula* and *Sporobolomyces*. The most commonly isolated hydrocarbon-degraders are Gram Negative rods such as *Pseudomonas* and *Alcaligenes* [1]. Further studies continue to expand the list of microbial species that have shown capable of degrading various components of petroleum [2]. However, it is the isolates, single species or consortia that can degrade both aliphatic and aromatic hydrocarbons that are highly desirable.

Mantanani Island, Kota Belud, Sabah was chosen as the study site because of the potential to be a resort and one of the requirements to become a sea resort is free from any contamination and harmful-free marine species as Mantanani Island also known for dugong's populations. Therefore, the purpose of this study is to identify and determine the amount of oil and grease contamination and also the bacterial population in the surrounding sea of Mantanani Island. The aim of this study is to evaluate the physical (temperature and pH), chemical (Total Oil and Gas and Total Petroleum

Hydrocarbon) and biological aspects (bacterial populations) and examining the relationship between TOG, TPH and populations of bacteria found and latter identify the dominant bacterial populations of surface and subsurface waters of selected stations.

2. Materials and Method

2.1. Study Area

This study was done entirely at Mantanani Island, Kota Belud, Sabah with the coordinates (6°4351.38" N, 116°21'13.56" E) and Environmental Microbiology Laboratory, Faculty of Science and Natural Resources, Universiti Malaysia Sabah. Seawater samples were collected at four sampling stations which were 1) Housing area of *Orang Ubian* (6°41.980" N, 116°21.251" E), 2) the end of the island next to the station 1 (6°42.298" N, 116°22.131" E), 3) rear area of the island and no settlement (6°43.206" N, 116°22.227" E), and 4) the Naval base area (6°42.533" N, 116°20.490" E).

2.2. Culture Medium

Culture media used in this study were Ramsay agar and broth. The preparation was according to Ramsay medium as described by Ramsay *et al.* (1983). The composition of Ramsay broth is of the same with Ramsay agar, only that 0.5gL^{-1} $MgSO_4.H_2O$ and 20.0gL^{-1} glucose were added to the broth after being autoclaved at 121°C for 15 minutes. The glucose and $MgSO_4.H_2O$ stocks were filter-sterilized under laminar floor before added to the culture media.

2.3. Sampling

The seawater samples were taken at depth of 10 cm and 25 – 60 cm from surface, for surface and subsurface seawater respectively. For oil and grease determination, seawater samples were taken using the blue cap bottle. Meanwhile, for the determination of enrichment technique, 2 mL of seawater samples was injected into the universal bottles. Samples were stored in a cool box containing ice cubes all the way to the laboratory for analysis. Preservation of the samples was done by adding sulphuric acid until the pH is less than 2 and cooled at 4°C. Acidification works to stop the biological activity in the sample and also converting the fatty acids from marine salt water into free acid which is not water soluble, increases the total amount of oil and grease.

2.4. Temperature and pH analysis

Temperature and pH of surface and subsurface waters at selected stations were tested during sampling using pH meter. Prior to use, pH meter should be calibrated to avoid errors.

2.5. Direct and Enrichment techniques

Direct technique is a technique in which the seawater samples are taken directly for bacterial isolation. Meanwhile, Enrichment technique is a technique in which the bacteria are supplied with fuel as a food source to increase the amount of bacteria present in the original sample. This technique need to be used as total bacterial populations in seawater are few, so through enrichment technique the number of bacteria were doubled and the isolation process is easily conducted. Seawater was injected into Ramsay broth using a syringe. Bacteria were allowed to grow in Ramsay broth and incubated for 24 hours at 30°C. After 24 hours, the bacterial isolation was performed on samples.

2.6. Isolation of bacteria

Isolation of bacteria was done for the purpose of breeding bacteria that were found in the samples. Two methods that have been used are the pour plate and serial dilution method. These methods are according to the Standard Method for the Examination of Water and Wastewater (1998). Serial dilution of the sample was performed at the series of 10^1,10^2,10^3,10^4,10^5,10^6,10^7,10^8 and 10^9. However, bacterial analysis was only performed at the series of 10^3,10^4 and 10^5 as for series before 10^3 and after 10^6 would not give accurate results. Each selected series of dilution were duplicated. The samples were then placed into the incubator for 48 hours at 30°C and 40°C. Bacterial colonies formed after 48 hours were calculated using Colony Forming Unit (CFU/mL).

2.7. Biochemical tests
Biochemical tests were carried out to identify key chemical compounds found in isolated colonies. The tests were also done to study the reaction of each bacterial colony towards the addition of chemicals. A total of four biochemical tests were carried out prior to characterization of bacteria which were oxidase, catalase, spores and motility.

2.8. Characterization of bacteria
Characterization of bacteria was done using Gram staining process for identifying populations of bacteria found in samples of surface and subsurface waters. Pure culture was used for Gram staining. The preparation of a pure culture was by getting a single culture for each colony formed and incubated for 24 hours at 30°C.

2.9. Gram staining
A loopful of the pure culture were transferred to the surface of a clean glass slide using wire loop and added few drops of distilled water, spread over a small area and heat fixed until dried. Then, the smear stained with crystal violet for 2 minutes, washed briefly with distilled water. Next, the slide was treated with iodine or lugol to fix the stain for 2 minutes, washed briefly with distilled water. After that, destained the slides with drops of alcohol (95%) and finally counterstained with safranin for 10 seconds, washed briefly with distilled water. All slides of bacteria were then examined under the oil immersion lens.

2.10. Analytical Profile Identification (API)
Identification of the species of bacteria present in the sample surface and subsurface waters was done by using the API test. API tools that were used in this study were API 20NE, API 50CHL, and API 50CHB / E. API test results were recorded after 24 hours and 48 hours using results table sheets provided with API box. The results data obtained then placed in the computer software APIweb as input data to identify the species of bacteria and the percentage accuracy of the bacterial species. Percentage of bacterial species identification accuracy APIweb was acceptable in between 88% to 99%.

2.11. Total Oil and Grease (TOG) and Total Petroleum Hydrocarbon (TPH)
Total oil and grease (TOG) were determined using gravimetric partition (APHA, 1998). Total petroleum hydrocarbon (TPH) was determined using the gravimetric method (silica gel treated n-hexane extractable materials @ SGT-HEM) similar to USEPA method 1664.

3. Results

3.1. Temperature and pH

As shown in Figure 1a and 1b, the temperatures of surface and subsurface of seawater are of similar ranges between 27.6°C to 29.8°C for Station 2, 3 and 4 where the lowest temperature of 27.8°C and 26.6°C was recorded at Station 1. Meanwhile, for pH value of surface and subsurface of seawater, Station 1, 3 and 4 also shows similar ranges from 8.01 to 8.31 where the lowest pH value of 7.23 and 7.74 was recorded at Station 2. Optimum temperature for marine water is from 10°C to 40°C in order for sustaining marine organisms in that particular environment. Based on the in-situ analysis of temperature in the study area, the obtained results are from 26.6°C to 29.8°C. This shows that the study area exhibits an optimum temperature.

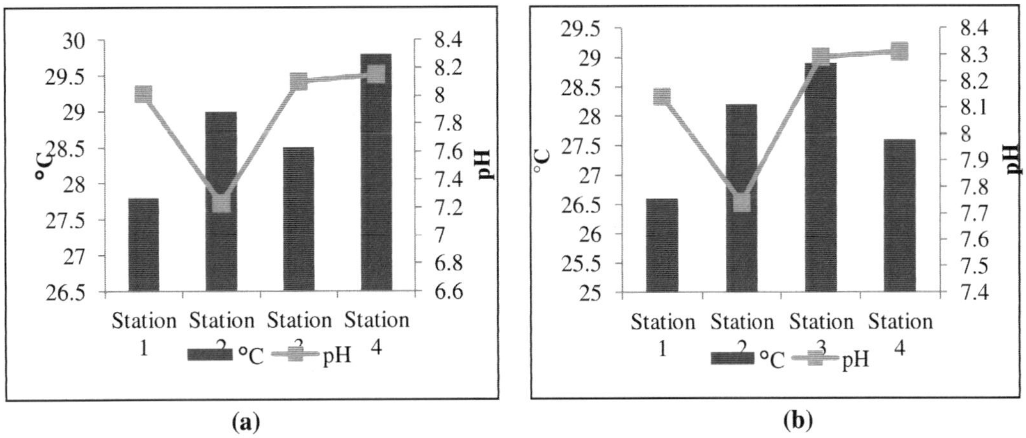

Figure 1 Temperature (°C) and pH of **(a)** Surface **(b)** Subsurface seawater in each sampling stations

Microbial growth rates in seawater are often associated with the surrounding temperature or optimum temperature. At temperature of about 20°C to 45°C, the microorganisms categorized as mesophilic. The optimum growth temperature is the temperature that is ideal for bacterial growth and rapid growth at this temperature [3]. The seawater temperature is also closely related to the concentration of petroleum hydrocarbons [4]. Evaporation rate of petroleum hydrocarbons from seawater into the air is influenced by water temperatures. In addition, water temperatures are correlated with the rate of degradation and microbial decomposition of petroleum hydrocarbons in water.

The range values of pH suitable for optimum marine conditions are from 7.0 to 8.5 meanwhile average range value of pH is form 7.9 to 8.3 [5]. Result shows that the range of pH values at study area are from 7.23 to 8.31, which is optimum and suitable for almost all aquatic organisms. A change of pH in the water column has an impact on aquatic life as aquatic organisms are very sensitive to changes in pH. Most microbes multiply rapidly in a neutral environment with a pH of 6.5 to 7.5, but based on the pH value recorded shows towards more alkaline environment. Thus, microbes in the environment can be categorized as Alkalophil. Alkalophil are of bacteria which tend to proliferate in the pH range of 8.5 to 9.0 [6].

3.2. Concentration of Total Oil and Grease (TOG) and Total Petroleum Hydrocarbon (TPH)

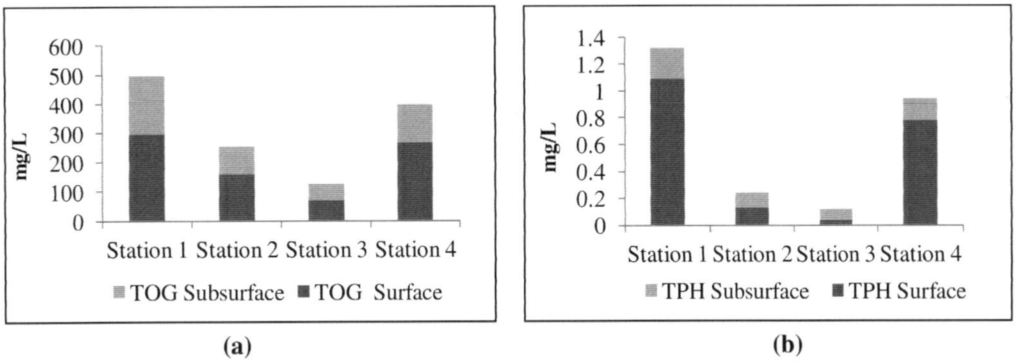

Figure 2 (a) TOG **(b)** TPH concentration of surface and sub surface seawater in each stations

Figure 2a shows TOG concentrations in surface and subsurface waters of selected stations. Highest reading recorded at Station 1 and the lowest is at Station 3, with a 4-fold difference. Meanwhile, highest TOG concentration for subsurface recorded at Station 1 and the lowest is at Station 3, similar to the surface seawater, only that the difference is a 3-fold.

Figure 2b shows the highest reading recorded at Station 1 and the lowest is at Station 3, with a 24-fold difference. Meanwhile, TPH concentrations for subsurface water range from 0.08 to 0.23 mg/L where the highest reading recorded at Station 1 and the lowest is at Station 3, similar to the surface seawater, only that the difference is almost 3-fold.

Based on the results, Station 1 recorded the highest concentration of TOG and TPH for both surface and subsurface seawater which may due to environmental factors that affect the concentrations of these compounds found in surface and subsurface waters. Through observation, Station 1 has a denser population distribution compared to the other three stations. Factors that may contribute to higher TOG and TPH concentrations are wastewater resulting from the daily activities of the villagers. Wastewater may contain a variety of contaminants such as food waste, cooking oil and many more. TOG also have poor solubility properties and tend to form a layer on the surface of the water [5]. TPH is one of the components in TOG. Higher concentration of TOG may contribute to the higher concentration of TPH. Among the major contributors to the high concentrations of TPH is from boating activity that uses gasoline as an energy source. Marine transportation is among the anthropogenic sources that introduced petroleum hydrocarbons into marine waters [7].

3.3. Bacterial population

Bacterial population for surface seawater using Direct technique has the range from 4.00×10^6 CFU/mL to 4.32×10^7 CFU/mL where the highest populations recorded at Station 4 with 4.32×10^7 CFU/mL and the lowest is at Station 3 with 4.00×10^6 CFU/mL, a 10-fold difference. Bacterial populations for Station 1 and Station 2 are 2.45×10^7 CFU/mL and 5.50×10^6 CFU/mL respectively. Meanwhile, bacterial population for subsurface seawater using Direct technique has the range from 4.00×10^5 CFU/mL to 1.57×10^7 CFU/mL where the highest reading recorded at Station 1 and the lowest is at Station 3, similar to the surface seawater, only that the difference is a 39-fold. Bacterial populations for Station 2 and Station 4 are 3.10×10^6 CFU/mL and 7.10×10^6 CFU/mL respectively.

Meanwhile, bacterial population for surface seawater using Enrichment technique has the range from 5.00×10^6 CFU/mL to 3.80×10^7 CFU/mL where the highest populations recorded at Station 1 with 3.80×10^7 CFU/mL and the lowest is at Station 3 with 0.05×10^8 CFU/mL, almost 8-fold difference. Bacterial populations for Station 2 and Station 4 are 9.00×10^6 CFU/mL and 2.00×10^7 CFU/mL respectively. As for bacterial population for subsurface seawater, the range is from 2.30×10^7 CFU/mL to 1.64×10^8 CFU/mL where Station recorded the highest population with 1.64×10^8 CFU/mL and Station 3 recorded the lowest population with only 2.30×10^7 CFU/mL, similar the surface seawater, only that the difference is a 7-fold. Bacterial populations for Station 2 and Station 4 are 3.20×10^7 CFU/mL and 1.50×10^8 CFU/mL respectively.

Based on the results, the bacterial population for surface and subsurface seawater has been identified and calculated using both Direct and Enrichment techniques. Bacterial populations were recorded high in surface waters using the Direct method, and using the Enrichment method showed high bacterial population in the subsurface.

3.4. Relationship of Total Petroleum Hydrocarbon (TPH) and Bacterial Population in Surface and Subsurface seawater

Bacterial population was recorded highest at Station 1, where TPH concentrations are also of the highest. As concentration of TPH decreases in Station 2 and 3, the bacterial population also declined and increased again at Station 4 when TPH concentrations increased. This suggests that the concentration of TPH and bacterial populations through Direct technique and Enrichment technique are directly proportional.

High bacterial population recorded at Station 1 shows the ability of the bacteria to tolerate the high concentrations of TPH. The presence of bacteria in the water column is important because microbes are the major organisms identified in the water column that are able to help in the decomposition of waste including hydrocarbons and other metabolites.

Typically, marine environment contaminated with petroleum hydrocarbons have an extensive bacterial population, which of population capable to degrade petroleum as compared to that population from uncontaminated areas. Bacteria that present on the surface and subsurface layer are highly resistant to UV radiation and chemicals. Resistance to pollution and the ability to degrade indicates that the bacteria isolated from surface and subsurface seawater are important in the process of bioremediation.

3.5. Isolation and characterization of surface and subsurface bacteria

3.5.1. Isolation of surface and subsurface bacteria. Morphological, physiological and biochemical identification of the isolates from surface and subsurface seawater samples were investigated. A selected number of the investigated sequences are shown in Table 1 and 2. A total of eight dominant bacterial colonies were identified (Table 1). Determination purpose of the dominant bacteria is to determine the ability of the bacteria to tolerate with high concentration of TPH. If the bacteria can survive and grow in a particular concentration of TPH, thus it can also decompose or degrade hydrocarbons in that particular area. Based on Table 1, there are two types of configurations for the selected colonies which are round and spread whilst the sides are wavy and smooth. All selected strains had similar colony colour of cream and white. All strains exhibits similar cell morphology which is rod-shaped. Through Gram staining, strains RMSF-C1 until RMSF-C5 identified as Gram-negative while RMSF-C6 until RMSF-C8 are Gram positive.

Biochemical tests performed on each strain to determine the APIs test that will be used for each selected bacterial strains. Almost the entire bacterial strains showed negative results for Spores Tests, where most of the bacterial strains do not produce spores. Instead, endospores are being produced by bacterial cells to survive. Endospore is a hard layer that enables the strain to be resistant to radiation and chemicals [4]. Catalase test showed that the eight strains showed positive results. Catalase is an enzyme containing iron. Meanwhile, Oxidase test showed that only strains RMSF-C6 and RMSF-C8 showed positive results. This indicates that all other strains are not of anaerobic bacteria. Most marine bacteria are facultative anaerobe which breeds well in the presence of oxygen.

Table 1. Morphological, physiological and biochemical identification of the isolated bacteria from surface seawater

Items	Isolates							
	RMSF-C1	RMSF-C2	RMSF-C3	RMSF-C4	RMSF-C5	RMSF-C6	RMSF-C7	RMSF-C8
Colony color	Cream	Cream	White	Cream	White	Cream	Cream	White
Colony surface	Smooth	Wavy	Smooth	Smooth	Smooth	Wavy	Smooth	Wavy
Shape of cells	Rod	Rod	Rod	Rod	Rod	Rod	Rod	Rod
Gram staining	-	-	-	-	-	+	+	+
Endospores	-	-	-	-	-	-	-	-
Catalase	+	+	+	+	+	+	+	+
Oxidase	+	+	+	+	+	-	+	-
Motility	+	+	-	-	-	-	-	-

Positive (+) Negative (-)

Table 2. Morphological, physiological and biochemical identification of the isolated bacteria from subsurface seawater

Items	Isolates					
	SSBR-C1	SSBR-C2	SSBR-C3	SSBR-C4	SSBR-C5	SSBR-W1
Colony color	Cream	Cream	Cream	Cream	Cream	White
Colony surface	Smooth	Wavy	Wavy	Irregular	Smooth	Smooth
Shape of cells	Rod	Rod	Rod	Rod	Rod	Rod
Gram staining	-	+	-	-	+	-
Endospores	-	-	-	-	-	+
Catalase	+	+	+	+	+	-
Oxidase	+	+	+	+	+	-
Motility	-	+	+	+	+	+

Positive (+) Negative (-)

However, the growth rate of marine bacteria is very slow compared to most soil bacteria [8]. Motility test only showed positive results for strain RMSF-C1 and RMSF-C2. Motility test using SIM aims to determine whether the bacteria are moving on its own or elsewhere during the growth occurs [9].

A total of six dominant bacterial colonies were identified (Table 2) from subsurface seawater. Based on Table 2, there are three types of configurations for the selected colonies which are round, spread and L-shaped, whilst the sides are smooth, wavy and irregular. All selected strains had similar colony colour of cream and white. All strains exhibits similar cell morphology which is rod-shaped. Through Gram staining, strains SSBR-CR1, SSBR-CR3, SSBR-CR4 and SSBR-W1 identified as Gram-negative while SSBR-CR2 and SSBR-CR5 are Gram positive.

Almost the entire bacterial strains showed negative results for Spores Tests, except for SSBR-W1 which show positive result by producing spore. Catalase test showed that the 5 strains showed positive results, only SSBR-W1 that show negative result which indicates that SSBR-W1 does not equipped with catalase, an enzyme containing iron. Meanwhile, oxidase test showed that only strains SSBR-W1 showed negative results. This indicates that all other strains are of anaerobic bacteria due to it exists in subsurface column of seawater. Motility test only showed negative results for strain SSBR-C1.

3.5.2. Characterization of surface and subsurface bacteria. API tests (API 20NE and API 50 CHL) are the test that showing results for characterization of surface and subsurface bacteria. API 20NE are used to identify the species of non-enteric bacteria of a Gram-positive with a rod-shaped. The API 50CHL are used to identify Gram positive bacteria with a rod-shaped and does not produce spores.

Table 3 shows the results of API test and percentage of identification of each bacterial species for surface seawater. Based on the API tests, there are three species of bacteria that have been identified in surface waters which are *Bacillus coahuilensis sp*, *Marinobacter hydrocarbonoclasticus sp* and *Alcanivorax borkumensis sp*. According to API *Web*, the percentage of identification of species for *Bacillus coahuilensis sp* for strain coded RMSF-C5, RMSF-C6 and RMSF-C8 are weak, while *Marinobacter hydrocarbonoclasticus sp* for strain coded RMSF-C1 and RMSF-C2 and *Alcanivorax borkumensis sp* for strain coded RMSF-C3, RMSF-C4 and RRMSF-C5 are good.

The percentage identification accuracy plays a very important role in determining the selection of bacterial strains. This is because only a percentage of accuracy approaching 100% will be accepted as a species of bacteria of the strain.

Table 3. Surface bacterial species identification using API and percentage of identification accuracy

Colony code	API	Identified species	Percentage of Identification (%)	Comment
RMSF-C1	20 NE	*Marinobacter hydrocarbonoclasticus sp*	96	Good
RMSF-C2	20 NE	*Marinobacter hydrocarbonoclasticus sp*	90	Good
RMSF-C3	20 NE	*Alcanivorax borkumensis sp*	90	Good
RMSF-C4	20 NE	*Alcanivorax borkumensis sp*	91	Good
RMSF-C5	20 NE	*Alcanivorax borkumensis sp*	89	Good
RMSF-C6	50 CHL	*Bacillus coahuilensis sp*	60	Weak
RMSF-C7	50 CHL	*Bacillus coahuilensis sp*	63	Weak
RMSF-C8	50 CHL	*Bacillus coahuilensis sp*	50	Weak

Meanwhile, Table 4 shows the results of API test and percentage of identification of each bacterial species for subsurface seawater. Based on the API tests, there are five species of bacteria that have been identified in subsurface waters which are *Burkholderia cepacia sp*, *Rhizobium radiobacter sp*, *Leuconostoc mesenteroides ssp cremois*, *Rhizobium radiobacter sp* and *Pseudomonas iuteola sp*.

According to API *Web*, the percentage of identification of all five species for *Burkholderia cepacia sp* for strain coded SSBR-C1, *Rhizobium radiobacter sp* for strain coded SSBR-C3, *Leuconostoc mesenteroides ssp cremois* for strain coded SSBR-C4, *Rhizobium radiobacter sp* for strain coded SSBR-C5 and *Pseudomonas iuteola sp* for strain coded SSBR-W1 are good.

Table 4. Subsurface bacterial species identification using API and percentage of identification accuracy

Colony code	API	Identified species	Percentage of Identification (%)	Comment
SSBR-C1	20 NE	*Burkholderia cepacia sp*	99	Good
SSBR -C3	20 NE	*Rhizobium radiobacter sp*	99	Good
SSBR -C4	50 CHL	*Leuconostoc mesenteroides ssp cremois*	78	Good
SSBR -C5	20 NE	*Rhizobium radiobacter sp*	97	Good
SSBR -W1	20 NE	*Pseudomonas iuteola sp*	99	Good

4. Conclusion

In this paper, it is examined that the highest bacterial population were obtained for surface and subsurface seawater using Direct technique and Enrichment technique. This suggests that the concentration of TPH and bacterial populations through direct technique and Enrichment technique are directly proportional. Morphological, physiological and biochemical identification of the dominant isolates from surface and subsurface seawater samples were investigated and results shows that the total of eight strains from surface seawater are showing almost similar morphological, physiological and biochemical characteristics with the total of six strains from surface seawater, although some are differ with another one. However, from API tests, the strains species for surface are of *Bacillus coahuilensis sp*, *Marinobacter hydrocarbonoclasticus sp* and *Alcanivorax borkumensis sp* while the strains species for subsurface are of *Burkholderia cepacia sp*, *Rhizobium radiobacter sp*, *Leuconostoc mesenteroides ssp cremois*, *Rhizobium radiobacter sp* and *Pseudomonas iuteola sp*. This shows that the differences in depth of water column for surface and subsurface do plays role in determining the species of bacteria identified. Ultimately, all of the results demonstrated by both surface and subsurface bacteria selected in our study may be applicable for further investigation on potential of biodegrading petroleum hydrocarbon of certain concentrations. Microbial degradation is one of the cost-effective treatments for hydrocarbon polluted environment. Nonetheless, periodic monitoring of TOG, TPH and bacterial population at Mantanani Island, Kota Belud is crucial to ensure the marine environment well maintained.

References

[1] Atlas R M and Bartha R 1972 Biodegradation of petroleum in seawater at low temperatures *Can. J. Microbial* **18** 1851
[2] Atlas R M 1981 Microbial degradation of petroleum hydrocarbons: an environmental perspective *Microbial Rev.* **45** 180
[3] Nester J M 2004 *Composition and fate of Petroleum and spill treating agents in the marine environment. Sea Mammals and Oil: Confronting the risks.* J R Geraci and D J St AUbin (eds.) Academic Press: New York **2** 1-33
[4] Ross D A 1995 *Introduction to Oceanography.* HarperCollins College Publisher, New York
[5] Radojevic M, Mohd Harun Abdullah and Ahmad Zaharin Aris 2007 *Analisis Air* Polar Press(M) Sdn Bhd
[6] Moat A G, Foster J W and Spector M P 2002 *Microbial Physiology.* Wiley-Liss Inc: New York
[7] Collins J F, Brown J P, Alexeff G V and Salmon A G 1998 Potencency Equivalency. Factors for some Polycyclic Aromatic Hydrocarbon Derivatives. *Journal of Regulatory Toxicology and Pharmacology* **28** 45
[8] Cevera E J, Karl D and Buckley M 2005 *Marine Microbial Diversity: The Key to Earth's Habitability.* The America Academy of Microbiology Report
[9] Moat A G, Foster J W and Spector M P 2002 *Microbial Physiology* Wiley-Liss Inc: New York

Acknowledgment

Authors wishing to acknowledge financial support from World Federation of Scientist (WFS) Malaysian National Scholarship Programme.

Review Paper on Cell Membrane Electroporation of Microalgae using Electric Field Treatment Method for Microalgae Lipid Extraction

C Joannes[1], C S Sipaut[1], J Dayou[2], S M Yasir[3] and R F Mansa[1]

[1] Energy and Materials Research Group (EMRG), Faculty of Engineering, Universiti Malaysia Sabah, Jln UMS, 88400 Kota Kinabalu, Sabah
[2] Energy, Vibration and Sound Research Group (e-VIBS), Faculty of Science and Natural Resources, Universiti Malaysia Sabah, Jln UMS, 88400 Kota Kinabalu, Sabah
[3] Seaweed Research Unit, Faculty of Science and Natural Resources, Jln UMS, 88400 Kota Kinabalu, Sabah

Email: costantinejoannes@gmail.com; rfmansa@ums.edu.my

Abstract. The paper reviews the recent studies on the lipid extraction of microalgae that mainly highlighted on the cell disruption method using variety of microalgae species. Selection of cell disruption method and devices are crucial in order to achieve the highest extraction percentage of lipid and other valuable intracellular (proteins, carotenoids and chlorophylls) from microalgae cell. Pulsed electric field (PEF) and electrochemical lysis methods were found to be potential for enhancing the extraction efficiencies either conducted in single step extraction or used as pre-treatment followed by conventional extraction method. The PEF technology capable to extract lipid as high as 75%. While, electrochemical lysis treatment capable to extract lipid approximately 93% using Stainless Steel (SS) and Ti/IrO$_2$ as the cathode and anode electrode respectively. PEF technology and electrochemical lysis are still considered to be a new method for microalgae lipid extraction and further investigation can still be done for better improvement of the system.

1. Introduction

The major concern of fossil fuel depletion has stressed out the importance of seeking an alternative renewable energy that fulfilled the criteria such as economic, sustainable and environmental friendly. The third generation of biofuel, which is biodiesel from microalgae had demonstrated to be the most potential biomass feedstock in replacing the other conventional crops like maize, jatropha and palm oil [1-3]. In comparison with the conventional crops based biodiesel, microalgae provide several benefits such as high growth rate [4; 5], high lipid content [6] and reduce in land requirements [7]. In addition, the issues regarding food versus fuel are less affected for microalgae crops, but yet still being used as an additional supplement in the human diet and animal nutrition such as *Spirulina* and *Chlorella* [7].

Fresh water microalga *Chlorella* species has been clearly stated to be a most favorable strain due to its high lipid content which rich with saturated fatty acids and suitable for biodiesel conversion [6; 8; 9]. Biodiesel yields are much more related to the percentage of lipid extracted from the microalgae cell. Hence, the selection of extraction methods and devices are totally dependable on the extraction efficiencies of lipid extracted from the microalgae biomass. Microalgae lipid can be extracted by traditional methods either via chemical, physical, enzymatic or a combination of the techniques. The methods included solvent extraction, supercritical fluid, microwave irradiation, osmotic shock, and homogenizer. Although these methods are still being used for lipid extraction, there are several drawbacks of using the traditional lipid extraction methods such as time consuming, energy intensive, potentially harmful and difficult to scale up [10].

Presently, microalgae lipid extraction using the concept of cell membrane electroporation or electropermebilization has gained interest lately. Pulsed electric fields (PEF) and electrochemical lysis methods had demonstrated to be able to increase the lipid extraction efficiency either used as pre-treatment or alone and potential for large scale application [11-20]. Besides, the methods have the advantages of eliminating dewatering process since the methods are applicable for direct wet biomass. The methods consume less energy and no additional chemicals required. Industrial scale application using PEF for microalgae lipid extraction [21] and algae wastewater treatment [22] had already been demonstrated.

To the best of our knowledge, the cell disruption related to the electroporation concept for microalgae lipid extraction has not yet been practiced in Malaysia country. The methods are relatively new for microalgae lipid extraction and only few literatures had been published so far. This paper focused on the potential application of PEF and electrochemical lysis treatment as the method to extract microalgae lipid for wet or dry biomass. The paper also includes the theory of electroporation mechanism. The methods can be an alternative approach for lipid extraction in biodiesel production specifically for algae based crop.

2. A cell membrane electroporation theory

In the middle of 1900, the electroporation concept had already been used in medical application. The concept then applied in food technology and processing to kill microbes and pathogens [23]. The concept is kindly unique, therefore, has attracted researchers to use it in biotechnology and engineering application specifically to extraction lipid and other valuable intracellular (e.g. Pigments and protein) from microalgae. Electroporation is a phenomenon which the cell wall exposed to high electric field strength in a certain period. During the electric field exposures the cell membrane is temporarily destabilized. Increasing the electric field intensity will cause the cell membrane to experience either reversible or irreversible electroporation [24]. Reversible electroporation is the condition which the cell membrane capable to recover or reseal back its membrane. While, irreversible electroporation is the cell membrane permanently damage and unable to recover its original form.

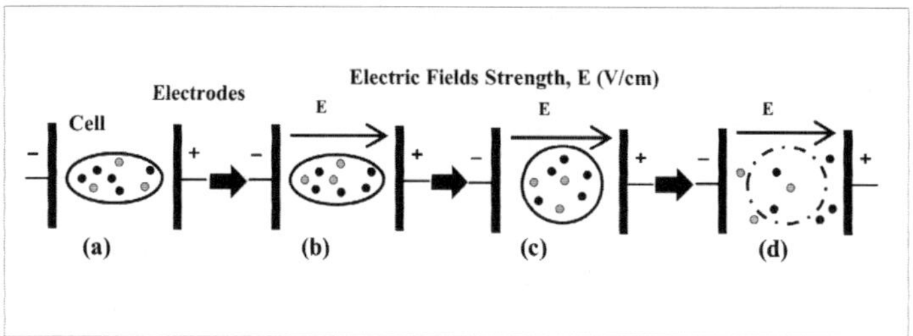

Figure 1. Electroporation Mechanism of a cell membrane: (a) At zero potential, (b) Osmotic imbalance, (c) Swelling, (d) Membrane ruptures. Modified from Ortega-Rivas [23].

Chemotherapy method for cancer treatment is using reversible electroporation to allow the injected drugs to attack the cell cancer after the cell membrane reseal back its initial form [25]. But from the engineering side of view, irreversible electroporation is preferable for the extraction of valuable components from the cell such as lipid and proteins. Figure 1 shows the illustration of the cell membrane electroporation mechanism. By placing a cell between two electrodes without introducing the electricity or at zero potential gives the cell in a static condition shown in figure 1a. The exposure of electric fields induced the cell causing an osmotic imbalance and swelling of the cell. At this moment, the cell membrane molecules that hold by the polar bonds were distracted by the electric field induction (figure 1b and figure 1c). In this state also, the formation of pores among the cell wall of membrane begin and the reversible electroporation may occur. The increasing of electric field strength will cause the cell membrane to break due to amount of pores increase. By maintaining the electric field strength above its threshold value or its critical strength, the cell wall will be electroporated [26]. The valuable components such as lipid, carotenoids, chlorophylls and protein will be released from the cell that shown in figure 1d. At this rate, the membrane will no longer be able to recover from its original form and this stated as irreversible electroporation [23]. The effectiveness of the membrane to be electroporated affected by several factors, for instance the size of the microorganism. Microorganisms that have smaller size required higher electric field strength for the cell wall to be broken down [27]. This is likely due to the thermal and mechanical fluctuation condition of the cell membrane [28].

3. PEF as the method for lipid extraction

Pulsed electric field (PEF) technology is a non-thermal method that is widely used in medical application for cancer treatment [25], in biotechnology for DNA transfer [29] and in food industries for microbial inactivation [23]. Lipid extraction from microalgae using PEF technology has received notable attention in the past few years. Generally, PEF system includes pulse generator, treatment chamber and control system. The technology is well known to be a promising method for microalgae lipid extraction due to low energy consumption, potential to be scaled up and economic [30]. The recent studies on microalgae lipid extraction in the application of PEF is discussed in this section. Table 1 shows the present studies done by several authors using the PEF as a single step extraction or as a pre-treatment followed by conventional method (i.e. Solvent extraction). From the table of summary, variety of microalgae species such as *Chlorella vulgaris*, *Spirulina*, *Nannochloropis sp.*, *Auxenochlorella protothecodies*, *Ankistrodesmus falcatus*, *Isocrysis sp.* and *Synechocystis PCC 6803* cyanobacteria a blue-green algae were used in their studies.

Flisar et al. [11] had studied the lipid extraction from *C. vulgaris* using PEF in a continuous flow system. Lipid content in microalga *C. vulgaris* is about 50 to 58% dry weight biomass. They fabricated the PEF treatment chamber using stainless steel as the electrodes and the housing act as an insulator was made from polytetrafluoroethylene material. The chamber designed with a co-field shape and the gap between the electrodes was fixed to 15 mm. The study found that, approximately 50% (0.43g of lipid/g of dry biomass) of lipid was extracted using electric field strength of 2.7 kV/cm for 21 pulses of 100 μs. According to Luengo et al. [13] microalga *C. vulgaris* that being exposed to 10 kV/cm electric field strength and 50 pulses for 3 μs was found to be at reversible electroporation state. But, increasing the field strength to 20 to 25 kV/cm and utilized lower pulses of 5 for 3 μs will cause the cell membrane at irreversible electroporation state. The electrodes distance used in their work was 2.5 mm which smaller than the distance used by Flisar and co-workers [11]. Foltz [18] had also conducted a study on *C. vulgaris* using PEF for cells lyse observation using smaller electrode distance of 0.25 mm. In his experiment, the cell can be lysed using the treatment condition of 4 kV/cm electric field strength and 60 pulses for 20 seconds. Based on the results reported from the three literatures, lower electric field strength and smaller distance between the electrodes can be used for the irreversible membrane permeabilization to occur. Longer time is required to lyse the cell membrane with lower electric field strength and larger space between the electrodes. Moreover, the size of microalga *C. vulgaris* is 2 μ to 10 μm which is very small thus, required higher voltage to break the cell wall.

Eing and co-workers had demonstrated that the percentage extracted lipid from *A. protothecoides* can be enhanced associated with solvent extraction after treating with PEF [15]. They constructed the PEF treatment chamber from polycarbonate and stainless steel electrodes with 4 mm electrode gap. Then the target sample was exposed to 35 kV/cm electric field strength for 1 μs duration but no lipid was detected. The result obtained had disagreed with previous literature that used lower field strength to extract lipid [11]. The best assumption that can be described, was the behavior of the cell wall of different microalgae species may permeable to certain water-soluble or ions but not specific to the lipid droplet [15].

Table 1. Summary of PEF application in microalgae.

Microalgae strain	Purpose of Study	Scale	Cell Disruption /Solvent Extraction	Lipid Extracted (% wt)	Electric Field Strength, E (kV/cm)	Reference
Chlorella vulgaris	Extraction of lipid	Lab	PEF	22	2.7	[11]
Spirulina	Inactivation of microalgae	Lab	PEF	NA	33.3-66.7	[12]
Chlorella vulgaris	Extraction of pigments	Lab	PEF	NA	20-25	[13]
Nanno-chloropsis sp.	Extraction of intracellular	Lab	PEF	NA	20	[14]
Auxenochlorella protothecoides	Extraction of lipid	Lab	PEF/ C_2H_5OH	22	35	[15]
Auxenochlorella protothecoides	Extraction of intracellular	Lab	PEF	NA	23-43	[16]
Ankistrodesmus falcatus	Extraction of lipid	Lab	PEF/H_2O, C_4H_8O and CH_3OH	6.1 mg/L	45	[17]
Dunaliella salina	Observation of cells lyse	Lab	PEF	NA	1.6	[18]
Chlorella vulgaris	Observation of cells lyse	Lab	PEF	NA	4	[18]
Synechocystis PCC 6803[a]	Extraction of lipid	Lab	FP[b]/ C_3H_7OH	25-75[c]	>35[d]	[19]
Isocrysis sp.	Extraction of lipid	Industrial	PEF	NA	20-30	[20]

[a] Cyanobacteria strain.
[b] Focused Pulsed (FP) adaptation of PEF technology.
[c] Extracted lipid as Fatty Acids Methyl Ester (FAME) in % wt.
[d] Treatment intensity unit in kWh/m^3.
NA (Data not available).

Besides lipid, other valuable components from *A. protothecoides* such as organic carbon, carbohydrates and proteins can be also extracted using PEF treatment. The work implemented by Goettel et al. [16] on the extraction of intracellular valuable from *A. protothecoides* and using the similar PEF treatment chamber with Eing and co-workers. The study used electric field strength of 23 to 43 kV/cm to disintegrate the cell wall to allow the intracellular to be drained out. It was observed that no lipid droplet is released from the cell which supported the result gained from Eing and co-workers. However, it was found that 1 MJ/kg_{dw} algae is required to lyse the cell wall of *A. protothecoides* for 100g dry weight over kg of suspension.

Qin et al. [12] had investigated the inactivation of *Spirulina* using the PEF method. Inactivation can be defined as the cell wall of microalga was damaged by the induction of electric field strength. The results indicated that the *Spirulina* can be lysed at field strengths of 33.3 kV/cm and impulses number of 100 to 500. The increasing of field intensity to 66.7 kV/cm will decrease the number of impulse to 50. Thus, the number of impulse is inversely proportional to electric field strength. Microalga *Spirulina* has lager size compared than that of *C. vulgaris*, *Nannochloropsis sp.* and *Isocrysis sp.* which is 0.5 mm in length. Different microalgae species and sizes can affect the performance of PEF to disintegrate the cell wall. Antezana-Zbinden et al. [17] had conducted a study on *A. falcatus* lipid extraction using PEF as the pre-treatment to disrupt the cell wall and followed by the solvent extraction. The solvents used were ethyl acetate, methanol and water. Ethyl acetate was selected instead of chloroform solvent due to less harmful to the environment. Using PEF prior to solvent extraction increased the lipid extraction. Approximately 90% of the cell algae had been electroporated at the condition of 45 kV/cm field strength in 100 ms. The PEF caused the disarrangement of the cell wall membrane molecules and using high field intensity attributed to irreversible electroporation. Then, allowing the solvent molecules to extract the lipid droplet and thus, this has enhanced the percentage of lipid recovery.

Certain of microalgae can be lysed easily with lower electric field strength as 1.6 kV/cm such as microalga *D. salina* [18]. This can be explained that some of the microalgae do not have a cell wall. Other than green microalgae, cyanobacteria (a blue - green algae) also been used for the test subject. Sheng et al. [19] reported that using PEF prior to solvent extraction increased the efficiency of lipid extraction from the dry or wet biomass. The study indicates that 87% of the cell being electroporated using PEF at treatment intensity of 35 kWh/m^3. It was found out that, it was no efficient to use isopropanol without the PEF treatment. But, with PEF treatment 25 to 75% of lipid recovery can be achieved using solvent to wet biomass ratio of 5. Although, pilot scale for microalgae lipid extraction using PEF has not yet been reported, the PEF application already had been patented in industrial scale. United State patent published by Kempkes and co-workers had designed a single equipment using the PEF concept to extract lipids from *Isocrysis sp.* [20]. The technology was provided by Diversified Technologies, Inc with the efficiency of power conversion up to 90%. The design was operated in continuous system and suitable with input flow rates of 10 to 100,000 liter per hour, hence suitable for large scale application. Additionally, the pulse only required 20 to 30 kV/cm for 1 to 10 microseconds to lyse the microalgae cell wall. In overall, PEF consumed less energy which is 4.6 times lower compared to conventional drying method for lipid extraction [15; 19].

4. Electrolysis as the method for microalgae lipid extraction

The electrolysis method utilizes electrically charged probes called an anode and a cathode electrode, which been submerged in salty water to separate the oxygen and hydrogen atoms. The technique uses two different electrodes to induce an electric field due to the different conductivities of the electrode materials supplied by lower voltage (typically from 10 to 30V). In comparison with PEF method that required a higher voltage range from 1 kV to 60 kV in a short period of time (nanosecond to microsecond). Lower voltage requires longer time of treatment and vice versa.

Electrolysis method is basically being used during the harvesting process in microalgae production to enhance the biomass concentration [31; 32]. A recent study conducted by Daghrir and co-workers using electric field formed by electrochemical method to lyse the cell wall and extract intracellular components of microalga *C. vulgaris* [33]. They used the electrochemical treatment terms as the method to extract lipid from the microalga in their studies. Treatment chamber was fabricated using plexiglass material and using 1 liter of microalga sample for the test subject. Both cathode and anode electrode plates were placed parallel with 1.0 cm distance and submerged into the electroporation chamber. Stainless steel was fixed at cathode mode while 4 types of electrodes were used at anode mode (table 2). Since, the voltage used was lower compared to PEF, thus longer treatment time is required. By fixing the treatment time at 60 minutes, Ti/IrO_2 at the anode electrode achieved the highest lipid and protein extraction of 3.27% $g_{Lip}/g_{dry\ wt}$ and 39.91 mg/L respectively. In addition, the lowest energy consumption also observed from Ti/IrO_2 as lower as 8.58 kWh/m^3. Table 2 shows the summary of the electrochemical treatment for *C. vulgaris* lipid extraction. Using Ti/IrO_2 as the anode electrode was better compared to other materials. This because, Ti/SnO_2 has high crystalline nature and Ti/PO_2 has weak interaction with the hydroxyl group [33].

Table 2. Summary of electrochemical treatment in microalgae.

Microalgae strain	Purpose of Study	Electrode		Lipid (% $g_{Lip}/g_{dry\ wt}$)/ Proteins (mg/L)	Electric Field Strength, E (V/cm)	Reference
		Cathode	Anode			
Chlorella vulgaris	Extraction of lipid and protein	SS	Ti	2.65/19.43	30.7	[33]
Chlorella vulgaris	Extraction of lipid and protein	SS	Ti/IrO_2	3.27/39.91	14.3	[33]
Chlorella vulgaris	Extraction of lipid and protein	SS	Ti/SnO_2	2.76/15.11	30.5	[33]
Chlorella vulgaris	Extraction of lipid and protein	SS	Ti/PbO_2	2.08/17.12	26.3	[33]

Note: SS represents stainless steel electrode, Ti represents Titanium electrode, IrO_2 represents Iridium (IV) Oxide electrode, SnO_2 represents Tin (IV) Oxide electrode, PbO_2 represent Lead (IV) Oxide electrode.

5. Conclusion

In conclusion, biodiesel from microalgae as an alternative feedstock for renewable energy is promising, not only it is eco-friendly and promotes clean energy, but also capable in reducing or eliminating carbon dioxide from the atmosphere. Nevertheless, extraction technology is one of the major challenges to obtain a high yield lipid from microalgae. Several conventional methods found to be efficient, but also have limitations. Cell electroporation of microalgae using PEF and electrochemical treatment methods are both demonstrated to have potential and promising results in increasing the extraction efficiency. Besides, both methods consume less energy and no additional chemicals are required for the system.

Acknowledgments

The authors wish to acknowledge the Ministry of Agriculture Malaysia a Seaweed Research Unit Grant GPRL 013 (Biofuel from Seaweed) and GPRL 017 (Biochemical Process of Seaweed for Industrial Product) for the financial support of the study.

References

[1] Gouveia L 2011 Microalgae as a feedstocks for biofuels *SpronerBriefs in Microbiology* (doi: 10.1007/978-3-642-17997-6_1)
[2] Rawat I, Ranjit-Kumar R, Mutanda T and Bux F 2013 Biodiesel from microalgae: A critical evaluation from laboratory to large scale production *Appl. Energ.* **103** 444–67
[3] Suali E and Sarbatly R 2012 Conversion of microalgae to biofuel *Renew. Sust. Energ. Rev.* **16(6)** 4316–42

[4] Mansa R F, Tahir A, Hua L M, Dayou J and Sipaut C S 2012 Design of a pilot scale outdoor photobioreactor for mass cultivation of local microalga *Int J. Eng and Phy Sci* 348–52.
[5] Suali E, Sarbatly R and Muhamad-Shaleh S R 2012 Characterisation of local *chlorella sp.* towards biofuel production *Int. Conf. on Applied Energy (ICAE)* (Suzhou, China, 5-8 July 2012)
[6] Mata T M, Martins A A and Caetano N S 2010 Microalgae for biodiesel production and other applications: A review *Renew. Sust. Energ. Rev.* **14** 217–32
[7] Brennan L and Owende P 2010 Biofuels from microalgae - A review of technologies for production, processing, and extractions of biofuels and co-products *Renew. Sust. Energ. Rev.* **14** 557–77
[8] Nashima K and Palanisamy A 2012 Biodiesel production by *chlorella sp.* and *oscillatoria sp.* *IJPI's J. Biotechnol. Biother.* **2(10)** 2229 – 6824
[9] Rasoul-Amini S, Montazeri-Najafabady N, Mobasher M A, Hoseini-Alhashemi S and Ghasemi Y 2011 *Chlorella sp.* : A new strain with highly saturated fatty acids for biodiesel production in bubble-column photobioreactor *Appl. Energ.* **88** 3354-56
[10] Lee A K, Lewis D M and Ashman P J 2012 Review disruption of microalgal cells for the extraction of lipids for biofuels: Processes and specific energy requirements *Biomass Bioenerg.* **46** 89-101
[11] Flisar K, Meglic S H, Morelj J, Golob J and Miklavcic D 2014 Testing a prototype pulse generator for a continuous flow system and its use for *E. coli* inactivation and microalgae lipid extraction *Bioelectrochemistry* (doi 10.1016/j.bioelechem.2014.03.008)
[12] Qin S, Timoshkin I V, Maclean M, Wilson M P, MacGregor S J, Given M J, Anderson J G and Wang T 2014 Pulsed electric field treatment of microalgae: Inactivation tendencies and energy consumption *IEEE Transactions on Plasma Sci* 1-6
[13] Luengo E, Condo´n-Abanto S, A´ lvarez I and Raso J 2014 Effect of pulsed electric field treatments on permeabilization and extraction of pigments from *chlorella vulgaris J. Membrane Biol.* (doi 10.1007/s00232-014-9688-2)
[14] Grimi N, Dubois A, Marchal L, Jubeau S, Lebovka N I and Vorobiev E 2014 Selective extraction from microalgae *nannochloropsis sp.* using different methods of cell disruption *Bioresource Technol.* **153** 254-9
[15] Eing C, Goettel M, Straessner R, Gusbeth C and Frey W 2013 Pulsed electric field treatment of microalgae-benefits for microalgae biomass processing *IEEE Transactions on Plasma Sci* **41(10)** 2901–7
[16] Goettel M, Eing C, Gusbeth C, Straessner R and Frey W 2013 Pulsed electric field assisted extraction of intracellular valuables from microalgae *Algae Research* **2** 401-8
[17] Antezana-Zbinden M D, Strum B S M, Nord R D, Carey W J, Moore D, Shinogle H and Stagg-Williams S M 2013 Pulsed electric field (PEF) as an intensification pretreatment for greener solvent lipid extraction from microalgae *Biotechnol. Bioeng.* **110** 1605–15
[18] Foltz G 2012 Algae lysis with pulsed electric fields *Master Thesis* California State Polytechnic University, San Luis Obispo Follow
[19] Sheng J, Vannela R and Rittmann B E 2011 Evaluation of cell disruption effects of pulsed electric field treatment of *synechocystis PCC 6803 Environ. Sci. Technol* **45** 3795–802
[20] Kempkes A K, Roth I and Gaudreau M P J 2011 Pulsed electric field (PEF) method for continuous enhanced extraction of oil and lipids from small aquatic plants US Patent 2011/0107655 A1, 12 May 2011.
[21] OriginOil 2009 *A breakthrough technology to transform algae into biofuel* National Algae Association Available from http://www.originoil.com/pdf/OriginOil_NAA_30_Apr_ 2009.pdf
[22] OpenCEL, *Focused Pulsed Technology* Accessed Date 25 June 2014 http://www.opencel.com/technology/focused-pulsed-technology/
[23] Ortega-Rivas E 2012 *Non-thermal food engineering operations* (*Food Eng. Series)* ed Barbosa-Cánovas (New York: Springer)

[24] Chang D C, Saunders J A, Chassy B M and Sowers A E 1992 *Guide to Electroporation and Electrofusion* ed Chang et al. (London: Academic Press Limited) pp 1-6
[25] Haberl S, Miklavcic D, Sersa G, Frey W and Rubinsky B 2013 Cell membrane electroporation – Part 2: The applications *IEEE Elec. Insul. Mag.* **29(1)** 29-37
[26] Dimitrov D S 1995 Electroporation and Electrofusion of Membranes *Handbook of Biological Physics* vol 1, ed by Lipowsky R and Sackmann E (Bulgaria: Elsevier Science) chapter 18 pp 851-895
[27] Zimmermann U, Pilwat G and Riemann F 1974 Dielectric breakdown of cell membranes *Biophys. J.* **14** 881-9
[28] Kotnik T, Kramar P, Pucihar G and Miklavicic D 2013 Cell membrane electroporation – Part 1: The phenomena *IEEE Electr. Insul. Mag.* **28(5)** 14-23
[29] Teissie J, Golzio M and Rols M P 2005 Mechanisms of cell membrane electropermeabilization: A minireview of our present (lack of ?) knowledge *Biochim. Biophys. Acta* **1724** 270 – 80
[30] Toepfl S, Mathys A, Heinz V and Knorr D 2006 Review: potential of high hydrostatic pressure and pulsed electric fields for energy efficient and environmentally friendly food processing *Food Rev. Int.* **22** 405–23
[31] Inman M, Taylor E J, McCrabb H, Skinn B, Kell J and Stuart B 2012 Electroconcentration and electroflotation for de-watering/water purification *Presented at 221st Electrochemical Society Meeting* (Seattle, WA, 6-10 May 2012)
[32] Misra R, Guldhe A, Singh P, Rawat I and Bux F 2014 Electrochemical harvesting process for microalgae by using nonsacrificial carbon electrode: A sustainable approach for biodiesel production *Chem. Eng. J.* **255** 327–33
[33] Daghrir R, Igounet L, Brar S K and Drogui P 2014 Novel electrochemical method for the recovery of lipids from microalgae for biodiesel production *J. Taiwan Inst. Chem. Eng.* **45** 153–62

A First Survey on the Abundance of Plastics Fragments and Particles on Two Sandy Beaches in Kuching, Sarawak, Malaysia

V James Noik[1] and P Mohd Tuah
Department of Environmental Science, Faculty of Science and Natural Resources, Universiti Malaysia Sabah, Jalan UMS, 88400 Kota Kinabalu, Sabah, Malaysia.

E-mail: victorjms8007@gmail.com, pmdtuah@ums.edu.my

Abstract. Plastic fragments and particles as an emerging environmental contaminant and pollutant are gaining scientific attention in the recent decades due to the potential threats on biota. This study aims to elucidate the presence, abundance and temporal change of plastic fragments and particles from two selected beaches, namely Santubong and Trombol in Kuching on two sampling times. Morphological and polymer identification assessment on the recovered plastics was also conducted. Overall comparison statistical analysis revealed that the abundance of plastic fragments/debris on both of sampling stations were insignificantly different ($p>0.05$). Likewise, statistical analysis on the temporal changes on the abundance yielded no significant difference for most of the sampling sites on each respective station, except STB-S2. Morphological studies revealed physical features of plastic fragments and debris were diverse in shapes, sizes, colors and surface fatigues. FTIR fingerprinting analysis shows that polypropylene and polyethylene were the dominant plastic polymers debris on both beaches.

1. Introduction

Ever since its discovery, man-made polymers of various types can be easily found virtually in any industrial materials and consumer products today thanks to its desirable performance. Commonly known as 'plastics' in layman's term, this man-made polymers are physically durable (i.e. tough, flexible, long-lasting), chemically and thermally stable, lightweight and can be easily molded into various shapes and sizes while addition of colour dyes improves visual attractiveness and functions of the end-products. With our 'addiction' to plastics and highly dependent on it economically, the high demand which led to lower production cost have seen our modern society reaching the so-called 'Plastic Age' due to the unprecedented amount of plastics that we produce, use and dispose [1].

By and large, plastics are designed either as a single- or extended-use product that eventually end up in the waste stream where it may be recycled or disposed off into solid waste treatment facilities. Apparently in most cases, irresponsible littering and indiscriminate dumping of solid wastes has made plastics unnaturally available ubiquitously in the environment. The adverse impacts exerted on the

[1] To whom any correspondence should be addressed.

ecosystem and marine wildlife by this synthetic anthropogenic debris is multitude [2] – [5]. Human health too is not spared by the potential dangers of plastics debris [6].

Disposed or littered plastics are commonly thought to resist physical and chemical breakdowns. This however is partially true as some plastics actually degrades albeit at a very slow rate depending on its type and chemical composition when left exposed to Nature's elements [7], [8]. Degraded plastics are fragile, discoloured and dull, brittle, flaky, powdery and fragmented in small tiny pieces or particles called microplastics in which it become an environmental contaminant [9]. The term microplastics itself is not uniformly defined by various authors, for example [9] – [12], but generally are plastics fragments and/or particles that range in the size of few millimeters to sub-millimeters (i.e. micrometers). For the sake of clarity, this paper will refer particles in the size of 1 to <5 mm as microplastics while those that measures between 5 to <20 mm are small fragments or meso-plastics and > 20 mm are large fragments or macroplastics [13] – [15].

The abundance of microplastics and plastic fragments in the environment are difficult to estimate largely due to spatial and temporal challenges and the heterogeneity of plastic debris [16], as well as the lack of standard sampling and analysis methods [17], [18], let alone monitoring them. Hence, the environmental risks and biological hazards of the contaminants are yet to be fully understood. Nevertheless, a number of studies e.g. [19] – [28] had shed some lights on the potential ill-effects and its roles on both environment and biota.

This paper presents the first ever findings of the abundance of micro-, meso and macro- plastics from two selected estuarine beaches in Kuching, located at Kampung Santubong (1°43.164'N, 110°19.165'E WGS84) and Kampung Trombol (1°42.191'N, 110°11.635'E WGS84) respectively. Kampung Santubong is situated about 30 km from Kuching City while the latter is about 35 km. Located at the river mouth of Santubong River, Kampung Santubong was once a well-known fishing village/town and public beach facing both the river and South China Sea. The inhabitants enjoy basic public amenities such as schools, treated water, electricity, well-maintained tar-sealed road and scheduled municipal solid waste collections. Kampung Trombol is located at the vicinity of Telaga Air town in Matang District. The population are sparse with roughly a couple dozen of household, mostly involve in coastal agriculture (i.e. coconut and small holdings of oil palm plantation). The village is connected from the main road via makeshift tar-sealed road and electricity supply is just recent. Treated water is not available at the time of sampling and is yet to receive scheduled solid waste collection services. The beach of Kampung Trombol faces South China Sea and its coastline stretches to the river mouth of Sibu River which flows through Telaga Air town. Unlike the beach of Kampung Santubong, the beach of Kampung Trombol is unknown to public hence low human activities. The objectives of this study were: 1) To estimate the spatial and temporal abundance of types and sizes of microplastics/plastic fragments, and 2) To assess the morphological features of the collected microplastics/plastic fragments.

2. Materials and Methods

2.1. Sampling station, sample collection and sample handling

Beach sampling works were carried out at Santubong (STB) and Trombol (TRB) stations in November 2013 and a revisit in August 2014. Sand samples were collected from three sampling sites (designated as S1, S2 and S3 respectively) on each station that was randomly selected. The sites were 100 m apart from each other, with a 50 m transect length along the strandline in each site. Sampling quadrats (20x20 cm) were placed 10 m apart within the site and marked with a handheld GPS receiver (Garmin GPSMap 60C) along the upper and lower strandline.

Sands were collected using a cleaned stainless steel scoop at a depth of 2 cm from the surface and transferred into a cleaned sanitized glass bottle with air tight cap. The bottles were then wrapped with aluminum foils and put into a cool box for transportation. All sampling activities were done during low tide.

Sampling precautionary measures were adhered for sample quality control and assurance to minimize any undesired contamination: Cotton fabric clothes were worn throughout sampling activities, the body of sample collector was opposite facing in relative to wind direction, and a field blank of washed and cleaned sand was used.

2.2. *Recovery of plastic fragments/particles*

Recoveries of plastics micro- meso- and macro-plastics were done at the Environmental Chemistry Research Laboratory, Faculty of Science and Natural Resources, Universiti Malaysia Sabah (UMS). All sand samples were dried in a drying oven at 40 °C until constant weight. Dried sands were then sieved through a 1 mm mesh size stainless steel sieve. Plastics fragments/particles that were retained by the sieve were visually pre-sorted to remove any unwanted coarse organic matter. The pre-sorted fragments/particles were then further visually sorted and identified from organic matters of similar size under a compound stereoscope for coarse fragments and microscopes for small particles, and subsequently measured for weights and size. All recovered plastics fragments were stored in a cleaned sanitized glass bottle container and kept in an auto-controlled drybox.

On the other hand, plastics fragments/debris that escaped the 1 mm sieve together with bulk of the sand were separated using a low-cost fluidized density separation system (FDSS) (Figure 1).

The FDSS used in this experiment was adopted and modified based on the system developed by [29], [30] and [31] by utilizing available laboratory glassware. Concentrated NaCl solution (ρ=1.2 $g{\cdot}cm^{-3}$) was prepared by dissolving a pure sea salt (food grade) in distilled water and subsequently filtered using Whatman filter paper No. 1. The solution was then poured into the glass column (I) until about 50% full. Sieved sand was then gently transferred into the glass column using clean metal spatula via a glass funnel. Sand and other particles denser than the solution will precipitate towards the bottom of the column, while those that are less dense will buoyed on the upper layer of the liquid column.

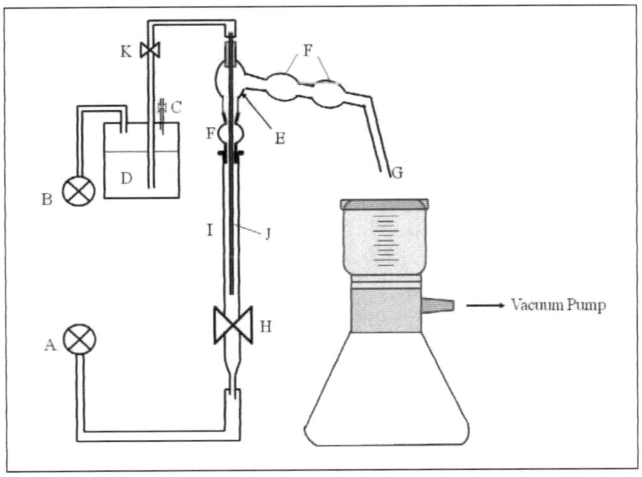

Figure 1. The set-up of the Fluidised Density Separation System (FDSS): **A** – Air Pump A, **B** – Air Pump B, **C** – Pressure release valve, **D** – NaCl solution reservoir, **E** – Overflow exit junction, **F** – Bulbous traps, **G** – Outfall, **H** – Air valve, **I** – Glass column, **J** – Glass pipette, **K** – Outflow valve.

Air Pump A (flow rate: 4 $L{\cdot}min^{-1}$) was then switched on and the air valve (H) is slowly opened, resulting in streams of air bubble in a controlled manner purging through the liquid column agitating the gravitated sand and buoying particles. This fluidisation of the column mixture was allowed to run for 10 minutes after which shutting sequence was then initiated beginning with the closure of the air valve (H) followed by the air pump. The mixture solution was then left for 30 minutes to allow dense particles gravitate, resulting a clear liquid supernatant. In some samples where silt and colloidal matter content are high, the gravitation was left overnight for complete settlement.

Pump B (flow rate: 3.5 L·min^{-1}) was then activated to pressurize the NaCl solution reservoir (D), forcing the solution into the column via a submerged glass pipette (J) for 2 minutes. The addition of solution causes an increase of the mixture's volume in the column until it overflows at the exit junction (E). Any sand grains or dense particles were retained at the bulbous trap (F). The overflow brought along flotsam and suspended solids in the column and get collected in the vacuum filter funnel. Shutting down sequence was initiated after 2 minutes beginning with the pump then closure of outflow valve (K) to prevent any possible backflow of the solution into the reservoir. The air release valve (C) was then quickly open to equalize the reservoir's internal pressure with that of atmospheric pressure.

Collected flotsam and suspended particles together with the saline solution were then vacuum filtered through a glass microfiber filter (Whatman Grade GF/B), whereas the filtered NaCl solution is transferred into the reservoir for reuse. The trapped flotsam and suspended particles on the glass microfiber filter were then rinse with distilled water and vacuum filter several times to remove residues of NaCl. The flotsam particles along with the microfiber filter were then dried in a drying oven for 40 °C until constant weight.

2.3. Removal of biogenics
Dried flotsam particles adhered to the glass microfiber filters were subjected to acid treatment for the removal of any biogenic matters present. The glass microfiber filters were then placed in a petri dish. Concentrated analytical grade hydrochloric acid (HCl) 37% v/v (Merck, Germany) and nitric acid (HNO_3) of 65% v/v (Merck, Germany) was then pipetted on the flotsams until saturated and left for 10 minutes. The glass microfiber filter was then gently lift up using a metal forceps and nozzle spray with NaCl solution to wash off any particles into the petri dish. The contained acidic saline solution in the petri dish was then gently poured into a beaker to collect any floating particles which was subsequently vacuum filtered and washed with distilled water for several times. The filter papers were then let dry in a drying oven at 40 °C for 2-days to affix any plastics particles on it.

2.4. Characteristics observation
Recovered plastics fragments/particles were observed under an inverted light microscope for physical features such as shapes and fatigues. Plastic particles that are loosely fixed on the filter paper were gently scrapped with a spatula onto a microscope slide and affixed firmly by sandwiching it with another slide and observed under normal light microscope. All microscopic images were captured using a 8 megapixels camera attached to the ocular lens. Images were enhanced using PhotoScape 3.6.3 photo editing software.

2.5. Fourier-Transform Infrared (FTIR) identification of plastic types
Mesoplastics with the size of 15 mm and above from all three sampling sites of each station were pooled together and randomly sampled for plastic type identification. A total of 100 samples (n=100) from STB and 132 samples (n=132) from TRB station were selected for identification. Samples were cut to a size of 10 mm and washed with distilled water to remove any dirt and residues. The samples were then placed and firmly clamped on the ZnSe crystal plate holder of the MIRacle™ Single-Reflection Horizontal attenuated total reflectance (ATR) accessory (PIKE Technologies, USA) installed on the Perkin-Elmer (Model: Spectrum 100) FTIR spectrometer. The samples were then scanned in the range of 4000 – 650 cm^{-1} wavenumber. Output from the scans in the form of spectrum peaks expressed as %T, were compared with established online polymer spectra databases and libraries.

2.6. Statistical analysis
Obtained data from the measured weight were statistically analysed using SPSS ver. 13.0 for descriptive statistics (Mean±SD). Comparison of temporal changes within station were analyzed using one-way ANOVA, while comparison between stations were analyzed using two-way ANOVA with

confidence level of 95% (significant difference, $p<0.05$) whereby values from the latter's analysis were assessed with Tukey's honesty significant difference (HSD) post-hoc test.

3. Results

3.1. Weight and size classification of recovered plastic fragments and particles

A total of 60 sand samples were collected from Kuching, with 30 samples each from Santubong and Trombol beach in November 2013 and August 2014 respectively. Weight can be use as a subjective indicator on the amount of plastics debris present on the sampling site since most of them were found in various sizes and shapes. Generally, STB recorded a mean±SD weight of 0.0358±0.062 g (November 2013) and 0.4108±0.610 g (August 2014), while the mean weight of sampled plastic fragments debris was 1.7343±2.173 g (November 2013) and 1.5419±2.291 g (August 2014). Table 1 shows the mean weight of plastic fragments and particles collected from both sampling stations at two different sampling times. Results from statistical comparison using one-way ANOVA between weights of paired same strandline with the respective sampling time showed that there were no significant difference ($p>0.05$) of fragments/particles collected, except for the upper strandline of STB-S2 which shows significant increase of the debris of interest.

The mean weight of sampled plastics from sites of both stations within the same month was compared using multivariate analysis (MANOVA). The analysis outcome indicate that there were no significant different, $F(4,54)=0.181$, $p=0.947$ i.e. $p>0.05$. Comparison of the August 2014 mean weight also yielded no significant difference, $F(4,54)=0.398$, $p=0.809$. Tukey's HSD post-hoc test affirmed that both the mean weight was insignificant different $p=0.73$ for the former month and $p=0.231$ for the latter's, respectively. It is however worthy to note that some of the collected individual fragment pieces were too light to be measured by analytical balance, hence the number of debris does not reflect by the weight in Table 1. Hence, both visual judgment from site sampling and laboratory sorting observation presented a more objective abundance.

Table 1. Mean weight of pooled plastics fragments/particles recovered from the respective sampling stations.

	Weight, g (Mean ±SD)			
	November 2013		August 2014	
Station, Site	Lower Strandline (n=15)	Upper Strandline (n=15)	Lower Strandline (n=15)	Upper Strandline (n=15)
STB-S1	0.0018 ±0.003	0.0496 ±0.031	0.0009 ±0.001	0.1794 ±0.223
STB-S2	0.0140 ±0.002	0.0963 ±0.110[a]	0.1883 ±0.419	1.2710 ±0.922[b]
STB-S3	Nil	0.0659 ±0.068	Nil	0.8254 ±0.885
TRB-S1	0.0049 ±0.011	4.0075 ±1.995	0.0273 ±0.033	3.3028 ±2.451
TRB-S2	0.0011 ±0.002	3.5698 ±2.569	Nil	3.6672 ±3.288
TRB-S3	0.0008 ±0.002	2.8219 ±0.491	Nil	3.2544 ±1.507

Superscript letters denotes significant difference ($p<0.05$).
Nil denotes no plastic found or recovered.

Table 2 shows the size of plastics recovered with apparently mesoplastics and macroplastics were the dominant size on the sampling stations particularly at the upper strandline.

Table 2. Size classification of recovered plastic fragments from upper and lower strandlines of the respective sampling stations.

Sampling period	Class (Size, mm)	Station: ΣSTB (1, 2, 3) Lower Strandline	Station: ΣSTB (1, 2, 3) Upper Strandline	Station: ΣTRB (1, 2, 3) Lower Strandline	Station: ΣTRB (1, 2, 3) Upper Strandline
November 2013	Microplastics				
	1 – 5	4	10	1	9
	Mesoplastics				
	6 – 10	1	11	2	18
	11 – 15	2	18	1	27
	16 – 20	2	17	0	38
	Macroplastics				
	> 20	0	25	2	37
Total:		**9**	**81**	**6**	**129**
August 2014	Microplastics				
	1 – 4	1	8	0	7
	Mesoplastics				
	5 – 10	0	10	0	3
	11 – 15	2	17	2	25
	16 – 20	4	22	1	37
	Macroplastics				
	> 20	2	23	2	28
Total:		**9**	**80**	**5**	**100**

3.3. Plastic polymer identification

Plastic fragments and particles were pooled together irrespective of strandline and the sampling month. Figure 2 shows the abundance of various plastic types identified using FTIR. Polypropylene (PP), polyethylene (PE), polyethylene terephthalate (PET) and polystyrene (PS) were the plastics identified and present most commonly on both beaches. PP and PE constitute the highest abundance on both beaches: 28.0% and 31.0% respectively on STB station while 37.12% and 30.30% respectively on TRB station. PP, PE, PET and PS are commonly found in a wide range of industrial and consumer products. Other types of polymers found and accounted for were nylon and polyurethane.

3.2. Observation of recovered plastics fragments and particles

Figure 3 shows some of the plastic debris and fragments that were recovered from STB and TBR sampling stations. Fragments were in the form of various shapes and sieze. Most of the fragments exhibit degraded form such as discoloration, torn, shrinked, fragmented, flaky, powdery, and fragile. Microscopic observations revealed that surfaces morphology of some plastics were cracked (Figure 4). Generally, most of the plastic fragments do not have any association with heavy industrial materials but of household as observed during site samplings. Items include broken kitchen and bathroom wares, motor oil containers, personal care and toiletries, children toys and derelict fishing gears and storage boxes.

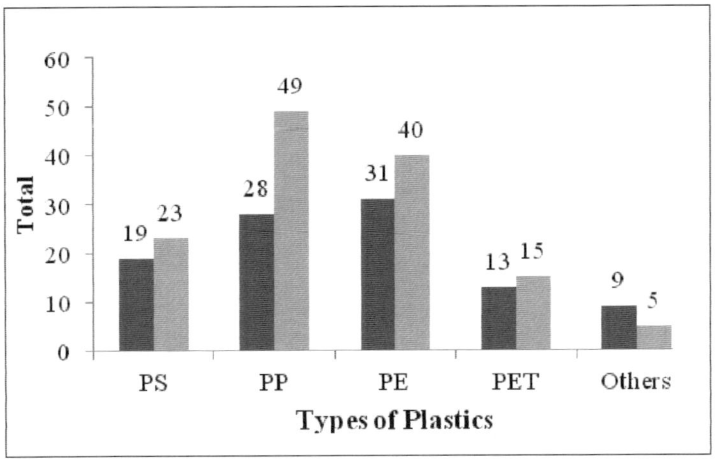

Figure 2. Total number of plastic fragments and debris pooled from Santubong (■) and Trombol (■) stations identified through FTIR analysis.

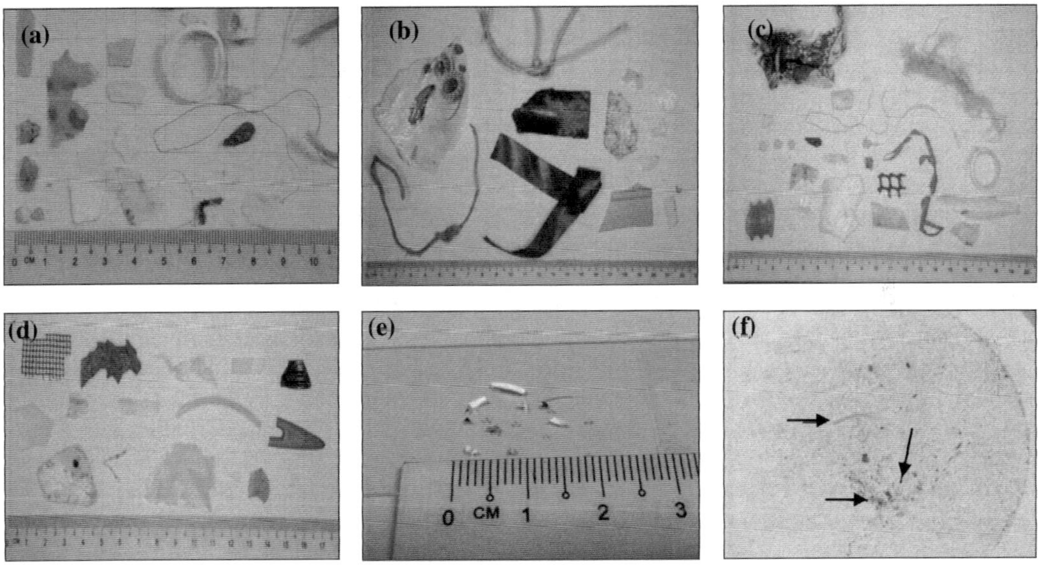

Figure 3. Some of the macro- and mesoplastics that were recovered from the 1 mm sieve, **(a)** – **(d)**; microplastic particles recovered via FDSS on a slide, **(c)**; colored microplastic indicated with arrowheads affixed on a filter paper after drying, **(d)**.

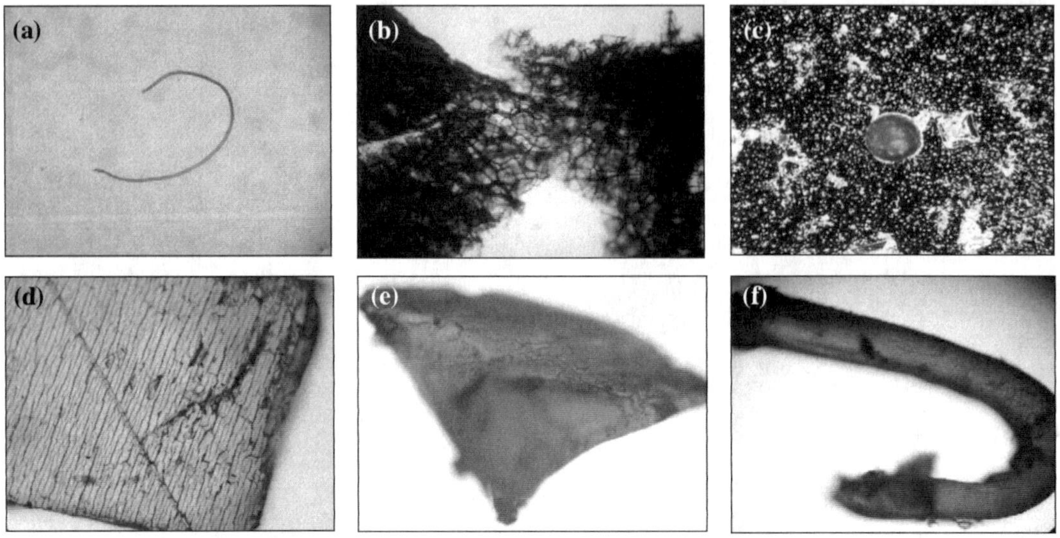

Figure 4. Microplastics viewed under light microscope. A strain of polymer, possibly a polyethylene fiber (100x mag.), **(a)**; Degraded polystyrene Styrofoam (100x mag.), **(b)**; A possible microbead with a salt crystal on its right (50x mag.), **(c)**; Cracked surface of polyethylene plastic (100x mag.), **(d)** and **(e)**; A fragment of possibly nylon debris (100x mag.), **(f)**.

4. Discussion

The present of plastic debris in the environment is not uncommon ever since humans have been using the polymer products for many years. Human factors are widely attributed to the sore sights of plastic being distributed and deposited in the environment. Most of the plastic were translocated to a great distance by wind, water and animals. Seas, oceans gyres, islands, beaches and rivers are the most common depository sites for these man-made polymers. Lagoons and estuarines are not even spared from the present plastic particles [32]. Besides marine environment, plastics are also invading freshwater ecosystem [33], [34]. With most of those ecosystems are biologically diverse, the present of plastics debris could potentially have great impacts on the well-being and safety of the animals inhabitants. What makes it even more of a concern is that those plastic debris gradually degraded over-time, making pieces of fragment present even far greater in numbers than the debris itself. This translates to greater translocation distance and into the food chain at the same time leaching out its additives compounds, plasticizers and adsorbing-desorbing hazardous chemicals along its way to the potential victim [35] – [40].

The results obtained from the present study shows that plastic particles present on the beaches of Santubong and Trombol. Most of the plastics particles were found on the upper strandline compared to the lower strandline. The upper strandline is the highest tide level on a beach nearest to land. Most debris and flotsams are washed to the uppermost strandline during the highest tide incident, leaving them there as when tide recede. Plastic debris might also be washed to the shore during storm events. Apart from being washed ashore by tides and rough seas, inhabitants in Santubong and Trombol and nearby locations might have explicitly disposed their household trash at the beach during high tide thinking that it would be drifted to the sea during tide recede. Observation of debris during site sampling found that most solid wastes stranded on the beach were not of industrial type. Additionally, both Santubong and Trombol are located at a river mouth hence high possibilities that those plastics have translocated from other populated places located upstream of the river. Plastics debris that were found on the lower strandline were mostly either partially embedded in the sedimented sand, got entangled on rocks or stranded tree branches, or stranded from fast receding tide.

The present findings also showed that there were slight decreasing in the total number of collected plastic debris on two sampling occasions. This can be explained from the weather/climate standpoint [41]. November is a wet month for Malaysia. Frequent heavy and prolonged downpour during the monsoon season in urban areas washes down plastic litters into storm drains that subsequently flow into rivers and seas. The incident of high tide during rainy months led to some locations experiencing flood. The receding floods brought along any flotsams and debris down to the sea and concentrated. The next tide might possibly transport the debris to beaches. Higher cloud coverage and cooler ambient temperatures during monsoon months, plastic debris experience slow onset degradation. The low present of humans and seaside animals on beach might also explain the high number of plastics at the shores during the wetter months. The incoming plastics debris from the sea were not being disturbed by humans nor animals, causing the amount of plastic debris increase.

The dry months in Malaysia normally begins in March until September or October. The dry and bright conditions favour degradation of plastics [9], [42]. Dry and windy seaside also causes some of the lighter plastics were blown away, and dry sand blown by wind burying the plastic debris underneath it. Some coastal plants, particularly those that creep on the sand, do not grow well during the wetter months but dry months. Their growth over some degraded plastics debris causes it hidden, and it grows denser, the roots began to push those plastics into the sand embedding them inside. The present of animals such as birds might have mistaken plastics debris for food, picking them up and translocate them elsewhere, while some littoral organisms such as shore crabs and insects might have translocate or ingested small fragmented plastics [43].

Microscopic observation on the recovered microplastic reveals its shapes and surface properties. Most plastics debris and particles recovered in this study were of degraded condition, hence the diverse shapes and sizes signifying that they were fragments of a larger item [44]. Spherical particles were of exception as it exhibit uneven surface. Some microplastics exhibit rough and cracked surface. The irregular surface with depressions and crevices provide higher surface area for biological, as well as chemical interaction due to the exposed polarity of the surface [45]. Microplastics of the polystyrene origin generally exhibit a ‘foam’ and ‘hollow’ character, making them lightweight less dense.

A combination of acids HNO_3 and HCl were employed to digest biogenic matter. Although both acids satisfactorily digested microscopic shells, spores, plankton and soft tissues present leaving residues to precipitate and easily remove any plastic debris later, the mixture however failed to digest small plant matter due to the present of fiber cellulose and lignin. Several published experiments demonstrated the effectiveness of hydrogen peroxide (H_2O_2) and to some extent hydrofluoric acid (HF), the strong oxidizing properties of both acids may affect plastic particles that are susceptible to acid corrosion and possibly digest it.

5. Conclusion

Plastic debris in the form of particles and fragments are getting more attention as an environmental contaminant and pollutant due to its hazardous potentials. The roles and fates of this emerging pollutant and contaminant are yet to be fully understood due to the nature of this debris and the challenges to sample, isolate and to quantify them. Thus far, only rough estimated from studies conducted in many places give us the idea on the abundance of this debris in the environment, as well as analyses on its chemical attributes.

This paper presents the first ever finding on the abundance of plastic particles on the shores of Santubong and Trombol in Kuching, Sarawak, Malaysia. The results from this study show that plastic particles were present in abundance on the selected sampling stations, with TBR had the significantly higher abundance of debris compared to STB. Analyzed data also showed that temporal fluctuations of plastic fragments were not significant for the sampling months. FTIR analysis showed that PE and PP were the dominant plastic types on both sampling stations, while morphological observations found that plastic fragments and particles exhibit diverse physical features.

This study also demonstrated the ability of FDSS to recover plastic particles. The use of saline solution to recover most plastic particles of lower density was satisfactory and at lower cost but was unable to recover higher density plastic particles. Heavier density salts such as sodium iodide (NaI) or sodium polytungstate (SPT) is recommended for full recovery however its price is a drawback for small scale research [17], [29], [46], [47].

The present of plastic debris mirrors the possibility of living organisms in the estuarine may face adverse threats in the long-term. It also signifies the condition of waste management in Kuching and the mentality of its inhabitants as a whole with regards to solid waste disposal especially plastic littering which need to be addressed.

6. Acknowledgment
This research was privately funded. I would to extend my since thanks and gratitude to the following individuals for their advice and guidance: supervising chairperson Associate Professor Dr. Piakong Mohd. Tuah, former supervisor Dr. Mahyar Sakari, and Professor Dr. Lau Seng of the Institute of Biodiversity and Environmental Conservation (IBEC), Universiti Malaysia Sarawak. Last but not least to the laboratory staff for their assistance.

7. References
[1] Thompson R C, Swan S H, Moore C J and vom Saal F S 2009 Our plastic age *Phil. Trans. R. Soc. B* **364** 1973 – 76
[2] Moore C J 2008 Synthetic polymers in the marine environment: a rapidly increasing, long-term threat. *Environ. Res.*108 131 – 9
[3] Teuten E L, Rowland S J, Galloway T S and Thompson R C 2007 Potential for plastics to transport hydrophobic contaminants *Environ. Sci. Technol.* **41** 7759 – 64
[4] Tanaka K, Takada H, Yamashita R, Mizukawa K, Fukuwaka M and Watanuki Y 2013 Accumulation of plastic-derived chemicals in tissues of seabirds ingesting marine plastics *Mar. Pol. Bull.* **69** 219 – 22
[5] Teuten E L *et al.* 2009 Transport and release of chemicals from plastics to the environment and to wildlife *Phil. Trans. R. Soc. B* **364** 2027 – 45
[6] Thompson R C, Moore C J, vom Saal F S and Swan S H 2009 Plastics, the environment and human health: current consensus and future trends *Phil. Trans. R. Soc. B* **364** 2153 – 66
[7] Andrady A L 1994 Assessment of environmental biodegradation of synthetic polymers: a review *J. Macromol. Sc.* **34** 25 – 76. 1994.
[8] O'Brine T and Thompson R C 2010 Degradation of plastic carrier bags in the environment *Mar. Pol. Bull.* **60** 2279 – 83
[9] Andrady A L 2011 Microplastics in the environment *Mar. Pol. Bull.* **62** 1596 – 1605
[10] Costa M F, Ivar do Sul J, Silva-Cavalcanti J S, Araújo M C B, Spengler Â and Tourinho P S 2010 On the importance of size of plastic fragments and pellets on the strandline: a snapshot of a Brazilian beach *Environ. Monit. Assess.* **168** 299 – 304
[11] Browne M A, Crump P, Niven S J, Teuten E, Tonkin A, Galloway T and Thompson R C 2011 Accumulation of microplastics on shorelines worldwide: sources and sinks. *Environ. Sci. Technol.* **45** 9175 – 79
[12] Ng K L and Obbard J P 2006 Prevalence of microplastics in Singapore's coastal marine environment *Mar. Pol. Bull.* **52** 761 – 7
[13] Frias J P G L, Sobral P and Ferreira A M. 2010 Organic pollutants in microplastic from two beaches of the Portuguese coast *Mar. Pol. Bull.* **60** 1988 – 92
[14] Ivar do Sul J A, Spengler A and Costa M F 2009 Here, there and everywhere. Small plastic fragments and pellets on beaches of Fernando de Noronha (Equatorial Western Atlantic) *Mar. Pol. Bull.* **58** 1236 – 38
[15] Lee J et al. 2013 Relationships among the abundances of plastic debris in different size classes on beaches in South Korea *Mar. Pol. Bull.* **77** 349 – 54
[16] Ryan P G, Moore C J, van Franeker J A and Moloney C L 2009 Monitoring the abundance of plastic debris in the marine environment *Phil. Trans. R. Soc. B* **364** 1999 – 2012

[17] Hidalgo-Ruz V, Gutow L, Thompson R C and Thiel M 2012 Microplastics in the marine environment: a review of the methods used for identification and quantification *Environ. Sc. Technol.* **46** 3060 –75
[18] Baztan J *et al.* 2014 Protected areas in the Atlantic facing the hazards of micro-plastic pollution: first diagnosis of three islands in the Canary Current *Mar. Pol. Bull.* 80 302 - 11
[19] Majer A P, Vedolin M C and Turra A 2012 Plastics pellets as oviposition site and means of dispersal for the ocean-skater insect Halobates *Mar. Pol. Bull.* **64** 1143 – 47
[20] Ugolini A, Ungherese G, Ciofini M, Lapucci A and Camaiti M 2013 Microplastic debris in sandhoppers *Estuar. Coastal Shelf Sc* **129** 19 – 22
[21] Farrel P and Nelson K 2013 Trophic level transfer of microplastic: Mytilus edulis (L.) to Carcinus maenas (L.) *Environ. Pol.* **177** 1 – 3
[22] Browne M A, Niven S J, Galloway T S, Rowland S J and Thompson R C 2013 Microplastics moves pollutants and additives to worms, reducing functions linked to health and biodiversity *Curr. Biol.* **23** 2388 – 92
[23] Wright S L, Rowe D, Thompson R C and Galloway T S 2013 Microplastic ingestion decreases energy reserves in marine worms *Curr. Biol.* **23** 1031 – 33
[24] Sanchéz-Avila J, Tauler R and Lacorte S 2012 Organic micropollutants in coastal waters from NW Mediterranean Sea: sources distribution and potential risk *Environ. Int.* **46** 50 – 62
[25] Reisser J, Shaw J, Hallegraeff G, Proietti M, Barnes D K A, Thums M, Wilcox C, Hardesty B D and Pattiaratchi C 2014 Millimeter-sized marine plastics: a new pelagic habitat for microorganisms and invertebrates *PLoS One* **9** 1 – 11
[26] Wright S L, Thompson R C and Galloway T S 2013 The physical impacts of microplastics on marine organisms: a review *Environ. Pol.* **178** 483 – 92
[27] Carson H S, Colbert S L, Kaylor M J and McDermid K J 2011 Small plastic debris changes water movement and heat transfer through beach sediments *Mar. Pol. Bull.* **62** 1708 – 1713
[28] Endo S, Yuyama M and Takada H 2013 Desorption kinetics of hydrophobic organic contaminants from marine plastic pellets *Mar. Pol. Bull.* **74** 125 – 31
[29] Nuelle M-T, Dekiff J H, Remy D and Fries E 2014 A new analytical approach for monitoring microplastics in marine sediments *Environ. Pol.* **184** 161 – 169
[30] Claessens M, van Cauwenberghe L, Vandegehuchte M B and Janssen C R 2013 New techniques for the detection of microplastics in sediments and field collected organisms *Mar. Pol. Bull.* **70** 227 – 33
[31] Imhof H K, Schmid J, Niessner R, Ivleva N P and Laforsch C 2012 A novel, highly efficient method for the separation and quantification of plastic particles in sediments of aquatic environment *Limnol. Oceanogr.: Methods* **10** 524 – 37
[32] Vianello A, Boldrin A, Guerriero P, Moschino V, Rella R, Sturaro A and Da Ros L 2013 Microplastic particles in sediments of Lagoon of Vanice, Italy: First observations on occurrence, spatial patterns and identification. *Estuar. Costal Shelf Sci.* **130** 54 – 61
[33] Eriksen M, Mason S, Wilson S, Box C, Zellers A, Edwards W, Farley H and Amato S 2013 Microplastic pollution in the surface waters of the Laurentian Great Lakes *Mar. Pol. Bull.* **77** 177 – 82
[34] Morritt D, Stefanoudis P V, Pearce D, Crimmen O A and Clark P F 2014 Plastic in the Thames: a river runs through it *Mar. Pol. Bull.* **78** 196 – 200
[35] Van A, Rochman C M, Flores E M, Hill K L, Vargas E, Vargas S A, and Hoh E 2012 Persistent organic pollutants in plastic marine debris found on beaches in San Diego, California. *Chemosphere* **86** 258 – 63
[36] Fisner M, Taniguchi S, Majer A P, Bícego M C and Turra A 2013 Concentration and composition of polycyclic aromatic hydrocarbons (PAHs) in plastic pellets: implications for small-scale diagnostic and environmental monitoring *Mar. Pol. Bull.* **76** 349 – 54
[37] Rios L M, Moore C and Jones P R 2007 Persistent organic pollutants carried by synthetic polymers in the ocean environment *Mar. Pol. Bull.* **54** 1230 – 37

[38] Lavers J L, Bond A L and Hutton I 2014 Plastic ingestion by Flesh-footed Shearwaters (Puffinus carneipes): implications for fledging body conditions and the accumulation of plastic-derived chemicals *Environ. Pol.* **187** 124 – 29
[39] Kim Y-J, Osako M and Sakai S-I 2006 Leaching characteristics of polybrominated diphenyl ethers (PBDEs) from flame-retardant plastics *Chemosphere* **65** 506 – 13
[40] Engler R E 2012 The complex interaction between marine debris and toxic chemicals in the ocean *Environ. Sc. Technol.* **46** 12302 – 15
[41] Browne M A, Galloway T S and Thompson R C 2010 Spatial patterns of plastic debris along estuarine shorelines *Environ. Sci. Technol.* **44** 3404 – 09
[42] Hidalgo-Ruz V and Thiel M 2013 Distribution and abundance of small plastic debris on beaches in the SE Pacific (Chile): a study supported by a citizen science project *Mar. Environ. Res.* **87** 12 – 8
[43] Watts A J R, Lewis C, Goodhead R M, Beckett S J, Moger J, Tyler C R and Galloway T S 2014 Uptake and retention of microplastics by the Shore Crab Carcinus maenas *Environ. Sc. Technol.* **48** 8823 – 30
[44] Dekiff J H, Remy D, Klasmeier J and Fried E 2014 Occurrence and spatial distribution of microplastics in sediments from Norderney *Environ. Pol.* **186** 248 – 56
[45] Fotopoulou K N and Karapanagioti K 2012 Surface properties of beached plastic pellets *Mar. Environ. Res.* **81** 70 – 7
[46] Corcoran, P L, Biesinger M C and Grifi M 2009 Plastic and beaches: a degrading relationship *Mar. Pol. Bull.* **58** 80 – 4
[47] Mathalon A and Hill P 2014 Microplastic fibers in the intertidal ecosystem surrounding Halifax Harbor, Nova Scotia *Mar. Pol. Bull.* **81** 69 – 79

IOP Conf. Series: Materials Science and Engineering **78** (2015) 012036 doi:10.1088/1757-899X/78/1/012036

Isolation, characterization and screening of rhizospheric bacteria of *Pittosferum resiniferum* Hemsl.

Hanirah R., Piakong M. T. and Syaufi L.
Faculty of Science and Natural Resources
University Malaysia Sabah

Email : hanirahramli@gmail.com

Abstract - The bacterial rhizosphere species of host plant, Petroleum Nut (*Pittosferum resiniferum*) were isolated and characterized morphologically. The isolates were designated as, TSArp- Cr2, TSArp- Cr3, TSArp- Cr4, TSArp- Cr5, TSArp- Cr6 and TSArp- Cr7. All of the species were tested on three different concentration of phenol (1mM, 3mM and 5mM). Only species TSArp- Cr4 and TSArp- Cr6 growth were detected. The highest growth is 6Log_{10}CFU/ml in 1mM by TSA-Cr4. The lowest reading was 3.6 Log_{10}CFU/ml in 3mM by TSA-Cr6. Species TSArp- Cr4 has higher tolerance on phenol compared to TSArp- Cr6

1.Introduction

The existence of anthropogenic (man-made) organic compounds in the environment is a serious issue to the world residents. The man-made organic compounds such as BTEX (benzene, toluene, ethyl-benzene, *o-, m-, p-* xylene),and phenol are important industrial raw materials for paints, pesticides, resins, fiber, glass unit, varnish, phenolic resin manufacture, textile unit, making of organic dyes and as solvents for rubber and plastic as well [1] [2] . These aromatic compounds has a high toxicity level and carcinogenic that commonly found as contamination relatable to human activities [3][4][5] [6][7].

The variation of microorganism undergo biodegradation ranging from *Pseudomonads* [8] [9][10][11],fungi [12][13][14] [15] yeast [16][17][18]. The source of these bacteria can be found at oil effluent water [19], contaminated oil mousse from beach simulator tank [20], Crude oil-polluted river [21], natural lake and soil [22], soil contaminated with Crude oil spills [23] and rhizosphere soil [24][25]. Rhizospheric soil is the zone that is surrounded by the root of plant and the area of biological and chemicals activity are influenced by the compounds discharged by the root [26], The most common rhizospheric bacterial can be found is *Pseudomonads*, *Bacillus* sp. and *Streptomyces* sp. [24] [25]. These isolated rhizospheric bacteria can degrade toxic compounds, especially petroleum compounds up until 13g/l [24] [25].

Metabolism of these compounds by microorganism is well known with physiological, biochemical and molecular research of the degradation report [27][28]. Microorganisms usually can be as pathogenic and harmful to human. But, apart from that, they can be beneficial to the environment. There are numerous type of pollutant that can be degraded by microorganism, phenol [29] n-alkane [30][31], TCE [32][33].

Since *Pittosferum resinisferum* (petroleum nut) can produce biofule, the hypothesis of rhizosheric bacteria of *Pittosferum resinisferum* has the potential to degrade hydrocarbon can be made. But currently there are no reports from the isolated *Pittosferum resinisferum* rhizospheric bacteria.

The aim of the study is to isolate and characterize the bacteria from rhizosphere soil and screening the bacteria with various concentration of phenol (1mM, 3mM and 5mM).

2. Material and methods

2.1 Sampling

Rhizospheric soil from *Pittosferum resinisferum* was collected in Kundasang, Sabah.

2.2. Physical properties of rhizospheric soil

The value of pH soil *Pittosferum resinisferum* was measured by using pH meter tool (Hanna Instrument, USA) (5.4). 1g of rhizospheric soil sample was mixed together with 3ml of distill water and the reading of the pH value was taken. Thermometer was used to determine the soil temperature (18°C).

2.3. Culture media preparation for isolation and screening of bacteria

For isolation of heterotrophic microorganism, Trypticase Soy Agar or TSA medium was prepared. The composition of TSA media consisting (g/L): 15.0g of Tryptone (Pancreatic Digest of Casein), 15.0g of Soytone (Papaic Digest of Soybean Meal), 5.0g of Sodium Chloride, 15.0g of Agar.

2.4. Isolation and Characterization of bacteria

10g of soil sample rhizospheric soil was taken for serial dilution series (10^{-1} until 10^{-9}) by using saline water (0.85%). The bacteria were originally isolated by direct technique on TSA agar and incubated for 24 hours at 37°C. The developed colonies was purified by streaking on nutrient agar for bacterial identification according to colony and cellular characteristics [34].

2.4.1. Colony morphological characteristics

The identification colony morphology based on colour, margin, elevation and configuration. The colony morphology was viewed and determined by under light microscope, Olympus 9800

2.4.2. Cellular morphological characteristics

Cellular morphology was done by doing Gram staining [35] and being viewed under light microscope, Olympus 9800.

2.5. Screening isolated rhizospheric bacteria with phenol.

The isolated bacteria were tested into 3 different concentration of phenol (1mM, 3mM and 5mM). The isolated rhizospheric bacteria were tested on 3 different concentration of phenol that is 1mM, 3mM and 5mM. The growth of isolated rhizospheric bacteria was cultivated in Ramsay broth for 24 hours under 37°C. After the incubation period, the dilution siries was done and the incubation condition is the same as the inoculum preparation.

3. Result and Discussion

3.1 Isolation and characterization of bacteria.

The rhizospheric bacteria were isolated and colony morphology and cellular morphology were identified. There are 5 isolated bacteria from the soil sample taken at the area of Kundasang Sabah namely : TSARp- Cr2, TSARp- Cr3, TSARp- Cr4, TSARp- Cr5, TSARp- Cr6 and TSARp- Cr7. The colony morphology of the isolated strains are Cream with the margins figure of smooth, wavy and branching. The elevations of the strains are round with scalloped margin, filamentous and round gram staining for each strain is positive. Table 1 shows the characteristics of the isolated bacteria. All of the isolated bacteria comprises of gram positive and negative bacteria.

Compared to other research, of the isolated rhizospheric bacteria was gram negative done by reference [36]. Other than that, there were gram negative and gram positive bacteria was isolated from rhizosphere soil [37]. So, it was agreeable that rhizospheric bacteria dominated by both of gram negative and gram positive bacteria. Those positive and negative bacteria belongs to Proteobacteria, Actinobacteria and Firmicutes [37

3.2 Physical and growth test

There were 6 species of isolated rhizospheric bacteria. Only 2 species were able to grow in three different concentration of phenol (1mM, 3mM and 5mM). The growth of the rest of the species were not detected. There were no colonies observed. Fig. 1 shows the trend of microbial growth (Log_{10}CFU/ml) of two different species.

Table 1 The colony morphology of selected isolates grown on TSA agar at 37^{0}C after 24 hours incubation.

Strains	**Colony characteristics**				**Gram Reaction**
	Colour	**Configuration**	**Margin**	**Elevation**	
TSARp-Cr2	Cream	Round with scalloped margin	Branching	Hilly	Positive
TSARp-Cr3	Cream	Round	Smooth	Convex	Negative
TSARp-Cr4	Cream	Round	Smooth	Convex	Negative
TSARp-Cr5	Cream	Filementous	Wavy	Hilly	Positive
TSARp-Cr6	Cream	Round	Smooth	Convex	Positive
TSARp-Cr7	Cream	Round	Smooth	Convex	Positive

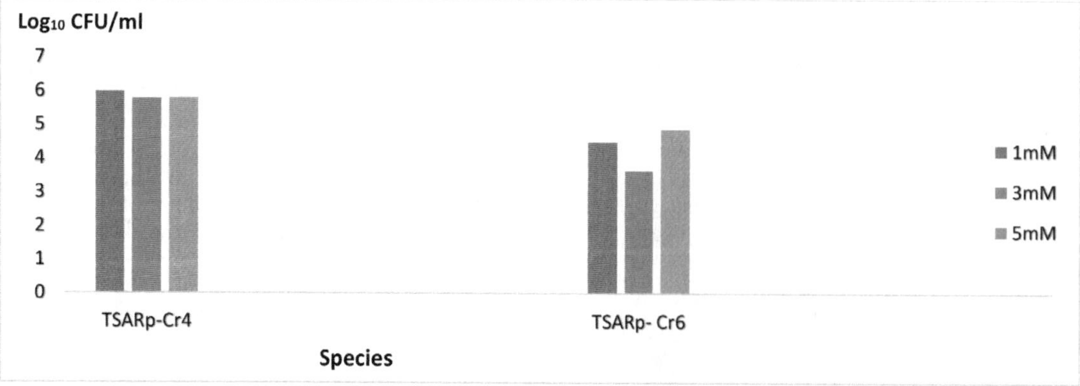

Figure 1 The growth pattern of isolated bacteria in different concentration of phenol.

The total number of culturable bacteria for TSARp-Cr4 ranging from 5.8Log_{10}CFU/ml to 6Log_{10}CFU/ml. While TSARp-Cr6 has the reading of total colony bacteria (Log_{10}CFU/ml) ranging from 3.6 Log_{10}CFU/ml to 4.9Log_{10}CFU/ml. Species TSARp-Cr4 growth on 3 different concentration did not vary significantly.as the value of Log_{10}CFU/ml in 3mM and 5mM is approaching 6Log_{10}CFU/ml. Whereas species TSARp-Cr6 growth on 1mM, 3mM and 5mM of phenol concentration shows fluctuation result.

Species TSARp-Cr4 has the optimum growth value (Log_{10}CFU/ml) at 1mM of phenol while TSARp-Cr6 has the optimum growth at5mM of phenol. The results of the growth species on 3 different concentration of phenol (1mM, 3mM and 5mM), TSARp-Cr4 able to grow greater than TSARp-Cr6.

4. Conclusion

Based on the screening result above, it illustrate that metabolically diverse and healthy community of microorganisms in the rhizosphere in different plants might be the 'hot spot' forhydrocarbon degrader [38]. The isolated rhizospheric bacteria could be highly potential of hydrocarbon degrader. This research could be continued for the bidegradation study on phenol by using these two species.

5. References

[1] Kumar A, Kumar S, Kumar S 2005 Biodegradation kinetics of phenol and catechol using *Pseudomonas putida* MTCC 1194 *Bio-chemical Engineering Journal* **22** 151–159

[2] Yeom, S-H and Daugulis, A J 2001 A two-phase partitioning bioreactor system for treating benzene-contaminated soil *Biotechnol. Lett* **23** 467–473

[3] ASTDR (Agency for Toxic Substances and Disease Registry), ASTDR-HE-CS-2001-0003, U.S Department of Health and Human Services. 1992

[4] Baun A. Reitzel L A, Ledin A, Christensen T H, Bjerg P L 2003 Natural attenuation of xenobiotic organic compounds in a landfill leachate plume (Vejen, Denmark) *Journal of Contaminant Hydrology* **65** 269-291

[5] Fang M Z, Shin M. K, Park K W, Kim Y S, Lee J W, Cho M H 2000 Analysis of urinary S-phenylmercapturic acid and trans, trans-muconic acid as exposure biomarkers of benzene in petrochemical and industrial areas of Korea.Scand *J. Work Environ. Health* **26** (1) 62-68

[6] Fraile J, Niairola J M, Olivella L, Figueras M, Ginebreda A, Vilanova M, Barcela D 2002 Monitoring of the gasoline oxygenate MTBE and BTEX compounds in groundwater in Catalonia (northeast Spain) *Scientific World J.* **8** (2) 1235-1242

[7] Kato M, Rocha M L, Carvalho A B, Chaves M E, Rana M C, Oliveira F C 1993 Occupational exposure to neurotoxicants: preliminary survey in five industries of the Camaçari Petrochemical Complex *Brazil Envrion.* **61**(1) 133-139

[8] Folsom B R, Chapman P J, Pritchard P H 1990 Phenol and trichloroethylene degradation by *Pseudomonas cepacia* G4. : Kinetics and interactions between substrate *Appl. Environ. Microbiol.* **56** (5) 1279- 1285

[9] Annadurai G, Ling L, Jiunn-Fwulee Y, 2007 Biodegradation of phenol by *Pseudomonas pictrom* on immobilizied with chitin *African J. Biotechnol.* **24** 541-547

[10] Kotresha D and Vidyasagar G M 2008 Isolation and characterization of phenol- degrading *Pseudomonas aeruginosa* MTCC 4996 *World J. Microbiol. Biotechnology* **24** 541-554

[11] Ahmed, A M Phenol degradation by *Pseudomonas aeruginosa* 1995 *J. Environ. Sci. Health* **30** 99- 103

[12] Adav S S, Chen M-Y, Lee D-J, Ren N-Q 2006 Degradation of Phenol by Aerobic Granules and Isolated Yeast *Candida tropicalis Biotechnology and Bioengineering* **96** 844 -852

[13] Jiang Y, Wen J, Lan L, Hu Z 2007 Biodegradation of phenol and 4-chlorophenol by the yeast *Candida tropicalis Biodegradation* **18** 719–729

[14] Varma R J and Gaikwad B G 2009 Biodegradation and phenol tolerance by recycled cells of *Candida tropicalis* NCIM 3556 *International Biodeterioration & Biodegradation* **63** 539–542, 2009.

[15] Weijian C, Li J, Zhang Z 2007 The characteristics and mechanisms of phenol biodegradation by *Fusarium* sp. *J. Hazard. Mater.* **148** 38 -42

[16] Ahmadi M, Vahabzadeh F, Bonakdarpour B, Mehranian M, Mofarrah E 2005 Phenolic removal in live oil mill waste water using loofah-immobilized *Phanerochaete chrysosporium World J. MiCrobiol. Biotechnol* **22** 119 – 127

[17] Begona P M, Aurelio H, Juan L S, Maria L S 2002 Degradation of phenol by *Rhodococcus erythropolis* UPV-1 immobilized on Biolite in a packed- bed reactor *J. Biotechnol* **97** 1-11

[18] Krallish I, Gonta S, Savenkova , Bergauer P, Margesin R, 2006 Phenol degradation by immobilized cold-adapted yeast strains of *Cryptococcus terreus* and *Rhodotorula Creatinivora Extremophiles* **10** 441- 449

[19] Piakong M T 2006 The performance of Phenol biodegradation by *Candida tropicalis* RETL-Cr1 using Batch and Fed-Batch Fermentation Techniques," Thesis for Philosophy Doctrate. UTM,

[20] Suntiknowati L I, 2007 Hydrocarbon Degrading Bacteria : Isolation and Identification **11** (2) 98- 103

[21] Koyode-Isola T M, Eniola K I T, Olayemi A B, Igunnugbemi A 2008 Response of resident bacteria of a crude oil-polluted river to diesel oil *American-Eurasian Journal of Agronomy* **1** (1) 06-09

[22] Malatova K 2005 Isolation and characterization of hydrocarbon degrading bacteria from environmental habitats in western New York state, Rochester, NY

[23] Mittal A and Singh P 2009 Isolation of hydrocarbon degrading bacteria from soils contaminated with Crude oil spills *Indian J. Exp. Biol.* **47** (9) 7650-7655

[24] Wang Z, Liu X, Zhang X, Hu X, Liu F 2011 Degradation of diesel with microorganisms in rhizosphere of *Carex phacota* Spr. *Second International Conference on Environmental Science and Technology* **6** 224-227

[25] Jussila M M, Jurgens G, Lindstrom K, Suominen L 2006 Genetic diversity of culturable bacteria in oil contaminated rhizosphere of *Galega orientalis Enviro. Pollut.***139** 244- 257

[26] Lines-Kelly, R. 2005. The rhizosphere : Soil Biology basics, Profitable and Sustainable Primary Industries. http//: www.dpi.nsw.gov.au. Accessed on 15 October 2015 at 3.22 PM

[27] Smith M R 1990 The biodegradation of aromatic hydrocarbons by bacteria *Biodegradation* **253** 111-118

[28] Pieper D H, Martins dos Santos V A, Golyshin P N 2004 Genomic and mechanistic insights into the biodegradation of organic pollutants *Curr. Opin. Biotechnol.* **15** 215- 224

[29] Sridevi V, Lakshmi M V V C, Manasa M, Sravani M, 2012 Metabolic Pathways for the biodegradation of phenol *International Journal of Engineering Science & Advanced Technology* **2**(3) 695 – 705

[30] Piasentier E, Bovolenta S, Malossini F, Suusmel P, 1995 Comparison of n-alkanes or chromium oxide methods for estimation of herbage intake by sheep *Small Ruminant Research* **18** 27- 32

[31] Hwang C W, Yano K, Takagi M 1991 Sequences 2 tandem genes regulated by carbon sources, one being essential for n-alkane assimilation in *Candida maltose Gene* **106** 61-69

[32] Morono Y, Unno H, Hori K, 2006 Correlation of TCE cometabolism with growth characteristics on aromatic substrates in toluene degrading bacteria *Biochemical Engineering Journal* **31** 173-179

[33] Kocamemi B A and Cecen F 2005. Cometabolicdegradation of TCE inriched nitrifying batch systems. *Journal of Hazardous Materials*. **125** : 260- 265

IOP Conf. Series: Materials Science and Engineering **78** (2015) 012036 doi:10.1088/1757-899X/78/1/012036

[34] EPA (Environmental Protection Agency). Method validation of U.S. Environmental Protection Agency Microbiological Methods of Analysis.Versar Inc.,Sprinfield, VA, 2009

[35] Gram C 1884 The differential staining of schizomycetes in tissue sections and in dried preparations **2** 185-189

[36] Bahig A E, Aly E A, Khaled, Amel K A 2008 Isolation, characterization and application of bacterial population from agricultural soil at Sohag Province, Egypt *Malaysian Journal of Microbiology* **4** (2) 42-50

[37] Nijhuis E H, Maat M J, Zeegers I W E, Waalwijk C, Van Veen J A 1993 Selection of bacteria suitable for introduction into the rhizosphere of grass *Soil Biol. Biochem.* **25** 885– 895

[38] Cunningham S D, Anderson T A, Schwab A P, Hsu F C 1996 Phytoremediatiocof soils contaminated with organic pollutants *Advances in Agron* **56** 114

Experimental investigation of wood fibre cement composite wall panel under axial loading

Mst. Sadia Mahzabin[1], Roszilah Hamid[2]

[1]Department of Civil Engineering;
Lee Kong Chian Faculty of Engineering and Science
Universiti Tunku Abdul Rahman (UTAR)

[2]Department of Civil & Structural Engineering
Faculty of Engineering and Built Environment
Universiti kebangsaan Malaysia (UKM)

sadia@utar.edu.my

Abstract. wood fibre cement (WFC) composite wall panels were cast and tested under axial load with 4/6 wood/cement ratio, 0.8 water/cement ratio, three chemical additives and horizontal and vertical reinforcement. Other panels with the same mix design proportion without reinforcement were also tested and compared with the commercially available WFC composite *Duralite* boards. An experimental result for the Duralite boards, the specimen showed quick failure with lower loading value and also with axial deformation. The WFC panel without reinforcement showed more brittle type of failure in that they were unable to sustain any more loading after reaching the maximum load. The failure for the WFC panel with reinforcement was gradual and this behaviour was attributed to the presence of steel as they act like bridges between cracks preventing sudden failure. The WFC panels without reinforcement results are higher than the theoretical value and also higher than the Duralite board panels.

Keywords-Wood Fibre Cement Composite; load bearing; wall panels; axial loading

1. Introduction

Wood fibre cement (WFC) composites products are lightweight (at low density of 400 kg/m^3) and can be applied for multi purposes in construction industry for both structural and non-structural element. WFC boards are suitable for non-load bearing wall structures. When it comes to economical aspects of WFC, there are two major aspects playing a role here (Frybort et al., 2008). First, the chemical additives (calcium chloride, calcium formate, soda ash and a number of organic materials) are essential in the production of WFC boards because the additives speed up the chemical reaction in the cement and water and accelerate the rate of setting and early gain in strength of cement. Thus, the type and dosage of chemical additives should be chosen carefully for compatibility with the type of wood fibre used. Secondly, the cement-wood ratio needs to be considered with the fact that wood is less expensive than cement. Wood as a naturally and almost everywhere grown material is cheaper than cement. Replacing an amount of cement (per volume) by wood or other lignocelluloses materials makes the product cheaper. Previous research with *Kelampayan* wood type had determined that the compressive strength of WFC mixture with chemical additives is greater than the strength of WFC matrix that was made without chemicals (2). At 28 days, the average strength of mixtures with wood-cement ratio of 40:60 that were made with chemical additives is the highest (increment of 62.7% in strength) than the WFC that is made with 50:50 and 30:70. Excess amount of cement in the WFC matrix with chemicals inclusion at wood/cement ratio of 30:70 had contributed negatively to the strength and had made the matrix brittle and failed at low strength.

Duralite (M) Sdn Bhd is one of the Malaysian owned companies. Duralite (M) Sdn Bhd was formed for the purpose of manufacturing and distributing their products under the name Duralite. Duralite panel (Dura) is manufactured accordance with BS 1105 or DIN 1101 standard. These boards are made from wood fibre that are chemically impregnated and cleverly bonded under pressure with Portland cement to form light weight and homogenises panels

For load bearing wall, its strength must be sufficient to carry load from the ceiling, roof or upper floor loads placed on it down to foundations, in addition to its own weight. Load bearing wall should be braced in the form of transverse walls directly connected to the load bearing wall (3). Wall also needs to support lateral loading such as wind. WFC composites wall panel can be used as external wall of an ordinary single story house. Masonry construction using clay brick, cement and concrete block is often used in load bearing wall of building up to four stories. By increasing the bending moment and compressive strength capacity, the light weight precast wall panel can be used as load bearing structural member. With a very little addition to the standard or specifically designed parameters, the WFC composite panels can easily be used as structural components and thus taking full advantage as concrete section. Several light weight alternatives to traditional precast concrete wall panels have been used in recent years. Dissanayake (4) investigated the behaviour of the slip formed load bearing wall panels made of cement and crusher dust subjected to two different support conditions (simple and continuous). It was concluded that the slip formed load bearing wall panels can sustain higher loads with crack free state than conventional masonry. Madsen (5) studied the experimental investigation of the plant fibre composites mechanical properties as load-bearing elements. The investigations included a number of relevant parameters: testing direction, matrix type, fibre volume fraction, process temperature and conditioning humidity. Mydin (6) verified the potential of using lightweight foamed concrete (LFC) in composite load-bearing wall panels in low-rise construction. The experiment investigated the performance of LFC panels of different densities whether the composite walling system had sufficient load carrying capacity, based on compression resistance at ambient temperature.

This research is to determine the capability of *Kelampayan* WFC composite by producing WFC composite wall with the optimized mix properties in previous research as described by Mahzabin et al. (2) with and without reinforcement to be used as load bearing wall. A laboratory test under concentric axial compression was conducted to investigate the axial behaviour of WFC load bearing wall panel. The tests were performed for all specimens to determine the ultimate load, load-deflection relationship and load-axial deformation relationship. A total of four wall panels were cast and tested, two WFC panel with reinforcement and another two WFC composite panel without reinforcement. The panels have been designated as similar to light weight concrete and were tested in one way action.

2. Experimental program

2.1. Sample preparation

The thickness of the WFC wall panels was chosen to be 100 mm to satisfy the fire resistance requirements for exterior bearing wall. It was the minimum code thickness for two hour fire resistive exterior bearing wall according to the international building code for the minimum thickness of cast in place or precast walls for various fire resistance ratings (7). The dimension of WFC composite for the application as load bearing wall panel were 600 mm for the width, 2000 mm for the height and 100 mm for the thickness (8). Two types of WFC wall panels (with two specimens for each type) were cast. The first was with 4:6 wood/cement ratio, 0.8 water/cement ratio, three chemical additives and horizontal and vertical reinforcement. The second panel was with same mixture except reinforcement.

The plane WFC wall panels without reinforcement were compared with the commercially available Duralite.

Two wall panels were reinforced with a single mesh consisting of 8 mm diameter bar with spacing of 94 mm for vertical reinforcement and 215 mm for horizontal reinforcement, placed in the centrally panel cross section. The vertical and horizontal reinforcements of panel were satisfying the minimum requirements of the British Standard BS 8110. The reinforcement layout is shown in Figure 1. This reinforcement was taken to keep the reinforcement to approximately the minimum level required by the code of practices.

2.2. *Test Setup*

The test frame was used to support the load testing mechanism and this frame consisted of two main steel columns each 4000 mm long. These columns were connected by complete fillet weld, to base plates. A series of channel members were used as cross beam were connected to the columns by bolts on each side. The test frame was used to support the independent hydraulic loading jack, of 500 KN capacities. The hydraulic loading jacks was attached to a strong mounted plate under the cross beam. The jack was required to transmit a uniform distributed load across the top through a loading beam. The bottom beam was identical to the upper beam and had small guide plate to prevent lateral movement. A schematic diagram of test setup is shown in Figure 2.

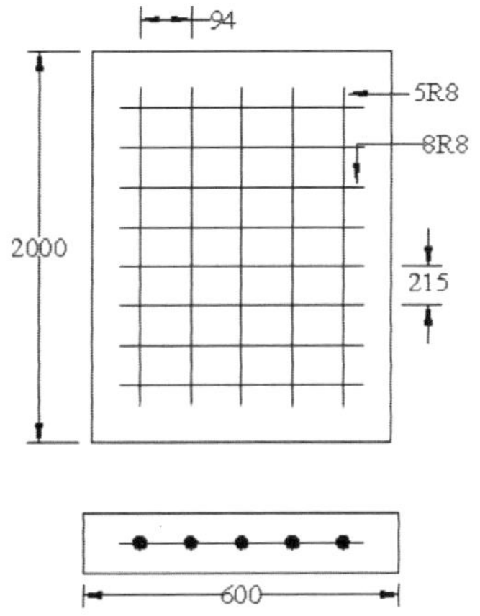

Figure 1. Arrangement of reinforcement in WFC wall panel (unit: mm).

2.3. *Test Procedure*

The walls were loaded with the crosshead of the machine was set to 0.5 kN/sec. With the load increments the deflection and the shortening of the panel were recorded. The linear vertical differential transducers (LVDT) were used to measure the deflection and axial deformation of the wall panel. They were placed at the centre of the panel (LVDT 2) and another one was placed on the top surface of the panels (LVDT 1).

The loading in the test program was a uniformly distributed nature along the length of the panel at the top and this distribution was achieved experimentally by transmitting the load from the fixed head of the testing device to the top through distributing with bearing plates. The detail of the test panel is shown in Figure 3.

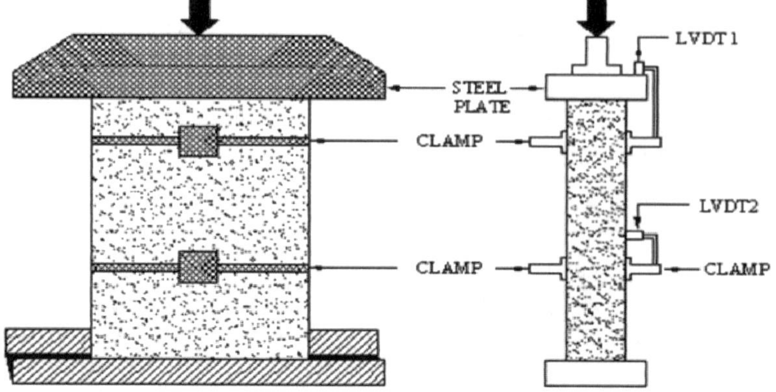

Figure 2. Schematic diagram of wall panel test setup

Figure 3. Testing setup of WFC composite wall panel

3. Results and Discussion

The wall panels were tested under concentric axial compression to determine the axial behaviour of the WFC panels as load bearing wall. All samples were tested in long direction in order to apply the WFC composite panel as load bearing slender wall. The ultimate axial load, mid span deflection and the axial deformations of the panels were determined.

The failure load for all panels tested are recorded and expressed in Table 1. The WFC panels without reinforcement results are higher than the theoretical value and also higher than the Duralite board panels. The overall test results were rather satisfactory except for the WFC panels with reinforcement results do not meet the theoretical value.

Table 1. Summary of Data of all WFC wall Panels

Panels	Experimental Ultimate load (kN)	Theoretical Ultimate Load (kN)
WFC Panel With Reinforcement	441 466.2	471.12
WFC Panel Without Reinforcement	345.2 357.8	275.17
Duralite Boards	98.2 110.9	-

The percent increase for the ultimate load of WFC panel with reinforcement is about 38.4 % and the percent decrease for the ultimate load of WFC boards without reinforcement is 27 % compared with Control panels (Duralite board). On the same time the mid span deflection of WFC panel without reinforcement is increased about 34.5 % and with reinforcement is decreased 2.11 % compared with control panels. The details of mid span deflection of all wall panels are mentioned in Table 2.

Table 2. Details of mid span deflection of all panels

Samples	Ultimate Deflection (mm)	Average Ultimate Deflection (mm)	Percent different (%)
Dura1 Dura2	15.6 12.44	14.02	Control
WP1 WP2	54.47 16.56	35.51	34.52
RWP1 RWP2	15.38 10.04	12.71	-2.11

For the Duralite boards the specimen showed failure with lower loading value at average 104.55 kN and also with small axial deformation 4.1 mm. For the WFC panel without reinforcement showed brittle type of failure in that they were unable to sustain any more loading after reaching the maximum load at 351.5 kN. The failure for the WFC panel with reinforcement was gradual and this behaviour was attributed to the presence of steel as they act like bridges between cracks preventing sudden failure. The percent increase for the ultimate load and deformation WFC panel without reinforcement is about 27 % and 48 % compared with control. However, the ultimate load of WFC panel with reinforcement is increased about 38.4 % and the deformation about 6.5 % compared with control panel. The details of axial deformation for all panels are mentioned in the Table 3.

Table 3. Details of axial deformation of all panels

Sam.	Ult. Load (kN)	Ave. ult. Load (kN)	Percent diff. (%)	Ult. Deform. (mm)	Ave. Ult. Deform. (mm)	Percent diff. (%)
Dura1	98.2	104.55	Control	3.79	4.1	Control
Dura2	110.9			4.41		
WP1	345.2	351.5	27	22.85	17.11	48
WP2	357.8			11.37		
RWP1	441	453.6	38.4	6.29	5.86	6.5
RWP2	466.2			5.44		

For the Duralite boards, the specimen showed quick failure with lower loading value and also with axial deformation. The WFC panel without reinforcement showed more brittle type of failure in that they were unable to sustain any more loading after reaching the maximum load. The failure for the WFC panel with reinforcement was gradual and this behaviour was attributed to the presence of steel as they act like bridges between cracks preventing sudden failure.

4. Conclusion

For the Duralite boards, the specimen showed quick failure with lower loading value and also with axial deformation. The WFC panel without reinforcement showed more brittle type of failure in that they were unable to sustain any more loading after reaching the maximum load. The failure for the WFC panel with reinforcement was gradual and this behaviour was attributed to the presence of steel as they act like bridges between cracks preventing sudden failure. The WFC panels without reinforcement results are higher than the theoretical value and also higher than the Duralite board panels. The WFC panels with reinforcement results do not meet the theoretical value.

References

[1] Frybort, S., Mauritz, R., Teischinger, A. & Muller, U. 2008. Cement bonded composites – a mechanical review, Bio Resources, 3(2): 602-626.

[2] M. S. Mahzabin, R. Hamid, W.H.W. Badaruzzaman, 2013. Evaluation of chemicals incorporated wood fibre cement matrix properties. Journal of Engineering Science and Technology, Vol. 8, No. 4 (2013) 385 – 398.

[3] Irwin, P.E.J.S. 1076. Considerations for Design of Precast Concrete Bearing Wall in Buildings to Withstand Abnormal loads. Prestressed Concrete institute (PCI), Vol. 21, No. 2, March-April.

[4] Dissanayake, T. A., Dissanayake, U. I., Ranaweera, M.P. & Kulasinghe, A.N. S. 2003. Behavoiur of low cost slip-formed load bearing wall panels under in-plane compressive loading. Annual Transactions of IESL, 46:53.

[5] Madsen Bo. 2004. Properties of Plant Fibre Yarn Polymer Composites. An Experimental study.

[6] Mydin, M. A. O. 2011. Potential of Using Lightweight Foamed Concrete in Composite Load-Bearing Wall Panels in Low-Rise Construction. *Concrete Research Letters*, 2(2): 213-227.

[7] IBC, 2000. International Building Code, International Code Council, INC. USA.

[8] Builder's Guide on Wood Wool Cement Board. Forest Products Research and Development Institute (FPRDI). Australian Centre for International Agricultural Research

The effects of Concentration and Salinity on Polymer Adsorption Isotherm at Sandstone Rock Surface

M Ali[1] and H Ben Mahmud[1]

[1]Petroleum Engineering Department, Curtin University Miri, Malaysia

E-mail: mohamed.a@curtin.edu.my and hisham@curtin.edu.my

Abstract. Adsorption of hydrolyzed polyacrylamide (HPAM) polymers on sandstone rock surface was studied by static adsorption experiments. Total of 10 Runs of static experiments were conducted in test tubes by mixing the desired solution with crushed rock sample, at temperature of 25 °C, and salinity range from 0-4 wt%. The results are in conformity with Langmuir's isotherm. Ten different isotherms were generated at each Run. The initial polymer concentration was varied from 0.3-2.1 g/l. The effects of salinity have been studied by observation on Langmuir adsorption coefficients (Y and K). The results show that the adsorption coefficient (Y) was found to have linear relationship with salinity. The adsorption coefficient (K) was found to be related to salinity by a quadratic relationship.

1. Introduction

The polymer used in the flood application in its hydrolyzed form [1]. Hydrolyzed polyarylamide (HPAM) was used in this study, which is a straight-chain polymer that has the acrylamide molecule as the monomer. This partial hydrolysis can occur in some of these monomers [2]. One of the most important properties which made polymer interested for EOR processes as mentioned earlier was the viscosyfing property. Another important parameter which may affect the feasibility study plan of a polymer flooding project is the adsorption of polymer in porous medium [3]. Adsorption may be defined as the interaction between the polymer molecules and the porous medium which leads polymer to be adsorbed. The adsorption of polymer will affect both, polymer and porous medium properties [4]. For example the viscosity of adsorbed polymer will be lower than the viscosity of the injected polymer.

Polymer adsorption is mainly physical interaction, electrostatic attraction due to the charges differences between the solid surface and the polymer molecules [5]. The difference in concentration of a chemical agent in the solution and rock surface is the main driving force for the adsorption [6]. Normally, adsorption is measured by what called isotherms. Isotherms are simply plots of the amount of adsorbed chemical per gram of solid or per surface area of solid vs. the equilibrium concentration at a constant temperature. Adsorption isotherm of polymer is influenced by many parameters; in this work only concentration and salinity are considered for investigation.

Among many adsorption isotherms, HPAM polymer was found to follow Langmuir isotherm [6]. The Langmuir adsorption isotherm equation was developed to describe the adsorption of gases onto clean solids and implies uniform adsorption sites and absence of lateral interactions. The Langmuir isotherm is expressed as follow:

$$Ad = \frac{YKC}{(1 + KC)} \tag{1}$$

where *Ad* is the amount adsorbed per unit mass; C is the equilibrium solution concentration; and Y and K are empirical coefficients [7]. However, the use of this model to predict adsorption in presence of salinity effects is not possible since it relates adsorption amounts to only equilibrium concentration. Static adsorption experiments in test tubes were conducted at 25 °C and different salinities. Adsorption isotherm was generated at each experiment (Run). The effect of salinity on the polymer adsorption has been studies by observing the values of adsorption coefficients (K and Y).

2. Materials and Methods

2.1. Materials

The polymer used in this study is the well-known partially hydrolyzed polyarylamide (HPAM). Polyarylamides are water soluble polymers which are produced by many manufactures in many ways for different purposes. For instance as enhanced oil recovery (EOR) agent. The monomer acrylamide is a compound derived from acrylic acid. The molecular weight of polycryamides is 14*106 Daltons. The size of the molecules is about 0.2 μm. By hydrolysis in a caustic water solution some of the $CONH_2$ groups react to form carboxyl groups (COOH). The degree of hydrolysis is an important parameter which determines the properties of polyacrylamide in aqueous solutions as used in EOR [8]. The polymer used here has hydrolysis degree about 25 % as provided by SNF group.

Rock samples used in this study are sandstone type which is a clastic sedimentary rock composed mainly of sand-sized minerals or rock grains. Most sandstone is composed of quartz and/or feldspar because these are the most common minerals in the Earth's crust. Like sand, sandstone may be any color, but the most common colors are tan, brown, yellow, red, gray, pink, white and black [9].

2.2. Static Adsorption Tests

It is simply, measuring the adsorption of solution on rock by mixing both in test tubes under static condition of flow. Samples of reservoir rock should be grounded for static measurements of adsorption so that the adsorptive has good access to the solid surface. As adsorption of EOR-agents is measured from a liquid phase, it may be difficult to measure directly (in the solid) the increase in weight, but it should be possible to measure this weight increase after the sample is dried. Total of 10 Runs of static adsorption tests were performed at different conditions of concentration and salinity. Each Run consists of ten solutions with different initial concentrations.

The adsorbed substance was determined from the decrease in concentration of the chemical solution after it has been brought into contact with the adsorbent and equilibrium has been reached, as in equation (2). The experiment was designed in such a way that the decrease in concentration is significant and can be easily measured; the mass of the adsorbent should not be too small and the volume of the solution not too big. For this purpose the solid to liquid (S/L) ratio of all static experiments was fixed at 0.1, which was found to be representative for the rock type we used (sandstone).

$$Ad = \frac{V}{M}(C_o - C_e) \tag{2}$$

where V is the volume (litre) of solution used, M is the weight of the solid sample (g), and C_0 and C_e are the initial and equilibrium concentrations (g/l), respectively of polymer solutions.

2.3. Concentration Measurements

To measure the concentration of polymer after adsorption a starch-triiodide method was used. This method gives a linear response in a concentration range of 0-3 g/l [10]. The samples had a higher concentration than this, so they were diluted first. Around 30 ml of the diluted sample was used. First of all 5 ml of a buffer solution (pH=3.5) is added. Secondly 1ml of saturated bromine water is added to the sample. After 15 minutes 5 ml of 1 wt. % sodium formate solution is added. At the end after 5 minutes 5 ml of a solution containing starch is added. The iodide is then reduced to iodine by a reaction product of bromine and the polymer. Starch forms a blue color with iodine and the adsorbance of the sample was measured using a UV-VIS spectrometer.

The adsorbance at peak wavelength, at around 568 nm, is used. It is necessary to have standard curve to measure the polymer concentration. The standard curve was generated using the same method as described above for known concentrations of polymer solution. The standard curve gives a linear relationship between the polymer concentration and absorbance of the solution.

3. Results and Discussions

3.1. Polymer Adsorption Isothermal

Ten solutions of polymer were used and were named P1 to P10, where the range of concentrations was between 0.3-2.1 g/l. The polymer is normally injected into reservoir with a concentration between (0.5-2 g/l) [21]. The adsorption isotherm of polymer is shown in Figure 1. Equation (3) was used to estimate adsorption coefficients for Langmuir isotherm.

$$\frac{1}{Ad} = \frac{1}{YKC} + \frac{1}{Y} \tag{3}$$

Figure 1 shows the adsorption of polymer as a function of equilibrium concentration for the solutions with no salinity. As it can be seen from Figure 1, at low polymer concentrations (below 1 g/l), polymer adsorption is directly proportional to the polymer concentration. Increasing concentration of polymer solution increases the numbers of polymer molecules per unit volume. Thus, it increases the probability of polymer adsorption on available adsorption sites in the rock surfaces. After concentration of the polymer solution reaches 1.5 g/l, polymer adsorption is basically concentration-independent (see Figure 1). The corresponding value of polymer adsorption at concentrations higher than 1.5 g/l was 1.87 mg/g. Based on the previous discussion, the maximum adsorption level of HPAM polymer solution (salinity = 0) on sandstone rock surface at 25 °C, and S/L ratio of 0.1 is 1.87 mg/g. From Figure 1, it can also be observed that the Langmuir model agrees better with the experimental data in the whole range of test concentration for polymer solution

3.2. Effect of Salinity

Many reservoirs suitable for chemical EOR have high salinities, i.e., brines with substantial hardness and having total dissolved solids (TDS) contents up to about 200 g/L [11]. Many physical and chemical properties of the injected chemical agents are affected by salinity, like viscosity, chemical activity, stability, and adsorption [12].

Adsorption, mainly physisorption of chemical molecule on the rock surface is exaggerated by changes in the ionic strength (salinity). Charge type (positive, negative, or neutral) of rock surface is disturbed in the presence of salinity, which leads to different behavior of the rock surface [12]. While the salinity (ionic strength) is an important factor that influences the adsorption of polymer to the rock surfaces. Salinity in this study reflects the concentration of Sodium Chloride (NaCl) in the system. Irrespective of the process and the nature of the polymer, the quantity of adsorbed polymer is increased by a factor of 2 to 5 in the presence of NaCI (100 g/l).

On the other hand, the maximum adsorption limit increases with the degree of polymer hydrolysis, with or without the presence of salt [13]. Some explanations of the general increase in the quantity of the polymer adsorbed in the presence of sodium chloride may be considered. First, a polyelectrolyte may have a coagulation effect on dispersed solids, due to a decrease in the electric double layer at the interface solution-dispersed solid particle [14], and a flocculation effect, resulting from polymer bridges between the mineral particles in suspension [15]. In addition, the decrease in thickness of the double layer of the dispersed particles and the surface potential due to polyelectrolyte results generally in polymer adsorption [16].

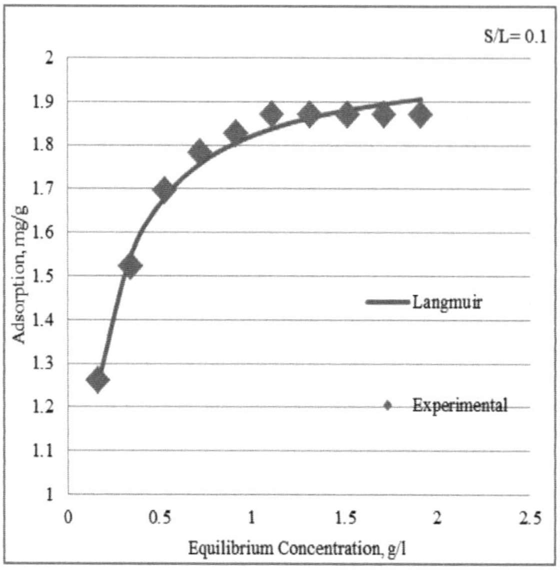

Figure 1: Polymer Adsorption Isotherm

The adsorption increase in the presence of NaCI may also be explained by a decrease in macromolecular bridges between the clay particles (viscosity decrease); the mineral particles are no longer connected, allowing a fixation of larger quantities of polymer (greater occupation of the adsorption sites) [18]. Different polymer adsorption isotherms were developed at different salinities (0-4 wt %) and are shown in Figure 2.

From Figure 2, it is clear that the adsorption of polymer was increased by adding (NaCl) to the polymer solution during the static adsorption test. The amount of absorbed polymer increased from 1.87 to 2.48 mg/g by adding 4 wt% salinity. This effect can be explained by the bridging of negatively charged polymer molecules to the negatively charged grain surface, and by coordination of negatively charged groups on the polymer molecule [17]. Moreover, at high salinity values, polymer molecules are more compact; this is due to electrolyte that leads to the reduction in the extension of the polymer molecule.

More compacted polymer molecules require smaller surface area than normal molecules to be absorbed on the rock surface, then the additional available area for adsorption (in case of compacted molecules) will attract more polymer molecules. As a result, adsorption of polymer increases with increasing salt concentration (salinity). The results shown in Figure 2 are in agreement with the results of the other researchers [17], [18].

To study the effect of salinity on adsorption of polymer on sandstone surfaces, four different sets of the polymer solutions with different levels of salinity ranging from 0 to 4 wt % were prepared. Adding salinity will change the values of Langmuir adsorption parameters Y and K. The parameter values were estimated by plotting (1/Ad) vs. (1/C) and then using the straight line analysis procedure based on equation (3). The estimated values of the parameters are given in Table 1.

From Table 1, it can be noted that when salinity is increased from 0 to 4 wt %, K increases from 9.648 to 29.673 l/mg, while Y_m values increases from 2.009 to 2.497 mg/g. To establish a mathematical relationship between the salinity and each of the parameters *Y* and *K*, *Y* vs. salinity and *K* vs. salinity were plotted in Figure 3 and Figure 4, respectively.

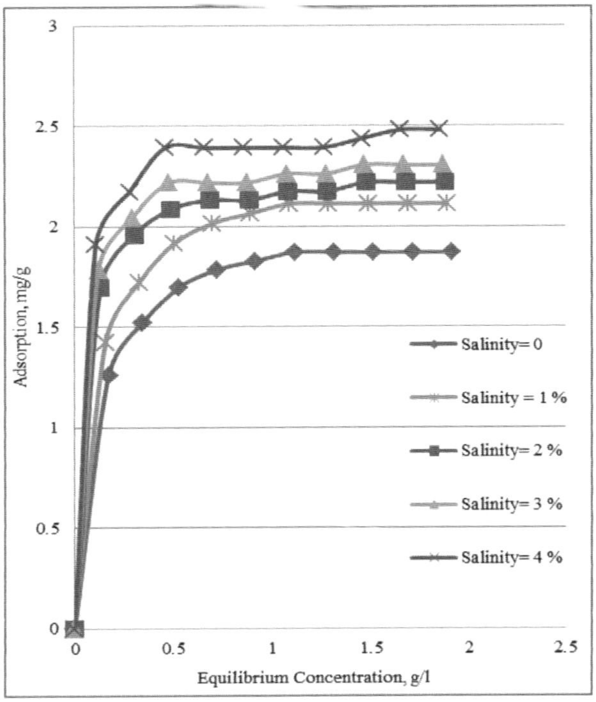

Figure 2: Effect of Salinity on Polymer Adsorption Isotherm

Table 1: Langmuir adsorption parameters for polymer with different salinities

Salinity, wt %	Y, mg/g	K, l/g
0	2.0094	9.6481
1	2.2533	10.7981
2	2.2626	22.6646
3	2.3468	25.7158
4	2.4964	29.6726

Figure 3 shows that Y changes with salinity in a linear manner. These findings are in agreement with the modified Langmuir model that have been used in some simulators to find the adsorption amount of chemicals species in presence of salinity effects [19], [20]. Whereas Figure 4 shows that the relationship between salinity and the adsorption coefficient K is of quadratic (see the equation given in the inset of Figure 4) in nature. The coefficient of determination for the fit of observed data of K with the quadratic model is very close to unity (0.996) which further improves the validity of the predicted quadratic relationship.

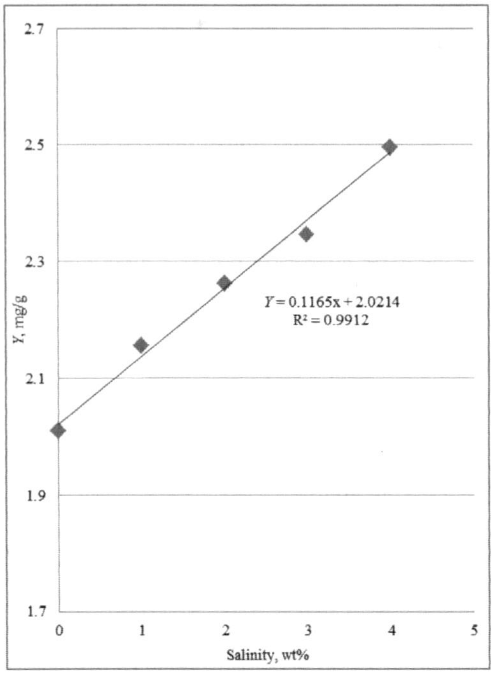

Figure 3. Relationship between salinity and *Y* for polymer

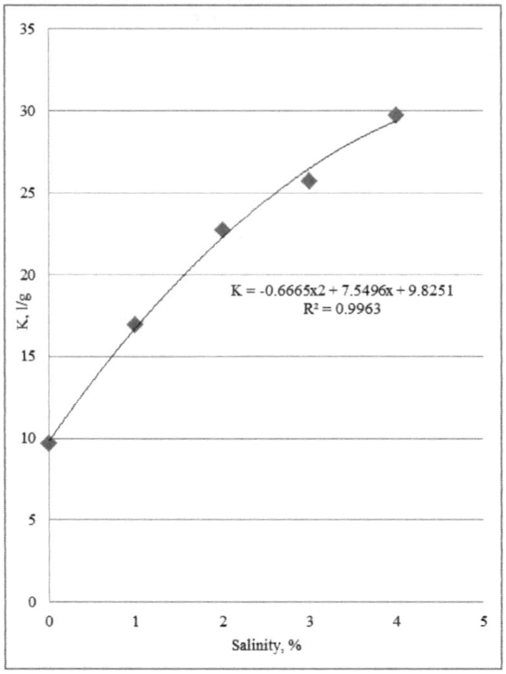

Figure 4. Relationship between salinity and *K* for polymer

4. Conclusions

- It was found that the adsorption data from the static experiments are best represented by the Langmuir model when compared to the Freundlich model.
- It was found that, adsorptions HPAM polymer increases when the salinity increases.
- It was found that for the HPAM polymer, adsorption coefficient *K* was bonded by a quadratic relation with salinity. While adsorption coefficient *Y* was found to be related by linear relation to the salinity.

5. References

[1] M. L. Taylor, G. E. Morris, P. G. Self, and R. S. C. Smart, "Kinetics of adsorption of high molecular weight anionic polyacrylamide onto kaolinite: the flocculation process.," Journal of colloid and interface science, vol. 250, no. 1, pp. 28–36, Jun. 2002.

[2] N. Arihara, T. Yoneyama, Y. Akita, and L. Xiang Guo, "Oil Recovery Mechanisms of Alkali-Surfactant-Polymer Flooding," SPE 54330, Proceedings of SPE Asia Pacific Oil and Gas Conference and Exhibition, Apr. 1999.

[3] H. Yang, C. Britton, P. J. Liyanage, S. Solairaj, D. H. Kim, Q. Nguyen, U. Weerasooriya, and G. A. Pope, "Low-Cost , High-Performance Chemicals for Enhanced Oil Recovery," Improved Oil Recovery Symposium, Tulsa, SPE 129978, 2010.

[4] G. Chauveteau, K. Denys, and a. Zaitoun, "New Insight on Polymer Adsorption Under High Flow Rates,"SPE 75183-MS, Proceedings of SPE/DOE Improved Oil Recovery Symposium, Apr-2002.

[5] M. J. McGuire, J. Addai-Mensah, and K. E. Bremmell, "Spectroscopic investigation of the adsorption mechanisms of polyacrylamide polymers onto iron oxide particles.," Journal of colloid and interface science, vol. 299, no. 2, pp. 547–55, Jul. 2006.

[6] C. Dang, Z. Chen, N. Nguyen, W. Bae, and T. H. Phung, "Development of Isotherm Polymer/Surfactant Adsorption Models in Chemical Flooding," SPE 147872, Asia Pacific Oil and Gas Conference and Exhibition, Jakarta, 2011.
[7] D. Chamovska, M. Cvetkovska, and T. Grchev, "The Kinetics of Polyacrylamide Adsorption on Polycrystalline Gold," Croatica Chemica Acta, vol. 81, no. 3, pp. 461–466, 2008.
[8] W. Littmann, Polymer Flooding, Development in Petroleum Science, ELSEVIER, 1988.
[9] P. L. Churcher, P. R. French, J. C. Shaw, and L. L. Schramm, "Rock Properties of Berea Sandstone, Baker Dolomite, and Indiana Limestone," SPE 21044, Proceedings of SPE International Symposium on Oilfield Chemistry, Feb. 1991.
[10] A. Kasal, M. Budesinsky, and W. J. Griffiths, "Spectroscopic Methods of Steroid Analysis," Springer Science-Business Media B.V. 2010.
[11] M. Puerto, G. J. Hirasaki, C. A. Miller, J. R. Barnes, and S. G. Solutions, "Surfactant Systems for EOR in High-Temperature , High-salinity Environments," SPE 129675, Improved Oil Recovery Symposium, Tulsa, Oklahoma, 2010.
[12] Elmofty and Shokir, "Applying Electrophoresis Technique to Study Adsorption of Surface Active Agents on Reservoir Rocks," SPE 85649, 27th Annual SPE International Technical Conference and Exhibition, Abuja, 2003.
[13] H. Moradi, "Experimental investigation of polymer flow through water- and oil-wet porous media," MSc. University of Stavanger, 1990.
[14] Zhou and Gunter, "The Nature of the Surface Charge of Kaolinite," Clays and Clay Minerals, vol. 40, no. 3, pp. 365–368, 1992.
[15] A. P. Black, F. B. Birkner, and J. J. Morgan, "The effect of polymer adsorption on the electrokinetic stability of dilute clay suspensions," Journal of Colloid and Interface Science, Apr. 1966.
[16] N. Dimov, "Adsorption of Ionic Surfactants on Solid Particles Determined by Zeta-Potential Measurements: Competitive Binding of Counter ions," Journal of Colloid and Interface Science, vol. 256, no. 1, pp. 23–32, Dec. 2002.
[17] L. Chiappa, A. Mennella, T. P. Lockhart, and G. Burrafato, "Polymer adsorption at the brine-rock interface : the role of electrostatic interactions and wettability," Journal of Petroleum Science and Engineering, 1999.
[18] P. Espinasse and B. Siffert, "Acetamide and polyacrylamide adsorption onto clays: influence of the exchangeable cation and the salinity of the medium," Clays and Clay Minerals, vol. 27, no. 4, pp. 279–284, 1979.
[19] Schlumberger, ECLIPSE Reservoir Simulation Software.Technical description, Version 2011.2.
[20] Computer Modelling Group Ltd., Advanced Process and Thermal Reservoir Simulator. Technical description, Version 2011.
[21] Z. Zain, "EOR – A New Look at an Old Field," Report, PETRONAS Research Sdn. Bhd, 2011.

Design and Development of Low-Cost Wind Tunnel for Educational Purpose

T H Yong and S S Dol[1]
Department of Mechanical Engineering, Curtin University
Sarawak Campus CDT 250, 98009 Miri, Sarawak, MALAYSIA

E-mail: sharulsham@curtin.edu.my

Abstract. The presence of wind tunnel is undoubtedly bringing infinite possibilities to studying and understanding complex fluid flows. However, commercial wind tunnel is expensive and only limited to highly-focus researchers or exclusive institutions. This paper discusses the design and development of a low-cost, educational-purposed, open-typed subsonic wind tunnel. In this work, an open-typed subsonic wind tunnel is designed with the aim of achieving turbulent intensity (in the working section) below or equal to 5%, within the budget of RM 1500 and a working speed of 6 m/s – 8 m/s to meet the Reynolds number in the order of 10^5. The conceptual design was studied using Ansys Fluent 14.5 and the optimal design was then developed and experimentally verified.

1. Introduction

The phenomenon of turbulent flow has been long studied and intrigued scientists and researchers. It was Dr. Price's [1] of Monash University belief that Leonardo Da Vinci was one of the first to be intrigued by turbulent flows. Due to his insightful observations, significant works were able to carry out easily and led to the development of the basic dynamics equations. Unfortunately, the knowledge of turbulent flows is too vast and is still fairly immature to date. There are limitations on numerical approach as it involves approximations and it lacks physical understanding to researchers. The invention of wind tunnels gave researchers an idea how flows behave in actual.

There are various different design types of wind tunnel for different applications as no single tunnel fits for all purposes. Wind tunnels can be classified based on air flow speed in test section or based on design. Open-typed wind tunnel and close-typed wind tunnel (see figure 1) are wind tunnels based on design while wind tunnels based on speed are subsonic, transonic, supersonic and hypersonic. An advanced commercial grade wind tunnel is very expensive and exclusively available only to research-focused universities.

[1] To whom any correspondence should be addressed.

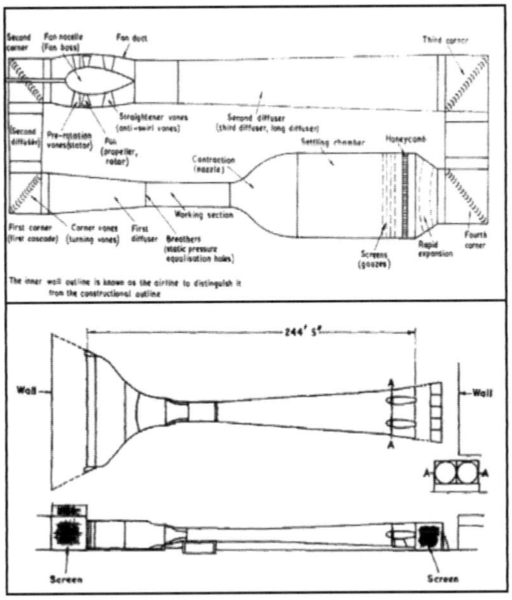

Figure 1. Closed Type Wind Tunnel proposed by National Physical Laboratory (NPL) (above) and Open Type Wind Tunnel of Indian Institute of Science (below) [2].

Thus, this project aims to design, develop, and construct a cost-effective open-circuit wind tunnel which is able to conduct small-scale experiments to visualize flow passing through fundamental objects for educational purpose. This wind tunnel mainly focuses on testing any scaled-down model, roughly giving a width based Reynolds number of the order of hundred thousand at approximately 5% blockage based on frontal area. A fourth to sixteenth times scale was deemed adequate for investigations of the underlying flow mechanisms and for basic research and study on simplified bluff and quasi-streamlined bodies. Turbulence intensity of less than 5% at working speed of 6 m/s to 8 m/s would be considered reasonable for testing without compromising the integrity of the results [2]. A 120 V gable fan was used to achieve the working speed. The budget for this project was allocated RM 1500.

2. Design methodology

This project can be divided into three components namely: Designing and Simulation, Construction, and Experimentation. The design of educational-purpose and cost-effective subsonic wind tunnel is verified and improved using Ansys Fluent 14.5 CFD simulation. In order to achieve the goal of designing a subsonic wind tunnel with working turbulent intensity of less than 5%, the relationship between the relevant parameters and components were studied. A detailed list of the parameters and the components are studied by [3].

All designs were simulated using Ansys Fluent 14.5 simulation software. The Computational Fluid Dynamics (CFD) model and setup can be referred to [3]. Table 1 show the final design of the subsonic wind tunnel which has a blockage ratio of 4.8% and a turbulence intensity of 2.72% - 3.39% at the flow speed of 6.28 m/s (see also figure 2).

Table 1. Dimensions of components for final conceptual design

Module	Width (m)	Height (m)	Length (m)	Area (m^2)	Area Ratio
Contraction Cone	0.81	0.81	0.610	0.656	7.28
Working Section	0.30	0.30	0.610	0.090	-
Diffuser	0.43	0.43	1.218	0.185	2.05

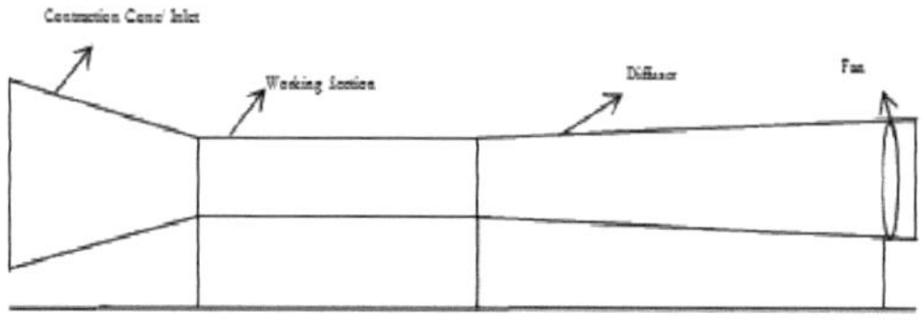

Figure 2. Schematic sketch of designed wind tunnel.

3. Wind tunnel construction

The wind tunnel was vastly of wood construction due to its affordable price compared to metal sheet or fibre glass of smoother surface but more expensive. Due to its nature of the wood, it is also relatively easy to handle for construction as well. The chosen material was standard 0.9 cm plywood. The working section had to be transparent to gain clear view of the internal of working section. Also, a working section is the most important part of the wind tunnel as experiment is carried out in the working section; thus, a smooth surface (see figure 3) was necessary to maintain the accuracy of results. A rough surface could induce unwanted development in boundary layer and stress and was experimentally demonstrated by Bradley [4]. At large surface roughness, the flow may be excessively clogged by the tunnel walls leading to increased pressure gradient and consequent acceleration of flow at the blockage. The blockage ratio of 5.0% - 5.3% is generally accepted as sufficiently low to avoid significant low wall interference effects [5].

Figure 3. Photograph of actual test section made from plexiglass.

3.1. Contraction cone

The contraction cone, with a contraction ratio of 7.28, which was in between the range of 6 and 9 proposed by Bradshaw [2] has a huge influence on working section turbulence intensity. A contraction cone is used to improve uniformity of the air flow and increases the mean velocity. The power factor contribution of screens in the settling chamber varies as $\frac{1}{c^2}$ [2 and 6]. However, if the contraction part of the wall along the flow direction is too large at certain points, the uniformity of velocity will be disturbed at the end of contraction cone [7]. It is therefore very important to achieve a perfect balance between the size of the test section and contraction ratio. The contraction cone was a perfectly straight without any curve as it was easier to construct with plywood. The trapezium shaped plywood were connected using 5 cm long, 0.4 cm diameter nails. The gap between joints (see figure 4) was filled with silicone glue to strengthen the structure and minimize exfiltration. The end which connect the cone to the test section was filled with rubber gasket and silicone glue (see figure 5 and 8).

Figure 4. Joint filled with silicone glue (top view).

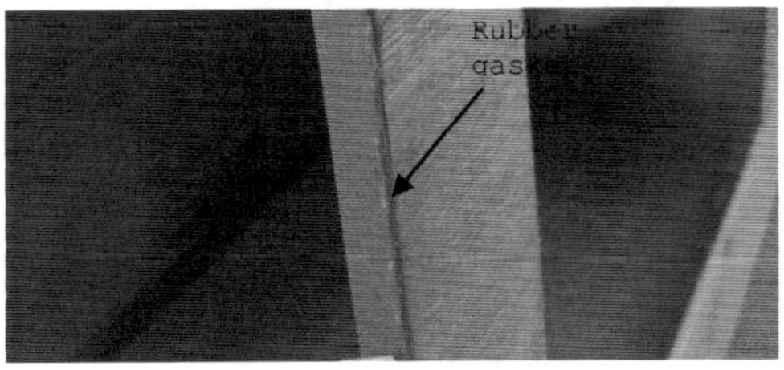

Figure 5. Rubber gasket adhere to the side of diffuser.

3.2. Honeycomb

A proper honeycomb is very expensive to acquire. The initial plan of improvising by using 1.2 cm diameter thick straw, which met the minimum requirements suggested by Metha [6] and Bradshaw [2] was time consuming to construct. Therefore, a wire mesh of 0.5 cm diameter was used as substitution (see figure 6) due to the consideration that a proper honeycomb was not accessible locally and would be expensive. The wire mesh was fitted to the contraction cone by stapler bullet. The performance of substituted honeycomb would be greatly reduced as it had no depth.

Figure 6. Contraction cone fitted with honeycomb made by mesh wire.

3.3. Diffuser

The diffuser is arguably the least important part in a suck down open circuit configuration because of its position at the downstream of test section. The diffuser (see figure 7) was a trapezium shaped with exit slightly larger than the entrance of an area ratio of 2.05 which was less than 2.5 [8]. The three components: contraction cone, test section and diffuser were assembled together by sitting on a frame (see figure 9) which provided a firm and stable structure for the wind tunnel. The assembly was as equally important as constructing the separated parts because if the structure was not strong and stable, the wind tunnel would vibrate when operating due to the vibration from the fan.

Figure 7. Trapezium shape diffuser made of plywood.

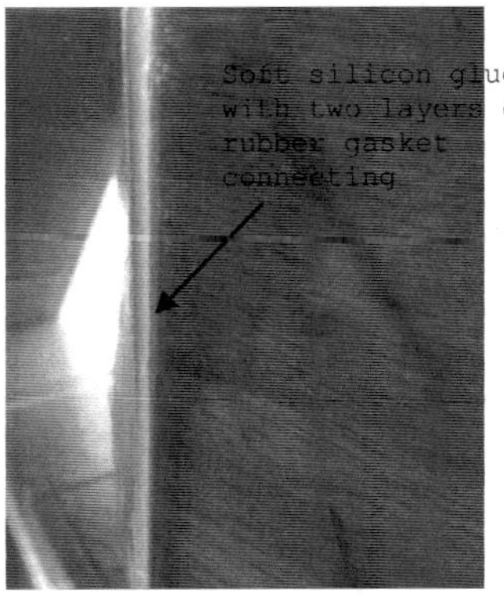

Figure 8. Joint between contraction cone and test section.

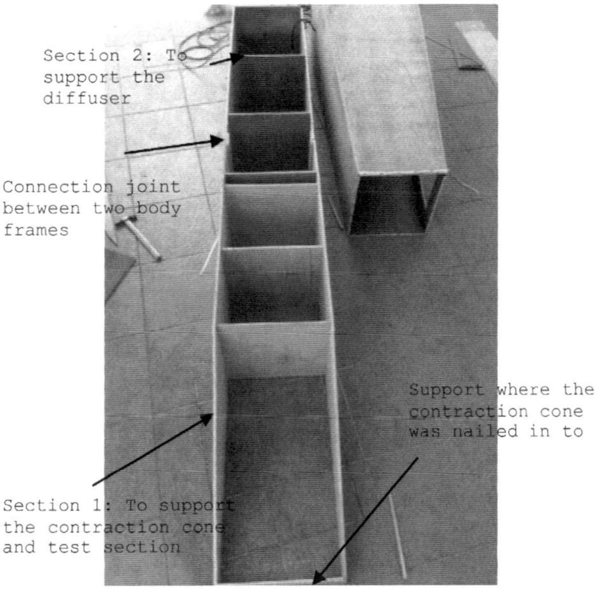

Figure 9. Body frame to support the wind tunnel.

4. Experiment

The prototyped wind tunnel had to be verified experimentally to know its specification. Due to the limited budget and equipment, a simple air flow anemometer accompanied with a data logger with RC-232C interface were used to check the oncoming wind speed. The air flow anemometer used has an accuracy of ± 3% at the range of 0.1 m/s - 45.0 m/s. The turbulent intensity is given in the equation below:

$$Tu = \frac{\sqrt{\overline{u^2}}}{U_{mean}} \qquad (1)$$

With the goal of achieving within 5% of turbulence intensity, the air flow anemometer was placed at two positions (see figure 10 marked 1 and 2) in the region of developed boundary layer over a consistent of 130 seconds.

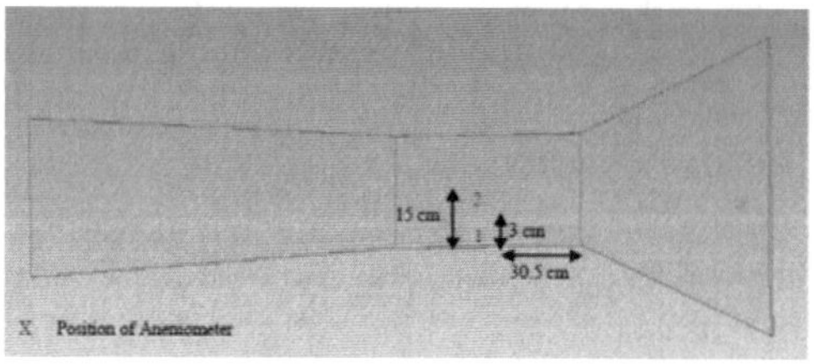

Figure 10. Position of anemometer.

Experiments were conducted for completed wind tunnel (figure 11) under 3 different conditions as shown in table 2.

Table 2. Detailed sets of experiments conducted

Experiments	Conditions	Positions (see figure10 for picture illustration)
1	Right after the fan was initiated	1
2	10 minutes after the fan was operated	1
3	10 minutes after the fan was operated	2

Figure 11. Completed wind tunnel.

5. Results and discussion

5.1. Simulation results

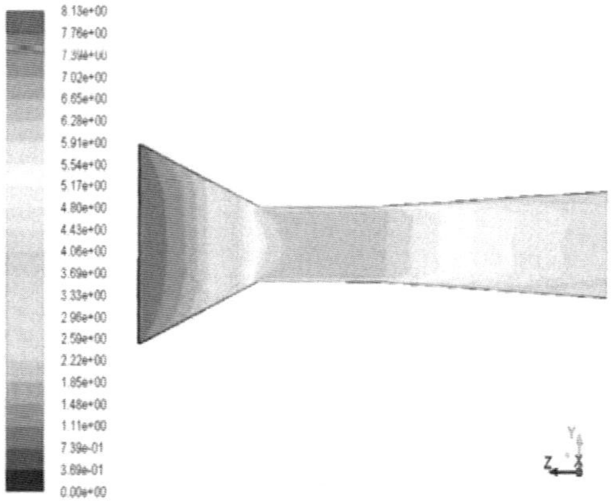

Figure 12. Contours of velocity magnitude for final conceptual design.

Figure 12 shows the simulation results of the final conceptual design, which simulation settings can be referred to [3]. The color bar indicates the velocity in m/s. It can be seen that the velocity at the working section has a uniform pattern peaked at 6.65 m/s and in the general region of 6.28 m/s. The velocity decreased gradually in the diffuser and satisfied the conversation of mass equation for subsonic flow.

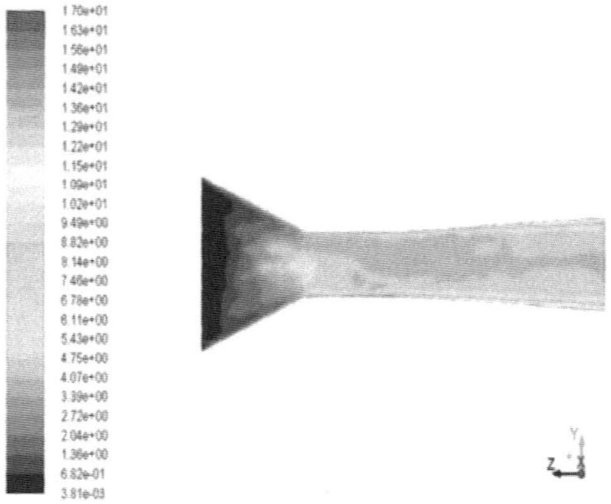

Figure 13. Contours of turbulence intensity for final conceptual design.

Figure 13 shows the simulation results of the final conceptual design. The color bar indicates turbulence intensity in percentage. The data shows the turbulence intensity within the working section ranged from approximately 2.72% - 3.39%. In consistence with the figure of velocity contours, the flow near wall was unstable due to the boundary layer thickness and skin friction. However, the usable area, which was approximately 3 cm off the working section floor, had a blockage ratio of 4.8% [3] in comparison to the 5.0% - 5.3% which was generally accepted as sufficiently low to avoid significant wall interference effects [5].

5.2. Experimental results

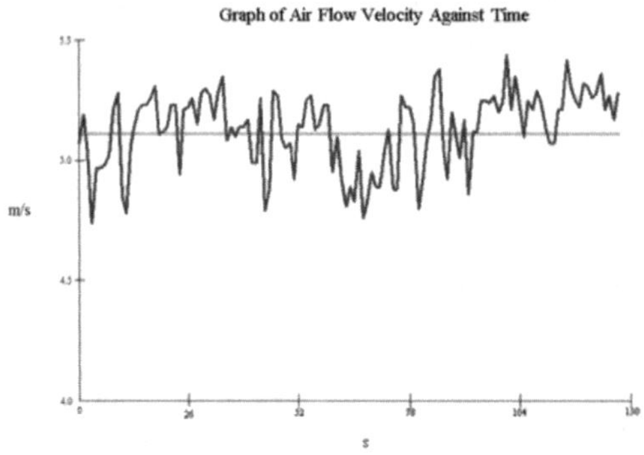

Figure 14. Graph of air flow velocity against time for experiment 1.

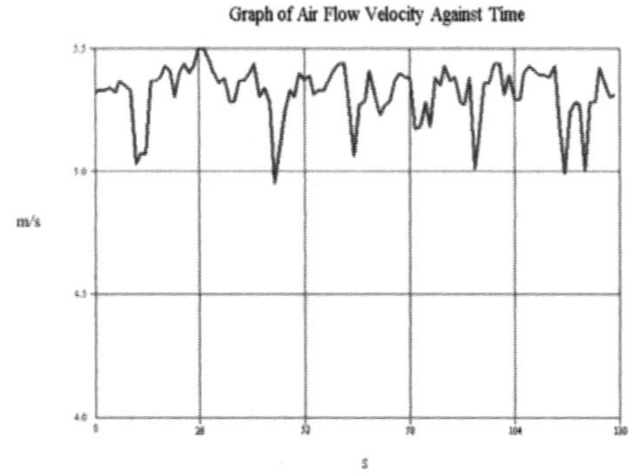

Figure 15. Graph of air flow velocity against time for experiment 2.

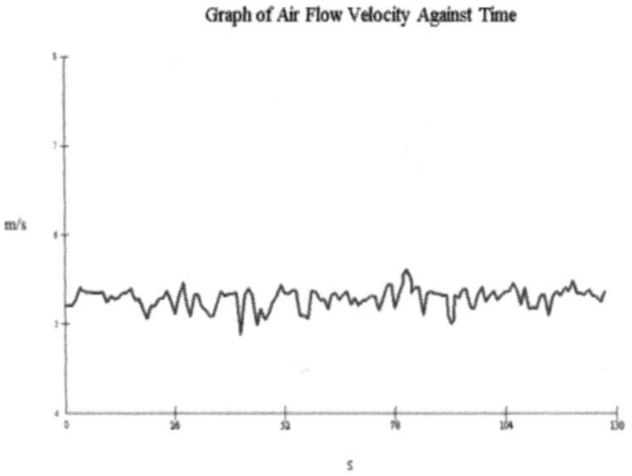

Figure 16. Graph of air flow velocity against time for experiment 3.

Figure 14 shows the graph of air flow velocity against time for experiment 1. The mean velocity was 5.12 m/s. The stream-wise turbulent intensity was 3.073 % at that position. Figure 15 shows the graph of air flow velocity against time at for experiment 2. The mean velocity was 5.32 m/s. The stream-wise turbulent intensity was 2.055% at that position. The mean velocity appeared to be slightly higher than in experiment 1 after the surrounding flow was steady. The flow inside the test section also appeared to be slightly more steady than in experiment 1 as the turbulence intensity of 2.055% was slightly less than 3.073%. Figure 16 shows the graph of air flow velocity against time for experiment

3. The mean velocity was 5.29 m/s. The stream-wise turbulent intensity was 2.028% at that position. The mean velocity appeared to be near the mean velocity in experiment 2 which was in position 1. The velocity appeared to be uniformly distributed in the boundary-layer-free region once the flow was steady.

Based on the experimental and simulation results, it is safe to say that they are close. The overall mean velocity in experimental data shows a 5.32 m/s in position 1 of condition 2 to 5.29 m/s in position 2. In contrast to that, the overall mean velocity generated by Fluent was 6.5 m/s (see figure 12). A velocity of ± 1.2 m/s was higher than experimental results. The reason could be due to the friction which was caused by wall surface roughness. Since the rough flow is dominated by the drag of the roughness elements over shear stress as compared to smooth flow where the shear stress is dominated by viscosity [9 and 10], the drag produced induces energy loss in the flow.

The turbulence intensity simulated by Fluent showed a region of 3.67% at the position 1 and a region of 2.93% at the upwards of position 2. The experimental results which were surprisingly lower than simulation results. This could be because the experimental results showed were only stream-wise turbulent intensity. The overall turbulent intensity should be added up from the longitudinal, transverse and vertical turbulence intensity instead of just stream-wise turbulent intensity as shown by the experimental results.

6. Construction cost

Table 3. Costs breakdown for materials bought

Materials	Price per unit (RM)	Total amount (RM)
Aluminum mesh wire 50 cm x 50 cm	15.00	15.00
Wington 12 inch gable fan	290.00	290.00
Silicone glue small	11.00	11.00
Wood glue small	10 .00	10.00
½ inch square x 3 ft PVC netting	13.00	39.00
Plexiglas 1cm thick (30 cm x 61cm)	155.00	310.00
Plexiglas 1cm thick (32 cm x 61cm)	155.00	310.00
Plywood 10 mm x 1 m x 2 m	95.00	745.00
1 inch adhesive double side tape	36.00	36.00
Total		1766.00

Table 3 shows the detailed cost of materials purchased to building the wind tunnel. An invoice of the purchase order can be referred to [3].

7. Conclusion

Several studies have provided brief guidelines on the relationships between the components and the reliability of the flow. The simulation model used was the *k-ε* model with the assumption of incompressibility. The materials used to construct the wind tunnel were mainly plywood and Plexiglas as the working section. The cost of the wind tunnel was relatively affordable despite the RM 266 more than the budget. The prototyped wind tunnel shows good agreement with the simulation results. The experimental results showed that the mean velocity was ± 1.2 m/s slower than simulation results but the turbulent intensity was as suggested by Fluent. However, due to the simplicity of the RC flow anemometer, only stream-wise turbulent intensity was tested. Overall this project has achieved its objective and was able to test small scale test specimen to study the flow around the object although if possible, is better to further refine the performance of the wind tunnel.

References

[1] Price D, Monaghan J, Valizadeh A and Tricco T 2014, Astrophysical Fluid Dynamics and MHD *http://moca.monash.edu/research/astrofluids.html*

[2] Bradshaw, Peter and Pankhurst RC 1964 The design of low-speed wind tunnels *Progress in Aerospace Sciences* **5** 1-69

[3] Yong T H 2013 Design and development of subsonic wind tunnel *Thesis (B.Eng) Curtin University* Malaysia

[4] Bradley E F 1968 A micrometeorological study of the velocity profiles and surface drag in the region modified by a change in surface roughness *Q. J. R. Meteorol. Soc.* 361-79

[5] Saltzman J and Theodore G A 1982 Review of flight-to-wind-tunnel drag correlation Edwin *Journal of Aircraft* **19(10)** 801-11

[6] Mehta R D and Bradshaw P 1979 Design rules for small low speed wind tunnels *Aeronautical Journal* **83(827)** 443-49

[7] Tsien H S 2012 On the design of the contraction cone for a wind tunnel *Journal of the Aeronautical Sciences* **10(2)**

[8] Arifuzzaman M and Mohammad M 2012 Design construction and performance test of a low cost subsonic wind tunnel *IOSR Journal of Engineering* **10** 83-92

[9] Benson J 2005 Boundary-layer response to a change in surface roughness *Thesis (Ph.D.) University of Reading* UK

[10] Dol S S and Nor M A M 2006 Flow visualization of the vortex shedding of a stationary circular cylinder by an improved smoke-wire technique *WSEAS Fluid Mechanics Transactions* **1(6)** 745-52

Design and Development of Low-Cost Water Tunnel for Educational Purpose

M Zahari and S S Dol[1]
Department of Mechanical Engineering, Curtin University
Sarawak Campus CDT 250, 98009 Miri, Sarawak, MALAYSIA

E-mail: sharulsham@curtin.edu.my

Abstract. The hydrodynamic behaviour of immersed body is essential in fluid dynamics study. Water tunnel is an example of facility required to provide a controlled condition for fluid flow research. The operational principle of water tunnel is quite similar to the wind tunnel but with different working fluid and higher flow-pumping capacity. Flow visualization in wind tunnel is more difficult to conduct as turbulent flows in wind dissipate quickly whilst water tunnel is more suitable for such purpose due to higher fluid viscosity and wide variety of visualization techniques can be employed. The present work focusses on the design and development of open flow water tunnel for the purpose of studying vortex-induced vibration from turbulent vortex shedding phenomenon. The water tunnel is designed to provide a steady and uniform flow speed within the test section area. Construction details are discussed for development of low-cost water tunnel for quantitative and qualitative fluid flow measurements. The water tunnel can also be used for educational purpose such as fluid dynamics class activity to provide quick access to visualization medium for better understanding of various turbulence motion learnt in class.

1. Introduction

Water tunnels are used for a variety of reasons such as testing a structure under flowing fluid, which affect the forces on the submerged body. Water tunnels are also used for preliminary measurements of prototypes (e.g. vehicles, buildings, towers and bridges) early design such as to conduct flow visualization around models. Water tunnels are generally preferred over wind tunnels for measurement like Particle image velocimetry (PIV). This is because flow visualization in wind tunnels is harder to control and it dissipates quickly [1].

However, a proper research-purposed and commercial water tunnels are too expensive and they are generally huge in size. For example, Model 1520 Water Tunnel (Hills Research Corporation) costs around hundred thousand USD and requires a large amount of lab space approximately 10 x 5 x 2 m^3 (length x width x height) as shown in figure 1 [2]. The velocities at test section vary from 0 up to 0.3 m/s and contain approximately 3785 litres of water. Due to the high cost of commercial water tunnel, it leads to the idea of fabricating a low-cost water tunnel with acceptable quality for the educational purposes.

[1] To whom any correspondence should be addressed.

Figure 1. Model 1520 Water Tunnel [2].

The present water tunnel is designed to focus on visualizing the effects of vortex-induced vibration (VIV) on the submerged structure. The structure, which is a circular cylinder experiences vortex shedding from turbulent flow. In order to achieve the goal of designing and constructing of an open-typed water tunnel for research and educational purposes, it is carefully built to provide a uniform flow at the test section for wide range of Reynolds numbers. Another aim is to minimize the cost of construction and recognize the capabilities of water tunnel in solving practical fluid flow engineering problems.

2. Water tunnel design

This section discusses on the design of open channel water tunnel that can be used to study the turbulent flow phenomena such as flow characteristics of VIV on immersed body. The water tunnel as shown in figure 2 is designed by using the SolidWork2012 and AutoCAD2012 softwares. The design is based on modifications of water tunnels from the previous studies [3-4]. The aim of this water tunnel is to provide a steady and uniform flow speed within data collection area. The term steady implies no change at a point with time while the term uniform implies no change with location over a specified region [5]. Besides, this water tunnel is designed to provide a similar flow speed of ocean current in the range of 0.5 to 1.2 m/s [6] for the VIV application.

2.1. Glass tank

In this water tunnel design, a rectangular glass tank which is made from 0.001 m thickness of glass plate with a dimension of 1.8 x 0.7 x 0.7 m^3 (length x width x height) is used to hold a large amount of water. Glass material is chosen due to its high hardenability and transparent surface. A transparent glass plate is important for flow visualization purpose.

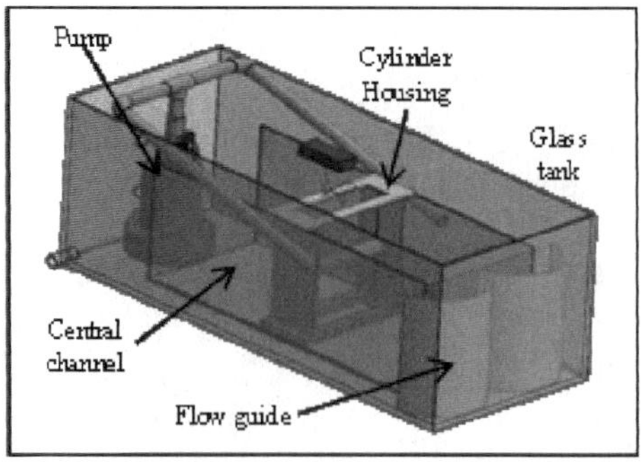

Figure 2. Open channel water tunnel designed using SolidWork software.

2.2. Central channel
A central channel, which is made from 1.3 x 0.5 m^2 Polyvinyl chloride (PVC) board plates, is placed at the centre of glass tank for data collection area. PVC is chosen because it is easy to construct and cheaper in price. Besides, its smooth surface can reduce friction of water flow and increase flow effectiveness. At the central channel entrance, screen which is made up from 1 cm^2 wire mesh is placed to provide uniform flow (figure 3). Screens have been used to minimize the intensity of the oncoming turbulence and improve the mean flow uniformity [7].

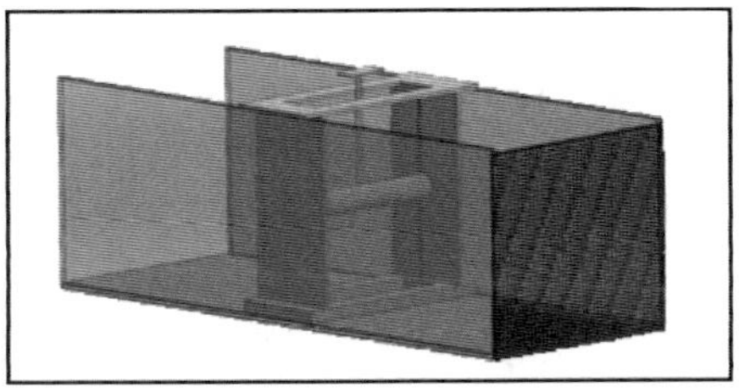

Figure 3. Screen positioned at the central channel entrance.

2.3. Flow guides
Two clear PVC flow guides with the radius of 17 cm as shown in figure 4 is placed at the end of the glass tank to join the water flow from both side of water tunnels and direct it through the central channel. The aim of this flow guides is to provide a steady flow within the central channel.

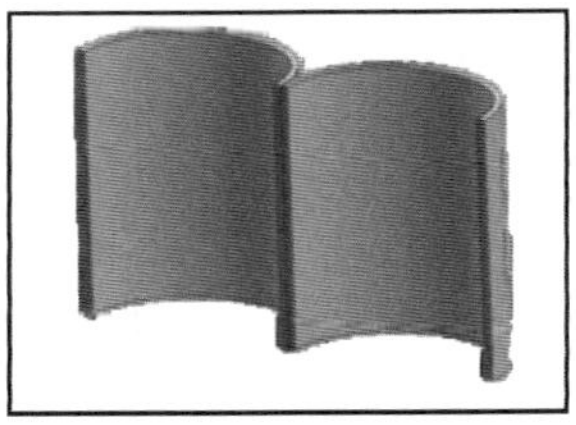

Figure 4. Flow guides.

2.4. Pump and piping systems
A proper water pump and piping systems (figure 5) is essential in designing water tunnel in order to produce adequate flow rates. This water tunnel is designed with a submersible pump as it can provide higher water flow rates. Other benefits of submersible pump are listed as the following:

- Submersible pump is completely watertight, and risks of leakages and electric contacts are eliminated.
- Submersible pump uses no suction to get the water through to the pipes thus the pressure is optimized.
- Submersible pump minimizes the pump cavitation problems compared to the centrifugal pump [8].

The piping system is designed to fit with the submersible pump. From the discharge of the pump, the water flows upward through a vertical PVC pipe with the diameter of 2 inch pipe. A valve is placed in the between of vertical PVC pipe to allow the adjustment of water flow rates. Then, the water flow is separated at the top using the t-shape pipe fitting and directs it along the side of water tunnel using 90o elbow.

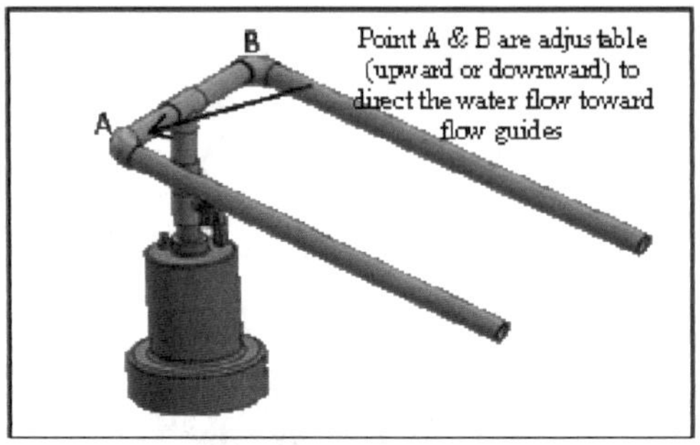

Figure 5. Submersible pump and piping systems.

2.5. Cylinder housing

The cylinder housing which is made from thick PVC plates is designed to place the testing cylinder at any location within the channel (figure 6). The top of cylinder housing is cut out to allow for cylinder visibility and space for measurement platform to oscillate. The cylinder housing will be placed within the channel and enable to move backward and forward along the channel to find a suitable data collection position. Hooks are connected with the cylinder housing to allow the attachment of springs and cylinder.

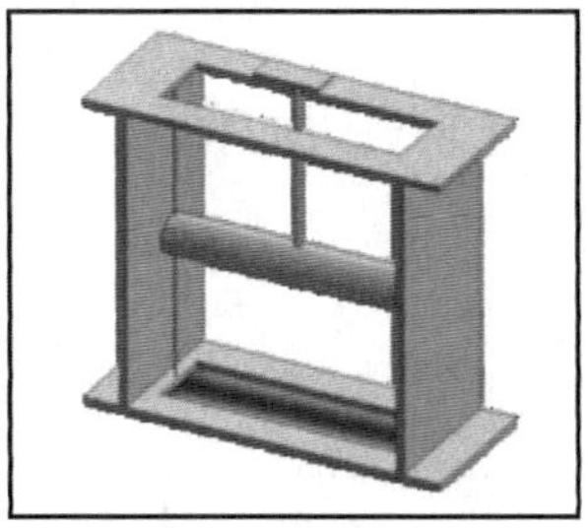

Figure 6. Cylinder housing.

The completed parts of water tunnel are assembled together for the clear view of complete water tunnel design as shown in figure 7.

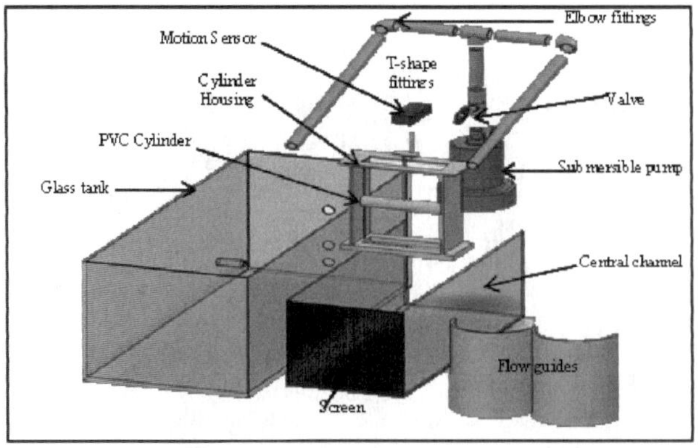

Figure 7. Exploded view of water tunnel.

IOP Conf. Series: Materials Science and Engineering **78** (2015) 012040 doi:10.1088/1757-899X/78/1/012040

3. Water tunnel construction

3.1. Glass tank

This glass tank has a rectangular shape and the most expensive part of the water tunnel. The glass tank plates are assembled using silicone sealant to prevent leaking on tank. For the safety reason, the edges of glass plates are sanded.

3.2. Central channel

The central channel which is made up from PVC boards is glued together to form a rectangular shape. Stapler bullets are used to attach the screen to the central channel as shown in figure 8. Near the edges of the central channel, the PVC board is drilled and the threaded steel rod is fixed inside of each holes. The threaded steel rod is bolted by the nuts to the board in order to ensure the position of central channel is placed centrally inside the glass tank.

Figure 8. Screen fitted at the central channel entrance.

3.3. Flow guides

Two flow guides are shaped from clear PVC plates to form semi-circular shape with the radius of 17 cm. These flow guides are placed at the end of the glass tank to join flow from both side channels and direct it through the central channel as shown in figure 9. Flow guides are attached to the glass tank using silicone sealant and dried in 2-3 days to ensure the strength of the flow guides to reflect the incoming flow from both sides of the tunnel to the central channel.

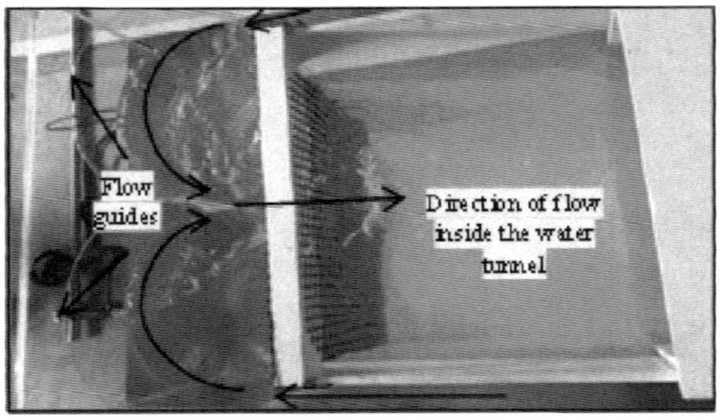

Figure 9. Flow guides are placed inside the water tunnel.

3.4. Pump and piping systems

In this project, submersible pump is used to provide the water flows inside the water tunnel. The pump flow rate is rated at 260 liter/minute with the output of 50 Watt. Thus, a single pump is sufficient to

IOP Conf. Series: Materials Science and Engineering **78** (2015) 012040 doi:10.1088/1757-899X/78/1/012040

provide the required flow rate 0.5 m/s within the data collection area. A valve is connected with the pump to allow the adjustment of the flow rates during measurement. Two elbows, a t-shape fitting, 2" and 1" inches pipes are used for the piping systems. All the fittings and valve are joined with pipes using the PVC glue, except on the elbow fittings to allow the adjustment of the flow direction toward the flow guides.

3.5. Cylinder housing

The cylinder housing with the height of 0.5 m is formed from four plates of thick PVC, joined together to form a square shape on the front view. It is constructed to place the testing cylinder at any location within the channel. On the top of cylinder housing, a 20 cm x 10 cm rectangular hole is cut to allow movement of measurement platform which is attached with the testing cylinder. The cylinder is oriented horizontally with four springs which is connected in this cylinder housing and created "H" shape. All the parts and equipment is assembled together as shown in figure 10.

Figure 10. Assembled open channel water tunnel.

4. Material and construction costs

One of the project objectives is to fabricate the water tunnel with the least cost possible so that the facility becomes affordable for all educational purposes without sacrificing its quality. Table 1 summarizes the overall materials and parts used in this project.

Table 1. Summary of material and construction costs

Items	Amount (RM)
1 cm thick glass tank	1500.00
PVC central channel	200.00
260 l/ min submersible pump	700.00
Piping systems (pipes, elbow, t-shape fitting)	100.00
Cylinder housing	150.00
Flow guides	50.00
Motion sensor	200.00
Springs (various size)	80.00
Others (steel rods, screws, bolts, glue, etc.)	200.00
Manufacturing cost	1000.00
TOTAL	4180.00

The total cost of this water tunnel is RM 4180.00 and it is relatively reasonable compared to the commercial water tunnel.

5. Flow velocity measurement

The flow within the data collection area needs to be steady and uniform for controlled condition measurement. For verification, flow measurement has been conducted. Flow meter is placed into the central channel, approximately 50 cm away from the flow guides and the flow sensor blade is parallel to the direction of flow. Open channel flow meter is used to determine the water velocity inside the central channel. Figure 11 and 12 below show the positions of flow velocity along the water depth which are measured inside the water tunnel.

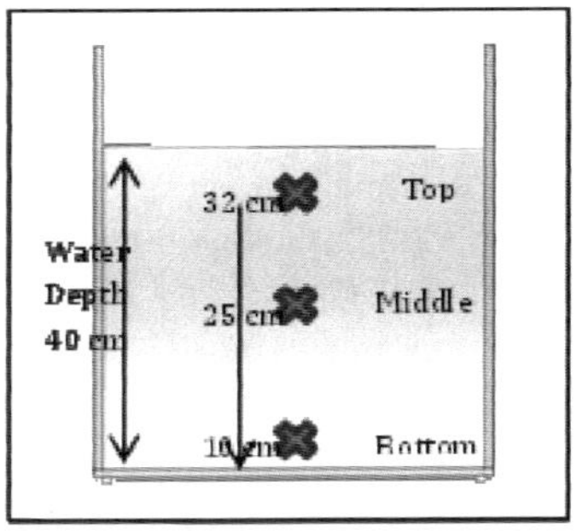

Figure 11. Position of measured water velocity.

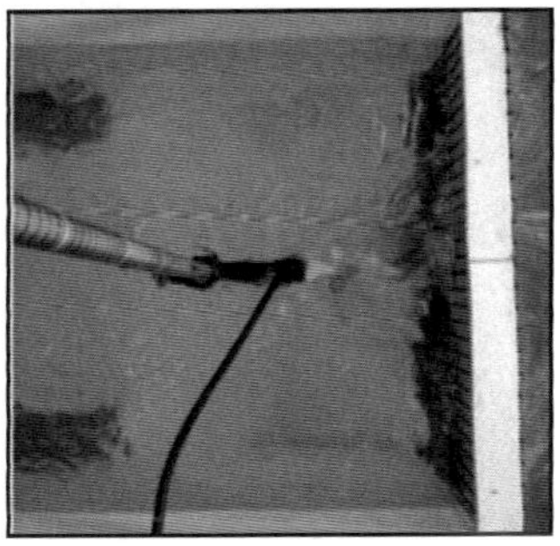

Figure 12. Open channel flow meter.

Figure 13 showed graphs of 10 samples of flow rate results. It shows the speed variation is small and within the acceptable region. Moreover, the flow characteristic of this water tunnel is similar to the characteristic of steady flow. This is due to the small different of flow speed changes with time for each trial. Top position gives a better result compared to the middle and bottom positions, indicating uniform flow. As the small different of flow speed along the length of the channel, it suggests the characteristic of this flow is nearly similar with the characteristic of uniform flow [5].

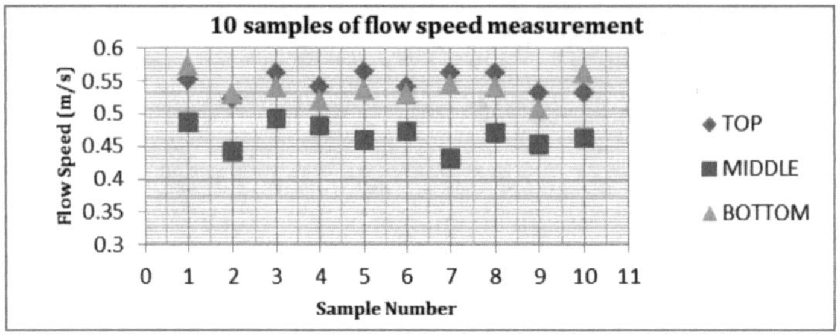

Figure 13. Graph of sample number versus velocity.

6. Conclusion

The flow measurement results show that the mean flow velocity is in the interest range of ocean current flow which is 0.5 m/s for the vortex-induced vibration measurements. The flow characteristic of this water tunnel is nearly similar with the characteristic of steady and uniform flow. The cost of this water tunnel is very reasonable, which is about four thousands Ringgit Malaysia (RM). Overall, this water tunnel design and development has achieved its objective and has a capability to study the flow around small scale of fundamental structures and behavior of vortex induced vibration (VIV).

References

[1] Dol S S and Nor M A M 2006 Flow visualization of the vortex shedding of a stationary circular cylinder by an improved smoke-wire technique *WSEAS Fluid Mechanics Transactions* **1(6)** 745-52

[2] Rolling Hills Research Corporation 2014 *Research Water Tunnels Specification* USA

[3] Hennessey T V 2005 The design and construction of an open channel recirculating water tank for the study of biological hydrodynamics *Thesis (B.S) Massachusetts Institute of Technology* USA

[4] Hall-Stinson A, Lehrman C and Tripp E 2011 Energy generation from vortex induced vibration *Thesis (B.S.) Worcester Polytechnic Institute* USA

[5] Cengel Y A and Cimbala J M 2006 *Fluid Mechanic: Fundamental and Application* (USA: McGraw-Hill International)

[6] Omar Y, Tengku T and Mohamad A 2006 Prospects for ocean energy in Malaysia *International Conference on Energy and Environment*

[7] Mehta R D 1985 Turbulent boundary layer perturbed by a screen *International Journal of American Institute of Aeronautics and Astronautics* **23(9)** 1335-42

[8] Brennen C 1978 The Bubbly flow model for the dynamic characteristics of cavitating pump *Journal of Fluids Mechanics* **89(2)** 223-40

The Effects of Spring Stiffness on Vortex-Induced Vibration for Energy Generation

M Zahari, H B Chan, T H Yong and S S Dol[1]
Department of Mechanical Engineering, Curtin University
Sarawak Campus CDT 250, 98009 Miri, Sarawak, MALAYSIA

E-mail: sharulsham@curtin.edu.my

Abstract. Vortex-induced vibration (VIV) is the turbulent motion induced on bluff body that generates alternating lift forces and results in irregular movement of the body. VIV-powered system seems a good idea in greening the energy sector and most importantly is its ability to take advantages of low current speed of water to generate electricity. This paper aims to investigate the effects of spring stiffness on the characteristic of VIV. The study is important in order to maximize these potentially destructive vibrations into a valuable resource of energy. Five cylinders with the range of 0.25 to 2.00 inch diameter are tested to study the behavior of VIV. Results from this experiment indicates that, the 2.0 inch cylinder gave the lowest error in frequency ratio which is 1.1% and have a high potential of lock-in condition to occur. In term of maximum amplitude, this cylinder gave the highest amplitude of oscillation motion that is equal to 0.0065 m.

1. Introduction

Renewable energy is an energy that originates from resources which is naturally replenished and continuously exists in the environment. Unlike the fossil fuels, renewable energy resources generate energy with no pollution and global warming emissions. It is a key in greening the energy sector by shifting the usage of fossil fuels to renewable energy in order to minimize the carbon-dioxide emissions and effects of global warming from burning of fossils fuels. However, alternative energy sources technologies exist currently such as solar, wind and hydro energy are unable to meet the demand of perpetually and ecological sources of energy. For instance, solar and wind energy are strongly reliant on the weather condition and work only where there is good exposure to sunlight [1]. Similarly, hydro energy can be generated only when there is abundant amount of flowing water. Besides, energy based on wind turbines does have an environmental cost as their effect on migratory paths of birds which could be killed by the rotating blades of wind turbines [2].

Ocean which is known as the world's largest storage of renewable energy seems an ideal alternative source of energy as it can offer an unlimited amount of clean energy [3]. However, the limitation of this source of energy is it low ocean current speed to generate energy. This is because ocean technologies based on tides and currents turbines can only work for current flow stronger than 2

[1] To whom any correspondence should be addressed.

m/s [4]. Thus, newer technology needs to be developed to efficiently convert this slow moving water into valuable energy.

One possible solution that meets these criteria is a hydroelectric power extraction system based on Vortex Induced Vibration (VIV) which can generate energy in low speed. VIV is a turbulence motions induced on bluff body that generates periodic irregularities lift forces on the body and pushing it up and down perpendicular to fluid flow [5]. Usually, engineers have been trying to suppress the presence of VIV because of it disruptive power on vibrating structures. A dramatic example of their disruptive power occurred on the Tacoma bridge that collapsed in 1940 due to the present of VIV. In this research, instead of suppressing the effect of VIV, these potentially destructive vibrations will be maximized and exploited in order to transform these vibrations into a valuable resource of energy [6].

Harnessing the phenomenon of vortex-induced vibrations (VIV) is originally designed by Professor Michael Bernitsas from the University of Michigan in 2005. The concept has been named Vortex-Induced Vibrations for Aquatic Clean Energy (VIVACE), and it works by mounting a cylinder horizontally in a fixed frame to allow an alternating movement of oscillation body in order to turn these potentially destructive vibrations into renewable and clean energy [6]. Based on previous study, VIV-powered system has an ability to generate power in the range of current speed 0.26 to 2.6 m/s [6]. Thus, this power generation system is suitable in generating alternative energy on slowing moving water as the typical current ocean are within the range of 0.41 to 1.18 m/s [7].

To the authors' best knowledge, the study still lacks fundamental reasoning as it focuses more on the application side. This paper aims to investigate the effects of spring stiffness on the behavior of VIV for further enhancement of energy generation system. When the vortex shedding frequency approaches the natural frequency, amplitudes of bodies' oscillation grow and start forming a strong interaction with the shedding mechanism in the flow [8]. This phenomenon is known as "lock-in" condition which is similar to linear resonance. Therefore, the stiffness of a spring is significant parameter in this research as it affects the natural frequency for controlling structural resonance. Thus, proper understanding on the effects of spring stiffness is studied.

2. Literature review

2.1. Reynolds number

The range of Reynolds number, *Re* targeted in this research are in the range of $300 < Re < 3\text{x}10^5$ which is known as "Fully Turbulent Vortex Street" regime for the natural low-flow [6]. *Re* can be determined using the equation (1):

$$Re = \frac{UD}{v} \tag{1}$$

2.2. Strouhal number

Another vital parameter that describes the vortex shedding frequency to the oscillating flow mechanism in this research is known as Strouhal number, *St* as shown in equation (2):

$$St = \frac{f_s D}{U} \tag{2}$$

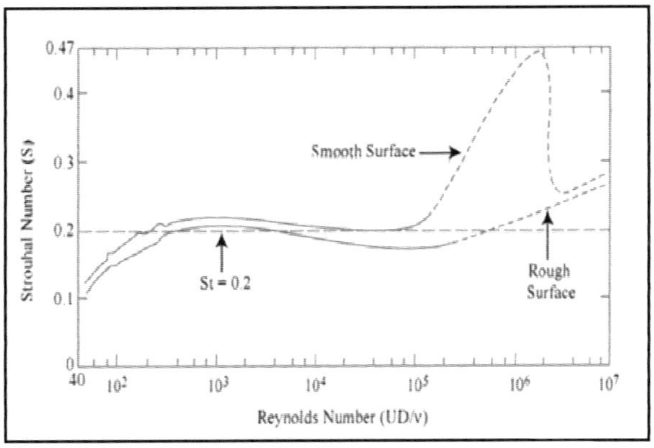

Figure 1. Strouhal number vs. Reynolds number [9].

Figure 1 shows that *St* is almost 0.2 for smooth surfaces within the range of $300 < Re < 10^5$, which matches well to the fully developed turbulent vortex street. Therefore $St = 0.2$ will be used as a constant value in any calculations for this research [9].

2.3. Lock-in phenomenon

In order to provide strong amplitude of the body oscillation, the natural frequency of the system must be synchronized with the shedding frequency. This phenomenon is known as lock-in condition. It is important to find the range of shedding frequency that matches with natural frequency, f_n to design energy harnessing device. The natural frequency of the body can be calculated using equation (3), where k is spring stiffness and m_{app} is mass applied to the cylinder:

$$f_n = \frac{1}{2\pi}\sqrt{\frac{k}{m_{app}}} \tag{3}$$

Mass applied, M_{app} for testing cylinder properties is determined using equation (4), (5) and (6):

$$\text{Volume} = \frac{\pi}{4} \times D^2 \times L = 8.596 \times 10^{-4}\ m^3 \tag{4}$$

$$M_{dis} = \rho_{water} \times \text{Volume} = 0.857\ kg \tag{5}$$

$$M_{app} = M_{pipe} + M_{dis} + M_{add} = 1.145\ kg \tag{6}$$

Then, the value of natural frequency obtained, the frequency ratio, f^* can be determined using equation (7):

$$f^* = \frac{F_s}{F_n} \tag{7}$$

According to [10], lock-in condition can occur when the value of frequency ratio is close to 1 and within the ± 30 % lock-in range. Another parameter that determined the range of lock-in condition is known as reduced velocity as shown in equation (8):

$$U^* = \frac{U}{D f_n} \qquad (8)$$

Shedding frequency can lock in and shift to match the natural frequency within the range of ($3 < U^* < 8$) [11].

3. Theoretical calculation

This section will express the result of spring stiffness for one testing cylinder based on theoretical calculation. The cylinder and flow parameters examined are listed in table 1. Example calculation is shown using cylinder E.

Table 1. Cylinder and water parameters

	Cylinder				
Label	A	B	C	D	E
Diameter [m]	0.0213	0.02683	0.0339	0.0486	0.0604
$Mass_{pipe}$ [kg]	0.0817	0.10357	0.1205	0.1944	0.2778
$Mass_{add}$ [kg]	0.0105 [Board apparatus]				
Length [m]	0.30 m				
	Water				
Velocity [m/s]	0.453 m/s (using data from water tunnel flow speed experiment) [12]				
Kinematic viscosity	$9.0 \times 10^{-7} m^2/s$				

The *Re* for cylinder E is given by equation (1):

$$Re = \frac{UD}{v} = 3.04 \times 10^5$$

For lock in phenomenon to occur it is an essential to match the vortex shedding frequency with natural frequency to enlarge the amplitudes of bodies' oscillation. The vortex shedding frequency, f_s will be determined by the flow condition and cylinder size as shown in equation (2):

$$f_s = \frac{StU}{D} = \frac{1.501}{s}$$

In order calculate the value of natural frequency, f_n of the oscillation body, the value of the spring stiffness which is equal to 100 N/m will be used to match the natural frequency to the shedding frequency. Mass applied, M_{app} for testing cylinder properties is determined using equation (4), (5) and (6):

$$\text{Volume} = \frac{\pi}{4} \times D^2 \times L = 8.596 \times 10^{-4}\ m^3$$
$$M_{dis} = \rho_{water} \times \text{Volume} = 0.857\ kg$$
$$M_{app} = M_{pipe} + M_{dis} + M_{add} = 1.145\ kg$$

Then, the value of natural frequency can be determined using equation (3):

$$f_n = \frac{1}{2\pi}\sqrt{\frac{k}{m_{app}}} = \frac{1.487}{s}$$

Based on the value of natural frequency obtained, the frequency ratio, f^* can be determined using equation (7):

$$f^* = \frac{F_s}{F_n} = \frac{1.501}{1.487} = 1.01$$

The frequency ratio obtained is equal to 1.01, which is well within the ± 30 % lock-in range [10].

In order to validate the range of lock-in condition on the testing cylinder, the value of reduced velocity is calculated using equation (8):

$$U^* = \frac{U}{D f_n} = 4.997$$

It showed that, shedding frequency for cylinder E can achieve the lock-in and shift to match the natural frequency as the value of reduced velocity falls within the range of $(3 < U^* < 8)$.

4. Experimental set-up

The goal of this paper is to study the effects of spring stiffness on vortex-induced vibration. The amplitude of testing cylinder is studied in order to validate the theoretical result obtained. In order to vary the mass applied on the cylinder, different diameters of cylinder are used in this experiment. All cylinders are made of PVC pipes with diameter of 0.25, 0.5, 1.0, 1.5 and 2.0 inch are labelled as A, B, C, D and E respectively. A constant value of spring stiffness which is 100 N/m will be used in this experiment in order to investigate the relationship between amplitude of cylinder and frequency ratio. The experiment is conducted in the low-turbulence free surface in the water tunnel constructed. In order to measure the amplitude of the oscillating cylinder, the Vernier displacement sensor is used to measure the displacement of the cylinder during experiment. Diameter of cylinder is measured using caliper and mass is measured using an electronic scale.

5. Results and discussion

5.1. Theoretical results

Based on calculation obtained, the results of theoretical values are tabulated in table 2.

Table 2. Theoretical results calculation

	Cylinder				
Label	A	B	C	D	E
Reynolds Number, *Re*	10710	13511	17068	24504	30401
Shedding frequency, f_s	4.265	3.381	2.677	1.864	1.501
Required spring stiffness, *k* [N/m]	142	128	113	104	102
Used spring stiffness, *k* [N/m]			100		
Natural frequency, f_n	3.609	3.022	2.540	1.843	1.487
Frequency Ratio, f^*	1.182	1.119	1.054	1.012	1.009
Reduced Velocity, U^*	5.908	5.594	5.269	5.058	4.997

Based on the results calculated in table 2, it showed that, natural frequency of the cylinder generally decreases with an increasing of mass applied on the cylinder [13]. A large mass causes the natural frequency to be small because of the larger inertia were produced on the system. This will increase the time of cylinder oscillation to return to the midpoint from a displaced position, so the period is increased and the frequency is reduced.

From the result calculated, all cylinders gave the values of frequency ratio within the lock in phenomenon range. Cylinder E showed the lowest error in frequency ratio which is 1.1 %. While, cylinder A showed the highest error in frequency ratio which is 15.37 %. The highest result on cylinder A is due to the high different of spring stiffness used which is 100 N/m compared to the required spring stiffness which is 142 N/m.

However, lock–in is most likely to happen for all of these cylinders as the frequency error is less than 30%. Besides, the entire cylinder tested were achieved at U^* between 3 and 8 which fulfill the lock-in theory. The relationship between frequency ratio and reduced velocity for each cylinder is shown (figure 2).

Based on the graph plotted, it clearly showed that as the frequency ratio closes to unity, the reduced velocity will be collected at the center of the lock-in range. As the frequency ratio is increased, the values of reduced velocity started to move away from the center of lock-in range. In this research, cylinder E most likely will create the large amplitude of oscillation due to the closeness of its natural frequency with shedding frequency which results in frequency ratio close to 1. Thus, it showed that, the potential of lock-in to occur is increased with the decrease of frequency error of the cylinder.

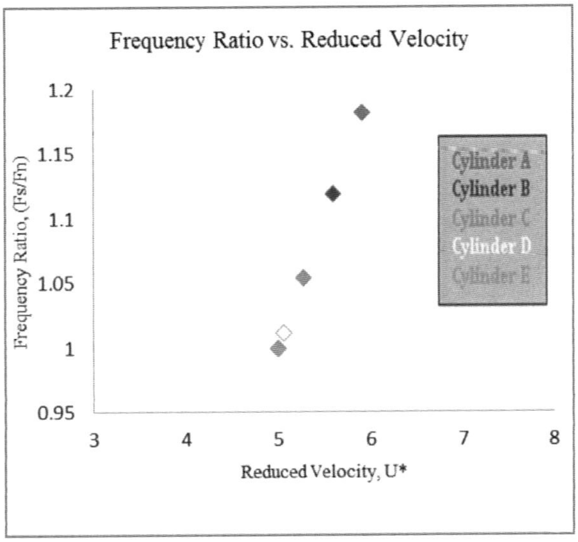

Figure 2. Frequency ratio vs. Reduced velocity.

5.2. Experimental results

In order to validate the results obtained in table 2, the corresponding cylinder amplitude is measured using motion sensor to give an overview of characteristic of VIV through stiffness of spring. The result of the cylinder amplitude was shown in (figure 3) [12].

Based on the figure 3, the maximum amplitude obtained by cylinder E is equal to 0.0065 m while Cylinder A gave the lowest maximum amplitude that is equal to 0.0035 m. The highest value of amplitude obtained by Cylinder E is due to the closeness of its natural frequency with shedding frequency which results in frequency ratio close to 1 and increase the possibility of lock-in condition to occur. Thus, it showed that the results obtained on the theoretical calculation were proven.

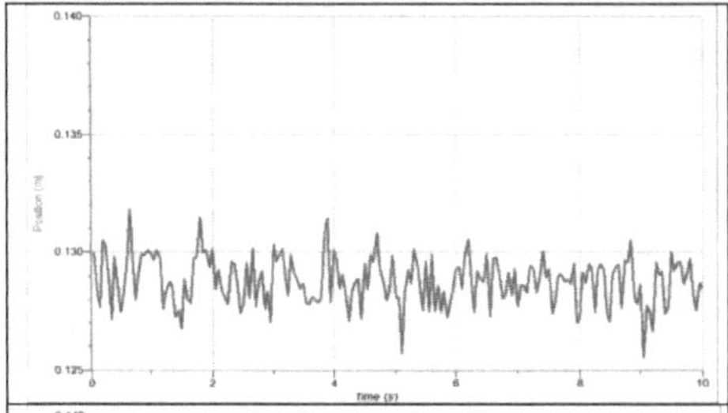	*Cylinder A* *Maximum Amplitude: 0.00315 m*
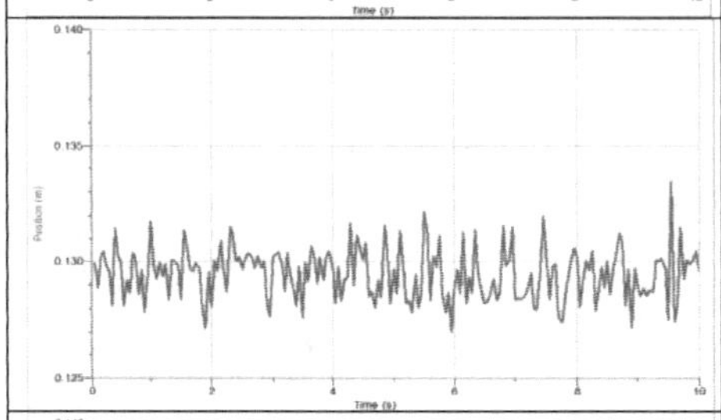	*Cylinder B* *Maximum Amplitude: 0.00325 m*
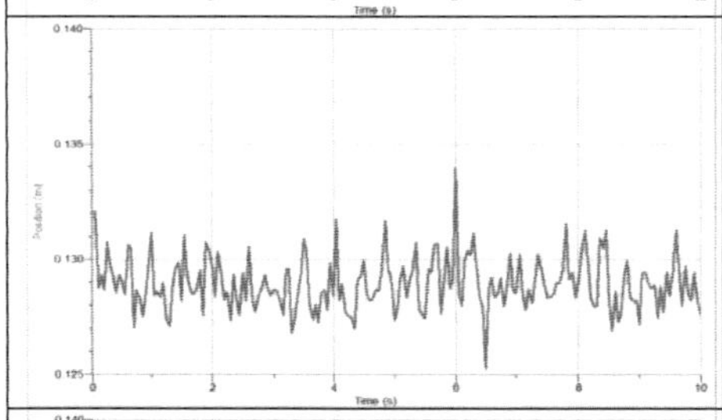	*Cylinder C* *Maximum Amplitude: 0.00430 m*
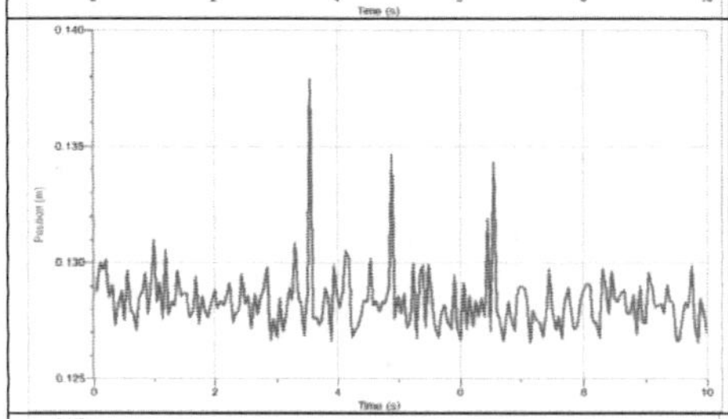	*Cylinder D* *Maximum Amplitude: 0.00570 m*
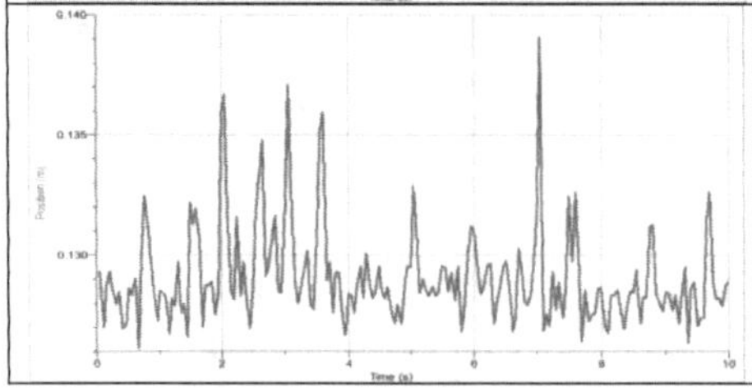	*Cylinder E* *Maximum Amplitude: 0.00650 m*

Figure 3. Amplitude profile of different sizes of cylinder

IOP Conf. Series: Materials Science and Engineering **78** (2015) 012041 doi:10.1088/1757-899X/78/1/012041

6. Conclusion

The VIV application in generating alternative energy is a viable solution of the current energy extraction problem for daily application. In this paper, the effects of spring stiffness were investigated theoretically and experimentally with the range Reynolds numbers (300 < *Re* < 300,000). Based on this research, it can be concluded that, the stiffness of spring is a vital parameter that affects the natural frequency of the oscillation body for the lock-in condition to occur. Results were shown on the Cylinder E which gave the highest maximum amplitude when it frequency ratio is close to 1. As the frequency ratio is close to 1, the possibility of lock-in condition to happen will increase. Then, amplitudes of bodies' oscillation grow and start forming a strong interaction with the shedding mechanism in the flow.

Nomenclature

Re	Reynolds Number
St	Strouhal Number
U	Current Speed, m/s
D	Diameter of Cylinder, m
v	Dynamic Viscosity, Pa.s
f_s	Shedding Frequency, Hz
f_n	Natural Frequency, Hz
U^*	Reduced Velocity
k	Stiffness of Spring, N/m
ρ_{fluid}	Density of Fluid, kg/ m^3
$\rho_{cylinder}$	Density of Cylinder, kg/ m^3
M_{app}	Mass of Application, kg
M_{pipe}	Mass of Cylinder, kg
M_{dis}	Mass of Fluid Displaced, kg
M_{add}	Additional Mass, kg
f^*	Frequency Ratio
g	Gravitational Force, m/ s^2

IOP Conf. Series: Materials Science and Engineering **78** (2015) 012041 doi:10.1088/1757-899X/78/1/012041

References

[1] Goffman E 2008 Why Not the Sun? Advantages of and Problems with Solar Energy *Journal of ProQuest Discovery Guides*

[2] Kuvlesky W P, Brennan L A, Morrison M L, Boydston K K, Ballarard B M and Bryant F C 2007 Wind energy development and wildlife conservation: Challenges and opportunities *Journal of Wildlife Management* **71(8)** 2487-98

[3] Sharma R C and Sharma N 2013 Energy from ocean and scope of its utilization in India *International Journal of Environmental Engineering and Management* **4**, 397-404

[4] Commission of the European Commission DGXII 1996 Wave energy project result: The exploitation of tidal marine current *Report EUR16683EN*

[5] Sarpkaya T 2004 A critical review of the intrinsic nature of vortex-induced vibrations *Journal of Fluids and Structures*

[6] Bernitsas M M, Raghavan K, Ben-Simon Y and Garcia E M 2006 VIVACE (Vortex induced vibration aquatic clean energy): A new concept in generation of clean and renewable energy from fluid flow *Journal of Offshore Mechanics and Arctic Engineering*

[7] Omar Y, Tengku T and Mohamad A 2006 Prospects for ocean energy in Malaysia *Journal of International Conference on Energy and Environment*

[8] Antiohos A A, Semercigii S E and Turan O F 2009 CFD design of a generic controller for vortex-induced resonance *Journal of Seventh International Conference on CFD in the Minerals and Process Industries*

[9] Techet A 2005 Vortex induced vibration *MIT OpenCourseWare* (USA: Massachusetts Institute of Technology)

[10] Hall-Stinson A, Lehrman C and Tripp E 2011Energy generation from vortex induced vibration *Thesis (B.S.) Worcester Polytechnic Institute* USA

[11] Williamson C H and Govardhan R 2004 Vortex-induced vibration *Journal of Fluid Mechanics* **36** 413-55

[12] Zahari M A 2014 Renewable energy harvesting using vortex induced vibration *Thesis (B.ME.), Curtin University* Malaysia

[13] Zahari M A and Dol S S 2014 Application of vortex induced vibration energy generation technologies to the offshore oil and gas platform: The preliminary study *International Journal of World Academy of Science, Engineering and Technology* **8(7)** 1331-34

IOP Conf. Series: Materials Science and Engineering **78** (2015) 012042 doi:10.1088/1757-899X/78/1/012042

The Feasibility of Wind and Solar Energy Application for Oil and Gas Offshore Platform

Y K Tiong, M A Zahari, S F Wong and S S Dol[1]
Department of Mechanical Engineering, Curtin University
Sarawak Campus CDT 250, 98009 Miri, Sarawak, MALAYSIA

E-mail: sharulsham@curtin.edu.my

Abstract. Renewable energy is an energy which is freely available in nature such as winds and solar energy. It plays a critical role in greening the energy sector as these sources of energy produce little or no pollution to environment. This paper will focus on capability of renewable energy (wind and solar) in generating power for offshore application. Data of wind speeds and solar irradiation that are available around SHELL Sabah Water Platform for every 10 minutes, 24 hours a day, for a period of one year are provided by SHELL Sarawak Sdn. Bhd. The suitable wind turbine and photovoltaic panel that are able to give a high output and higher reliability during operation period are selected by using the tabulated data. The highest power output generated using single wind energy application is equal to 492 kW while for solar energy application is equal to 20 kW. Using the calculated data, the feasibility of renewable energy is then determined based on the platform energy demand.

1. Introduction

Nowadays, most of the offshore platforms are operated by burning of fossil fuels to generate electricity. The gas turbines and diesel generators on most platforms are used to drive pumps and compressors on board, powered up by combustion of conventional fuels. Continuous burning of these conventional fuels will generate about 80% of the total CO2 and NOx emissions from offshore installations [1]. Thus, offshore platform are facing difficulties in term of operating their activities in an environmentally manner.

Currently, great efforts have been taken in greening the energy sector by shifting the usage of fossil fuels to renewable energy in order to minimize current rate of fossil fuel usage and their ensuing effects of climatic changes. Renewable energy such as wind energy, solar energy and ocean energy are highly focused in the feasibility test of replacing conventional fuels to operate an offshore oil and gas platform. Some of the offshore platform has already started to exploit a renewable energy sources to generate power supply to lessen consumption of fuels. For example, a new offshore platform in the Southern North Sea as shown in figure 1 is operated by using their own energy generated through solar panels and wind turbines [2].

---

1 To whom any correspondence should be addressed.

Figure 1. Wind turbines and solar panels on offshore platform [2]

Wind and solar seems like ideal alternative sources of energy as it can provide an infinite amount of clean energy for offshore application. This paper will focus on the capability of renewable energy available around the offshore platform in order to determine the possibility of such utilization to meet the demand needed for these platform activities. Out of all renewable energies available these days, wind energy and solar energy are highly focused in this feasibility test since data collected predicted a rather promising result. Electrical power consumption of a platform is within the range from 10 MW to 50 MW, while the power consumption for smaller unmanned varies from 6 MW to 7 MW [1]. The targeted platform for this study is carried out at SHELL Malaysia Oil and Gas Sabah Water Platform, with the rated power consumption is approximately 10 MW. Based on the data collected from the targeted platform, the rated power output are determined in order to meet the demand of power consumption of this platform.

2. Literature review

2.1. Wind energy

In this feasibility study, SIEMENS SWT-4.0-120 is selected due to its suitable cut-in speed as low as 2.9 m/s. The wind turbine selected is a direct drive horizontal axis wind turbine, with the latest patented Quantum Blade technology used by Siemens, and able to generate rated power output as high as 4 MW with rated wind speed of 13 m/s [3].

In order to determine the rated power output, P_{rated} for specific month, equation (1) is used. Where,A_r is the swept area of the rotor; PD is the power density by the wind; η is the efficiency of the selected wind turbine stated by the manufacturer [4].

$$P_{rated} = A_r \times \mathrm{PD} \times \eta \qquad (1)$$

The swept area, A_r by the rotor of wind turbine can be determined by using equation (2) where, D is the diameter of the rotor for the selected wind turbine.

$$A_r = \pi \times \left(\frac{D}{2}\right)^2 \qquad (2)$$

The power density, PD of the wind is determined by the equation (3) where, ρ_{air} is the air density at 50 m above sea level and v is the wind speed.

$$PD = \frac{1}{2} \times \rho_{air} \times \left(\frac{v^3}{1000}\right) \quad (3)$$

For Siemens SWT-4.0-120, the efficiency for the selected wind turbine is 32%, giving the value of η is equal to 0.32 [3].

2.2. Solar energy

For the solar energy, the selected photovoltaic panel used for this feasibility test is LG290 N1C-G3 MonoX, with the panel area of 1.64 m^2 and peak power output of 290 W [5]. The power output for photovoltaic panel can be determined via the equation (4):

$$P = A_p \times r \times H \times \mathrm{PR} \quad (4)$$

where, A_p = area of solar panel in m^2, r = solar panel yield (%), H = average solar radiation on panels (W/m^2) and PR = Performance ratio.

Previous studies suggested that the performance ratio is set as 0.75 as default when solar panel is tilted at 45° of the direction of irradiation [6]. This value varies depending on the shadings present around the solar panel, as well as the cleanliness of the solar panel itself which is referring to the dust accumulation on solar panel.

Solar panel yield, r can be determined by the equation (5) where, PE is the electrical power of solar module selected; $A_p$is the area of the solar module.

$$Solar\ Panel\ Yield, r\ (\%) = \frac{\mathrm{PE}}{10A_p} \quad (5)$$

3. Data

3.1. Wind energy

In this research, wind speed data surrounding the SHELL Sabah Water Platform is provided by Sarawak SHELL Bhd. With the tabulated data, the average monthly wind speed hitting around the platform 50 m above sea level are plotted as in figure 2.

Based on figure 2, the wind speeds around the SHELL Sabah Water Platform are within a range of 3.22 m/s to 6.07 m/s. A sample calculation to determine the rated power output is shown using the average wind speed on January (6.07 m/s). The wind turbine specification and wind parameter are listed in table 1.

The power output for January is calculated using the equations discussed before. The swept area by the rotor of wind turbine is given by equation (2):

$$A_r = \pi \times \left(\frac{D}{2}\right)^2 = \pi \times \left(\frac{120}{2}\right)^2 = 11310\ \mathrm{m}^2$$

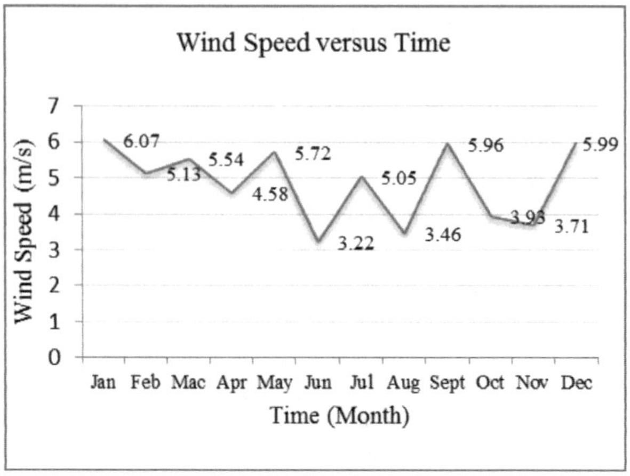

Figure 2. Average Wind Speed collected on SHELL Sabah Water Platform in 2008

Table 1. Wind turbine and wind parameters

Wind Turbines (SIEMENS SWT-4.0-120)	
Rotor diameter, D [m]	120
Blade length, L [m]	58.5
Efficiency, η [%]	32
Wind	
Air density, ρ_{air} [kg/m^3]	1.225 at 50 m above sea level

The power density, PD by the wind onto the wind turbine is determined by the equation (3):

$$\text{PD} = \frac{1}{2} \times \rho \times v^3 = 0.5 \times 1.225 \times 6.07^3 = 137 \text{ kg/s}^3$$

Then, the power output for specific wind turbine for January can be determined via the equation (1):

$$P_{rated} = A_r \times \text{PD} \times \eta = 11310 \times 137 \times 0.32 = 495830 \text{ W} = 495 \text{ kW}$$

3.2. Solar energy

The solar irradiation available in the surrounding of SHELL Sabah Water Platform is provided by Sarawak SHELL Bhd. With the tabulated data, the average monthly solar irradiation hitting around the platform are plotted as in figure 3.

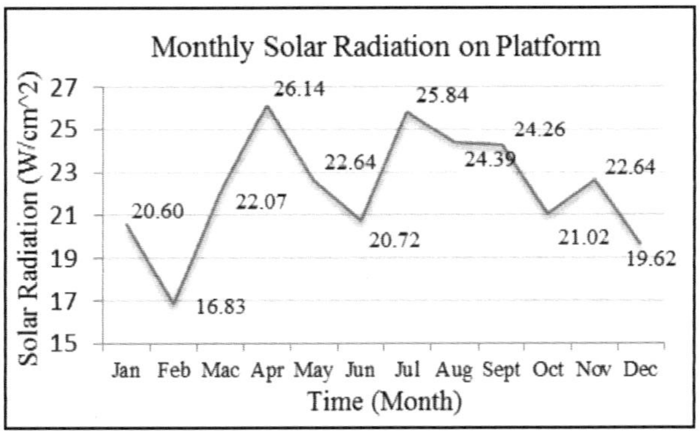

Figure 3. Average monthly solar radiation on SHELL Sabah water platform in 2008.

However, as the energy based on solar is dependent on the availability of the sun, additional graph is plotted based on daily solar irradiation to determine the peak hour of solar irradiation. With the collected data, the average daily solar irradiations hitting around the platform on are plotted as in figure 4.

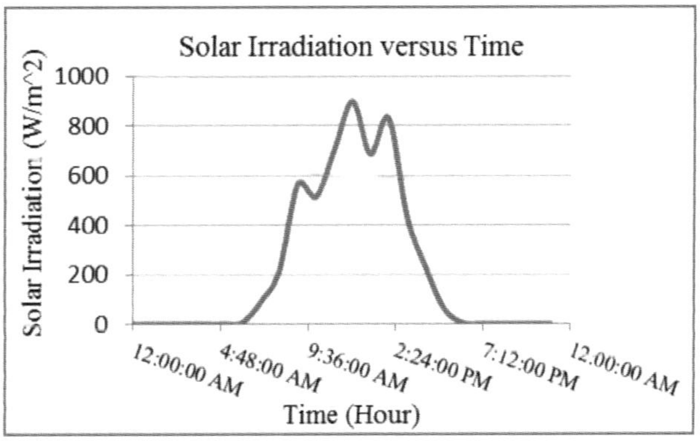

Figure 4. Daily solar irradiation peak hour on SHELL Sabah water platform.

Based on figure 3 and 4, the solar irradiation on the SHELL Sabah water platform varies within a range of 16.83 W/cm^2 to 26.14 W/cm^2 with the peak hour starting from 6 am to 6 pm. A sample calculation to determine the rated power output is shown using the average solar irradiation collected on April (26.14 W/cm^2). The selected photovoltaic panel used for this feasibility test is LG290 N1C-G3 MonoX, with the panel area of 1.6 4m^2 and peak power output of 290 W [5]. While its performance ratio is set as default which is 0.75 based on the previous study [6]. The power output for April is calculated using the equations discussed before. Solar panel yield, r can be calculated using equation (5):

$$r\ (\%) = \frac{\text{PE}}{10A_p} = \frac{290}{10(1.64)} = 17.68\ \%$$

The average solar irradiation collected in April is converted from 26.14 W/cm^2 to 94.10 kWh/m^2. Then, the power output for photovoltaic panel can be calculated via the equation (4):

$$P_{rated} = A_p \times r \times H \times \text{PR}$$
$$= 1.64 \times 0.1768 \times 94.1 \times 0.75 = 20 \text{ kWh}$$

4. Result and discussion

The rated power output calculated for each month based on selected wind turbine for wind energy application are plotted as shown in Figure 5.

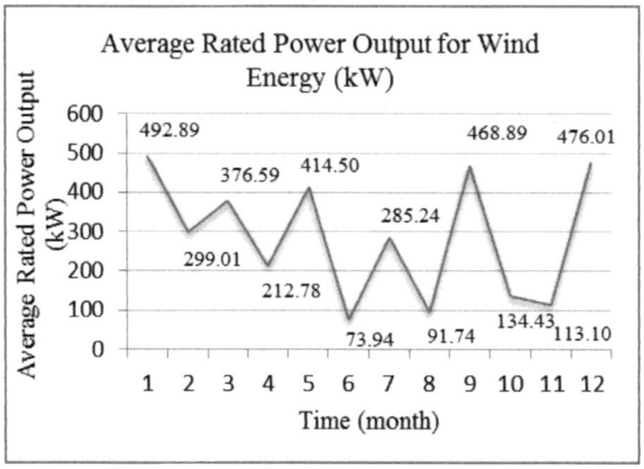

Figure 5. Monthly average rated power output for SIEMENS SWT4.0-120

Based on the figure 5, the monthly average rated power output generated by selected wind turbine SIEMENS SWT4.0-120 varies between the ranges of 73.94 kW to 492.89 kW. The highest power output was generated in January while the lowest month of power generated by this turbine was in Jun.

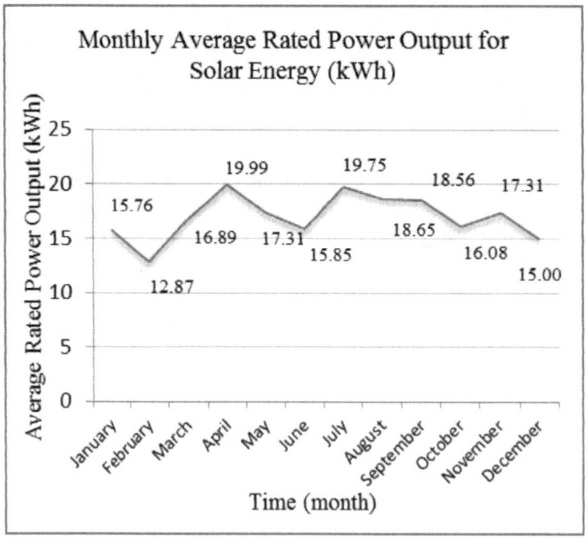

Figure 6. Monthly average rated power output for LG290 NIC-G3 MonoX

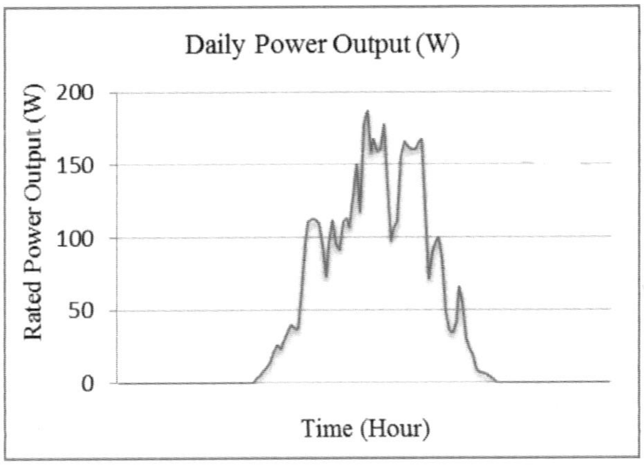

Figure 7. Daily average rated power output for LG290 NIC-G3 Monox

Based on the figure 6, the monthly average rated power output generated by solar panel LG290 NIC-G3 MonoX varies between the ranges of 12.87 kWh to 19.99 kWh. The poorest month of solar irradiation falls on February with the power output generated 12.87 kWh while the highest month of solar irradiation falls in April with the power output generated 19.99 kWh.

Based on daily power output in figure 7, at time between 12:00 a.m. and 6:00 a.m. as well as between 6:00 p.m. up till the next morning, the solar irradiation recorded is zero, indicating that the system has its downtime during this period of time. This is due to the availability of the sun [7]. The peak hour of daily solar irradiation falls at time between 11 a.m. to 3 p.m.

Based on the results discussed before, these wind turbine and solar panel can be installed in multiple units to form an array system in order to provide sufficient power supply to the selected offshore platform (figure 8). In order to accommodate the rated power output of 10 MW, the solar irradiation investigated should be from the poorest out of all months and days of all data collected to form an array. Moreover, an integrated system should be considered using both of these wind turbine and solar panel applications [9]. This is due to the complementary power output generated by these two forms of energy under seasonal variant in weather conditions. The solar panel plays a major role in generate highest power output in the summer months when the output from the wind turbine is at the lowest and vice versa during the monsoon season.

Figure 8. Multiple units of wind turbines and solar panels installed on offshore platform (conceptual design from [8])

5. Conclusion

The solar and wind energy application in generating energy is a viable solution of the current energy extraction problem for offshore application. In this paper, it showed that the highest power output generated for wind energy application is equal to 492 kW while for solar energy application power generated is equal to 20 kW. As a single unit of solar or wind turbine system is insufficient in order to provide power supply to offshore platform application, these applications can be installed in multiple units in order to generate sufficient power supply. Both of these systems can also be integrated to each other due to their different capabilities in generated power output based on weathers and climates condition.

Nomenclature

P_{rated}	Rated Power Output, W
P	Peak Power Output, W
PD	Power Density, W/m^2
D	Diameter of Rotor, m
L	Blade length of Rotor, m
r	Solar Panel Yield, %
H	Average Solar Radiation on Panel, kWh/m^2
PR	Performance Ratio
PE	Electrical Power of Solar Module, W
η	Efficiency, %
A_r	Swept Area of Rotor, m^2
A_p	Area of Solar Panel, m^2
v	Wind Velocity, m/s
ρ_{air}	Density of air above sea level, kg/ m^3

I.

Acknowledgement

The current speed and solar irradiation data on the SHELL Sabah Water Platform are provided by SHELL Sarawak Sdn. Bhd. and dated back in year 2008.

IOP Conf. Series: Materials Science and Engineering **78** (2015) 012042 doi:10.1088/1757-899X/78/1/012042

References

[1] He W, Uhlen K, Hadiya M, Chen Z, Shi G and Rio E 2013 Case study of integrating an offshore wind farm with offshore oil and gas platforms and with onshore electrical grid *Journal of Renewable Energy* 1-10

[2] Rosebro J 2006 Fossil-fuel platform runs on renewable energy *http://www.greencarcongress.com/2006/04/fossilfuel_plat.html*

[3] Siemens 2014 Siemens 4.0MW offshore wind turbine *http://www.energy.siemens.com/*

[4] NPower nd Wind turbine power calculations *The Royal Academy of Engineering*

[5] Tecnospot 2013 MonoX Neon LG290NiC-G3 *http://www.lg.com/uk/solar*

[6] Hanid M, Ramzan M, Rahman M, Khan M, Amin M and Amir M 2012 Studying power output of PV solar panels at different temperatures and tilt angles *ISESCO Journal Science and Technology* **8** 9-12

[7] Goffman E 2008 Why Not the Sun? Advantages of and Problems with Solar Energy *Journal of ProQuest Discovery Guides*

[8] Tiong Y K 2013 The renewable energy application for oil and gas offshore platform *Thesis (B.ME.) Curtin University* Malaysia.

[9] Zahari M A and Dol S S 2014 Application of vortex induced vibration energy generation technologies to the offshore oil and gas platform: The preliminary study *International Journal of World Academy of Science, Engineering and Technology* **8(7)** 1331-34

AUTHOR INDEX

Abdulmunem, Abdulmunem R 167
Abdul-Rani, Ahmad Majdi 209
Alam, A K M Moshiul 191
Ali, M 311
Alkali, Adam Umar 209
Al-Kayiem, Hussain H 148
Al-Qrimli, Haidar F 197
Al-Wakeel, Haitham B 148
Alzorqi, Ibrahim S 197
Balaji, P S 1
Balakumar, S 38
Beg, M D H 191
Benard, L D 270
Bhuiyan, M A S 88
Bing, Chua Han 53
Chan, H B 337
Chan, Y S 38
Chandrasekaran, K 80
Chong, H Y 225
Chong, Kit Wee 216
Dayou, J 279
Dol, S S 318, 329, 337, 347
Ezhumalai, P 94
Ginta, Turnad Lenggo 209
Gorin, A 176
Hadi, S S 134
Hamid, Mohamed Ali 53, 62
Hamid, Roszilah 305
Hanirah, R 299
Harno, H G 161
Harno, Hendra G 241
Hussien, Hashim A 167
Ibrahim, Raafat 250
Ibrahim, Raheek I 46
Irawan, S 70
Islam, Sumaiya 250
Ismail, Firas B 197
Jaison, J 38
Jamian, N 270
Jayakumar, M 11
Jayamani, E 94
Jessica, M S 263
Joannes, C 279
Karim, Z A Abdul 148
Khondoker, Noman 250
Khur, Wee Siaw 216
Kiu, G Q 115
Lau, H H 1
Law, M C 182
Lee, H V 161
Lee, V C C 161, 182
Liang, G V Y 182
Mahdi, Fadhil A 197
Mahmud, H Ben 311
Mahmud, Hisham Khaled Ben 62
Mahzabin, Mst Sadia 305
Mansa, R 30
Mansa, R F 23, 279
Mansuit, H 30
Minhad, K N 88
Mohammad, A W 46
Moussa, Leblouba 1
Mungkasi, Sudi 257
Nabilla, A F 128
Nagarajan, Ramasamy 53
Nasir, Meseret 216, 225, 233
Noik, V James 287
Numan, Ali H 167
Ofei, T N 70
Osman, Sabtuni 216, 225
Palanichamy, C 105, 115, 233
Pao, W 70
Piakong, M T 263, 299
Rahman, L F 88
Rahman, M E 1
Raja, C Ashok 38
Ramachandaramurthy, V K 80
Ramakrishnan, M 94
Raza, Arshad 53
Reaz, M B I 88
Ruzanna, R 263
Samsuri, M D C 30
Shafiq, Mian Umer 62
Shazryenna, D 263
Sipaut, C S 23, 30, 279
Soon, Lee Chee 11
Suadin, E G 270
Syaufi, L 299
Tholhath, S 140
Tiong, T C 128, 134, 140
Tiong, Y K 347
Tuah, P M 270
Tuah, P Mohd 287
Veeramani, S 233
Wee, A Y L 176
Wee, S K 182
Wong, S F 347
Wong, Y C 105
Wong, Z H 46
Woo, K C 161
Woon, Raymond Song Kiin 241
Yasir, S M 30, 279
Yee, C F 30
Yong, T H 318, 337
Yugis, A R 23
Yunus, Rosli Mohd 191
Zahari, M 329, 337
Zahari, M A 347

Institute of Physics
Dirac House, Temple Back
Bristol BS1 6BE UK

CURRAN ASSOCIATES INC
proceedings
.com